Advances in
FISH AND WILDLIFE ECOLOGY AND BIOLOGY

— Vol. 7 —

The Editors

Dr. Bansi Lal Kaul (b. 1942) Former Professor of Zoology and Principal, Government S.P.M.R. College of Commerce and Management, University of Jammu is a noted teacher, researcher, and author. He has to his credit more than hundred articles on functional anatomy and ecology of fishes, wildlife and Himalayan environment. Besides having edited titles on Himalayan biodiversity, ecology and environment he is editing the series titled, "Advances in Fish and Wildlife Ecology and Biology". The present volume is the seventh in the series.

Dr. Kaul is the recipient of several awards including the prestigious F.A.O (U.N) World Food Award. He is a fellow of the Linnean Society and Zoological Society of London. He is a founder member of Indian Society for Popularization of Science and Jammu Consumer's Council, Jammu. He is also associated with solutions exchange programme of the United Nation Development Programme (UNDP) in India.

Prof. (Dr.) P. L. Koul (b.1950) has worked on morphology, histology, histopathology and biochemical composition of Acanthocephala. He taught zoology, biotechnology, ecology, wildlife, fisheries and parasitology at UG level for more than thirty years. He also worked as Principal for more than eight years in Govt. Degree and PG Colleges in J&K State. He is credited with starting PG courses in various colleges of the state. Dr. Koul worked as Senior member of the committee constituted by the Govt. of Jammu and Kashmir to frame new higher education policy. He also introduced some ad-on (skill development) courses in various colleges. Dr. Koul has many research publications on Acanthocephla, Protozoology, Histology, Pathology and Histopathology to his credit.

Dr. Anil Verma (b.1963) Associate Professor and Head, Faculty of Sciences, Government Post Graduate College of Education, Jammu, J&K State did his M.Sc. Zoology in 1986. He was awarded M.Phil and Ph.D degrees by the University of Jammu, Jammu for his work in the field of animal reproduction. Dr. Verma was elected a Fellow of the Linnean Society of London in 2006. He has to his credit more than 70 research papers and review articles published in reputed journals.

Dr. Verma is the recipient of Rashtriya Gourav Award of the India-International friendship society and the top 100 scientists/professional of the year 2012 by American Biographical Institute, Washington.

Advances in
FISH AND WILDLIFE ECOLOGY AND BIOLOGY

— Vol. 7 —

– Editor-in-Chief –
Dr. B. L. Kaul
*Former Professor of Zoology and Principal,
Government SPMR College of Commerce and Management,
University of Jammu, Jammu – 180 001, J&K, India*

– Editors –
P. L. Koul
*Former Professor of Zoology & Principal,
Government Degree College
Akhnoor, J&K, India*

A. K. Verma
*Associate Professor and Head,
Faculty of Sciences, Government Postgraduate College of Education
Jammu – 180 001, J&K, India*

2018
Daya Publishing House®
A Division of
Astral International Pvt. Ltd.
New Delhi – 110 002

© 2018 EDITORS

ISBN **9789387057517** (*International Edition*)

Publisher's Note:

Every possible effort has been made to ensure that the information contained in this book is accurate at the time of going to press, and the publisher and author cannot accept responsibility for any errors or omissions, however caused. No responsibility for loss or damage occasioned to any person acting, or refraining from action, as a result of the material in this publication can be accepted by the editor, the publisher or the author. The Publisher is not associated with any product or vendor mentioned in the book. The contents of this work are intended to further general scientific research, understanding and discussion only. Readers should consult with a specialist where appropriate.

Every effort has been made to trace the owners of copyright material used in this book, if any. The author and the publisher will be grateful for any omission brought to their notice for acknowledgement in the future editions of the book.

All Rights reserved under International Copyright Conventions. No part of this publication may be reproduced, stored in a retrieval system, or transmitted in any form or by any means, electronic, mechanical, photocopying, recording or otherwise without the prior written consent of the publisher and the copyright owner.

Published by : **Daya Publishing House®**
A Division of
Astral International Pvt. Ltd.
– ISO 9001:2015 Certified Company –
4736/23, Ansari Road, Darya Ganj
New Delhi-110 002
Ph. 011-43549197, 23278134
E-mail: info@astralint.com
Website: www.astralint.com

Editorial Board

1. **Dr. B.L. Kaul**
   Former Professor of Zoology and Principal,
   Government SPMR, College of Commerce and Management,
   Canal Road, Jammu – 180 001, J&K, India
   Res: 186, Upper Laxmi Nagar, Sarwal Jammu
   E-mail: blkaul@gmail.com

2. **Dr. S.S.S. Sarma**
   Professor, Universidad Nacional Autonoma de Mexico, Campus Iztacala.
   Av. de los Barrios #1, Col. Los Reyes, Iztacala,
   Tlalnepantla, State of Mexico. C.P. 54090, Mexico
   E-mail: ssssarma2008@gmail.com

3. **Dr. Ashwani Wanganeo**
   Professor and Head,
   Department of Limnology, Barkatullah University,
   Bhopal, M.P, India
   E-mail: profwanganeoa@gmail.com

4. **Dr. S. Chellappa**
   Professor, Department of Oceanography and Limnology,
   Centre of Bio-Science, Universidade Fedral do Rio,
   Grandodo Norte (UFRN), Natal, Brazil
   E-mail: chellappa.sathyabama63@gmail.com; chellappa@dol.ufrn.br

5. **Dr. P.L. Koul**
   Professor of Zoology and Principal (Retd.),
   37/2A, Roop Nagar Enclave, Jammu, India
   E-mail: pearaykoul@gmail.com

6. **Dr. Anil K. Verma**
   Associate Professor & Head, Faculty of Sciences,
   Government (P.G.) College of Education, Jammu, J&K, India
   E-mail: anilverma.ak@gmail.com

Professor Dina Nath Fotedar: A Tribute

Professor Dina Nath Fotedar was born at Srinagar, Kashmir on 17th October 1923. He did his B.Sc from Punjab University, Lahore and M.Sc. Zoology with distinction from Aligarh Muslim University, Aligarh. He also did his M.Sc. in Parasitology from London School of Hygiene and Tropical Medicine and Ph.D. from the University of London.

Prof. Fotedar was a legendary figure in Jammu and Kashmir. He acted as Head, Department of Zoology, Dean, Faculty of Sciences and also as acting Vice Chancellor of the University of Kashmir from time to time. He was a member and fellow of a number of scientific and research bodies in India and abroad and an editor and member of editorial boards of several research journals. It was during his tenure that the research in Parasitology was initiated at the University of Kashmir. He guided research on vide range of parasites of vertebrate, animals, pathogenic plant parasitic nematodes of vegetable, cereal and fruit crops in Kashmir. He guided 15 Ph.D. scholars and published more than 150 research papers in leading national and international research journals. For his commendable research in the field of Parasitology, he was honoured with award of Bhalerao Gold Medal.

Prof. Fotedar edited more than 10 books; he translated research papers and books into English from other languages mostly Russian, which proved immensely useful to researchers in the field of Parasitology. He worked as s coordinator in "the Himalayan Eco development" Project and as chief investigator in the project "Wildlife Ecology and Conservation" of the Ministry of Environment and Forests, Govt. of India. Prof. Fotedar relinquished office as Head of Zoology Department in 1985 but continued his association with the University upto 1988. By that time Department of Zoology was fully developed with specializations in Parasitology, Ichthyology, Entomology and Wetland and Wildlife Ecology. Prof. D.N Fotedar was a highly respected academic. He died on 18th April 2011 and his death was condoled in the Zoology Department of Kashmir University, Srinagar by teaching and non-teaching staff, students, research scholars, Ex-students and well wishers.

Society for Popularization of Science also organized a condolence meeting at Jammu which was attended by his Ex-students and Teaching fraternity of Jammu University. Glowing tributes were paid to him at the meeting. His contribution to advancement of Zoology will be remembered for a long time to come. By writing this tribute to him I myself feel honored.

Prof. Dr. B.L Kaul

Preface

It is a well established fact that there is a link between climate change and biodiversity. Climate change that we have been witness to especially during the last two or three decades is not new. Throughout the history of our planet the climate has been changing affecting ecosystems. As a consequence ecosystems have been changing and resulting in changes in the species. Rapid climate changes due to anthropogenic factors have affected the ability of the species to adapt and loss of some of them.

The United Nations Global Biodiversity Outlook 3, in May 2010 thus summarized some concerns that climate change will have on ecosystems as follows:

"Climate change is already having an impact on biodiversity and is projected to become a progressively more significant threat in the coming decades. Loss of arctic sea ice threatens biodiversity across the entire biome and beyond. The related pressure of ocean acidification, resulting from higher concentrations of CO_2 in the atmosphere is also already being observed.

Ecosystems are already showing negative impacts under current levels of climate changes……..which is modest compared to future projected changes…….. In addition to warming temperatures, more frequent extreme weather events and changing patterns of rainfall and drought can be expected to have significant impacts on Biodiversity."

Nothing can be more prophetic than the above lines. In India we are currently tasting the effects of climate change. Drought and water scarcity is adversely affecting the lives of people and animals in many states like Maharashtra, Gujarat, Madhya Pradesh, Chhattisgarh, Jharkhand, Odisha, Tamil Nadu and Andhra Pradesh. In many states excessive and untimely rains have brought havoc. In Tamil Nadu, Jammu and Kashmir we have witnessed floods, cloud bursts and flash floods.

Landsides and avalanches in the Himalayas have become routine and adversely affected humans, live stock and wildlife alike. Likewise climate change has impacted life in the oceans lakes, rivers, wet lands and threatened coral reefs, lizards, birds, terrestrial animals, forests, water sources and last but not the least food security. It is heartwarming that a global agreement was finalized in Paris Conference on Global Climate in December 2016. The deal unites all the world's nations in a single agreement on tackling climate change for the first time in history.

Coming to a consensus among nearly 200 countries on the need to cut green house gas emissions is regarded by many as an achievement in itself and is being hailed as "historic." The Kyto Protocol of 1997 set emission cutting targets for a handful of developed countries, but the US pulled out and others failed to comply.

However, scientists point out that the Paris accord must be stepped up if it is to have any chance of curbing dangerous climate change. Pledges thus far could see global temperatures rise by as much as 2.7°C, but the agreement lays out a road map for speeding up progress.

The measures in the Paris agreement included the following:

1. To peak green house gas emissions as soon as possible and achieve a balance between sources and sinks of greenhouse gases in the second half of this century.
2. To keep global temperature increase well below 2°C (3.6°F and to pursue efforts to limit it to 1.5°C.
3. To review progress every five years.
4. US Dollar 100 billion a year in climate finance for developing countries, with a commitment to further finance in future.

If emissions of greenhouse gases are legally and voluntarily cut the world will be a better place to live in. With the cut of green house gases global climate will improve and threat to biodiversity will be reduced.

The practice in earlier volumes of having two sections namely Fish and Limnology and Wildlife has been followed in this volume as well. The current volume is dedicated to a legendary Zoologist from Kashmir Prof. Dina Nath Fotedar. He was a noted Helminthologist and Parasitologist and I have included two chapters on parasites in this volume to honour his legacy.

It is hoped that the present volume will be well received like the earlier volumes. I must thank the contributors without whose cooperation the volume would not see light of the day. I must also thank the board of Editors especially Dr. Anil Verma for their help with suggestions and also in editing. My thanks are also due to my wife Promila for her continuous support. Thanks are also due to Mr. Anil Mittal the publisher and his supporting staff for bringing out the volume in the shortest possible time.

Prof. Dr. B.L Kaul

Contents

Editorial Board	v
Professor Dina Nath Fotedar: A Tribute	vii
Preface	ix
List of Contributors	xv

Section I: Fish and Limnology

1. **Review on the Reproductive Strategies of Marine Fish Species from Northeastern Brazil**   3
   Sathyabama Chellappa, Monica Rocha de Oliveira and Naithirithi T. Chellappa

2. **Biodiversity Status and Measures for Conservation of Threatened Indian Fishes: An Update**   25
   A.K. Pandey, Rehana Abidi, Madhu Tripathi and P. Das

3. **The Effect of Endocrine Disruptors on Aquatic Animals: A Short Review**   63
   B.K. González-Pérez, S.S.S. Sarma, M. E. Castellanos-Páez and S. Nandini

4. **Management of Catchment Area of River Ganga to Conserve it for Posterity**   77
   Ashwani Wanganeo and Rajni Wanganeo

5. Review on Reproductive Tactics and Strategies of Marine Fish Species from the Coastal Waters of Rio Grande do Norte, Brazil — 89

   Sathyabama Chellappa and Naithirithi T. Chellappa

6. Macrobenthic Community Structure along the Temporal Shift from Non-paddy Cultivating (Flooded Phase) to Paddy Cultivating (Paddy Phase) Phases in Kole Paddy Fields, Vembanad Kole Wetland, India — 105

   S. Vineetha, S. Bijoy Nandan and K.P. Rakhi Gopalan

7. Demography of the Ostracod *Heterocypris incongruens* (Ramdohr, 1808) Fed Alga and Organic Wastes — 121

   Marissa F. Juárez-Franco, S.S.S. Sarma and S. Nandini

8. Benthic Faunal Diversity in Indian Mangroves — 135

   Philomina Joseph, S. Sreelekshmi, Rani Varghese, C.M. Preethy and S. Bijoy Nandan

9. Water Bugs as Forage Base in World Fishery — 167

   P. Venkatesan

10. Induced Spawning and Seed Production of *Pangasianodon hypophthalmus* in Three different Types of Hatcheries under Agro-climatic Conditions of Raipur (Chhattisgarh), India — 181

    C.S. Chaturvedi, Rashmi S. Ambulkar, R.K. Singh and A.K. Pandey

11. Allelopathic Interactions in Freshwater Ecosystems with Special Reference to Zooplankton — 195

    S.S.S. Sarma and S. Nandini

Section II: Wildlife

12. Breeding Ecology of Yellow-Wattled Lapwing *Vanellus malabaricus* in the Kole Wetlands of Thrissur, Kerala — 225

    P. Greeshma and E.A. Jayson

13. A Taxonomic Review on the Genus *Pareumenes* de Saussure (Hymenoptera : Vespidae : Eumeninae) from the Indian Subcontinent — 233

    P. Girish Kumar and P.M. Sureshan

14. Responses of Serum Luteinizing Hormone (LH) and Testosterone (T) Levels as Correlated with Testicular Morphology of Albino *Mus norvegicus* Induced by Sublethal Heroin Administration — 247

    Kaminidevi K. Bhoir, S.A. Suryawanshi and A.K. Pandey

15. **Wildlife Diversity of Odisha, India and their Conservation**   261
    Sudhakar Kar

16. **Ecology and Conservation of Mammals of Oak Forest of Central Himalaya, India**   273
    Aisha Sultana and Mohammad Shah Hussain

17. **On Two New Species of *Balantidium* Claparede and Lachmann, 1858 emend. Stein, 1867 from *Rana (Dicroglossus) cyanophlyctis* Schneider and First Report of Two Species from Kashmir Valley, India**   287
    Kanwar Narain, M.K. Raina and P.L. Koul

18. **A Review on Diversity and Distribution of Nematodes Associated with Paddy in West Bengal, India**   305
    Viswa Venkat Gantait, Suresh Mandal, Paromita Roy and Soumendranath Chatterjee

19. **Observations on Feeding and Breeding Biology of Red-Vented Bulbul *ycnonotus cafer* (Linnaeus 1766) in Jammu, J&K State, India**   329
    B.L. Kaul and A.K. Verma

    Previous Volumes   337

    Index   353

List of Contributors

Abidi Rehana, ICAR–National Bureau of Fish Genetic Resources, Canal Ring Road, Lucknow – 226 002, Uttar Pradesh, India.

Ambulkar Rashmi S., ICAR–Central Institute of Fisheries Education, Versova Mumbai – 400 061, Maharashtra, India.

Bhoir Kamini Devi K., Department of Zoology, Institute of Science, 15 Madam Cama Road, Mumbai – 400 032, Maharashtra, India.

Bijoy Nandan S., Department of Marine Biology, Microbiology and Biochemistry, School of Marine Sciences, Cochin University of Science and Technology, Fine Arts Avenue, Cochin – 682 016 Kerala, India.

Challappa Naithirithi, Department of Oceanography and Limnology, Centre of Bioscience, Universidad Federal do Rio Grande do Norte, Natal, R.N. Brazil

Challappa Sathyabama, Department of Oceanography and Limnology, Centre of Bioscience, Universidad Federal do Rio Grande do Norte, Natal, R.N. Brazil

E-mail: challappasathyabama63@gmail.com

Chatterjee Soumendranath, Department of Zoology, the University of Burdwan, Burdwan – 713 104.

Chaturvedi, C.S., ICAR–Central Institute of Fisheries Education Versova, Mumbai – 400 661, Maharashtra, India.

Das P., Ex Director, NBFGR (ICAR), A-8/4 Indralok Estate, Paikapara, Kolkata – 700 002, West Bengal, India.

Gantait Viswa Venkat, Zoological Survey of India, M- Block, New Aliproe, Kolkata – 700 053, West Bengal, India

E-mail: v.gentait@rediffmail.com

Girish Kumar P., Western Ghat Regional Centre, Zoological Survey of India, Kozhikode – 673 006, Kerala, India.

E-mail: kpgiris@gmail.com

Gonzalez-Peraz, B.K., Programme do Doctorado Unversidad Autonoma Metropolitana, Unidad Xochimilco, Alzada del Huesco, No. 11 00, Ville de Quietud, Mexico city, Codigo Postal-04960, Mexico.

Gopalan, K.P. Rakhi, Department of Marine Biology, Microbiology and Biochemisty, School of Marine Sciences, Cochin University of Science and Technology, Fine arts avenue, Kochi – 682 016, Kerala, India.

Greeshma P., Wildlife Department, Kerala Forest Research Institute, Peechi – 680 653 Thrissur, Kerala, India.

E-mail: greeshmap@kfri.res.in

Hussain Mohammed Shan, Department of Wildlife Sciences, Aligarh Muslim University, Aligarh – 202 002, Uttar Pradesh, India.

Joseph Philomina, Department of Marine Biology, Microbiology and Biochemistry, School of Marine Sciences, Cochin University of Science and Technology, Cochin –682 016, Kerala, India.

Juaraz-France Marissa F., Postgrado enCiencias Biologicas, National Autonomous University of Mexico Ciudad Universitaria Circuito De Posgrades, CP04510, Mexico City, Mexico.

Jysen, E.A., Wildlife Department, Kerala Ferest Research Institute, Peechi –680 653, Thrissur, Kerala, India.

E-mail: jaysen.58@gmail.com

Kaul, B.L., 18-6 Upper Laxmi Nagar, Sarwal, Jammu – 180 005, J&K, India.

E-mail: blkaul@gmail.com

Koul, P.L., 37/2-A, Reep Nagar Enclave Jammu, J&K India.

E-mail: pearaykaul@gmail.com

Mandal, Suesh, Zoological Survey of India, M–Block, New Alipore, Kolkata – 700 053, West Bengal, India.

Nandini, S., Universidad National Autonomie de Mexico, Campus Iztacala, Av. de los Barrios # 1, Col. Los Reyes, Iztacala Tlalnepantla, State of Mexico, C.P. 54090, Mexico.

Narain Kanwar, ICAR–Regional Research Centre, North-Eastern, ICMR, Post Box No. 105, Dibrugarh – 786 001, Assam, India.

Pandey, A.K., National Bureau of Fish Genetic Resources, Canal Ring Road, Lucknow – 226 002, Uttar Pradesh, India.

Preethy C.M., Department of Marine Biology, Microbiology and Biochemistry, School of Marine Sciences, Cochin University of Science and Technology, Cochin – 682 016, Kerala, India.

Raina, M.K., 174/5 Trikuta Nagar, Jammu, J&K, India.

Roy Paromita, Parasitology and Microbiology Research Laboratory, Department of Zoology, The University of Burdwan, Burdwan – 710 104, West Bengal, India.

Sarma, S.S.S., Universidad National Autonomie de Mexico, Campus Iztacala, Av. de los Barrios # 1, Col. Los Reyes, Iztacala Tlalnepantla, State of Mexico, C.P. 54090, Mexico.

E-mail: sarma@unam.mx

Singh, R.K., Chhattisgarh State Fisheries Department, Raipur – 492 001, Chhattisgarh, India.

Sreelakshmi, S., Department of Marine Biology, Microbiology and Biochemistry, School of Marine Sciences, Cochin University of Science and Technology, Cochin – 682 016, Kerala, India.

Sultana Aisha, Biodiversity Parks Programme, Centre for Environmental Management of Degraded Ecosystems (CEMDE), Department of Environmental Studies, University of Delhi, Delhi – 110 007, India.

E-mail: aishasultana28@yahoo.com

Sureshan, P.M., Western Ghat Regional Centre, Zoological Survey of India, Khozhikode – 673 006, Kerala, India

E-mail: pmsuresh43@gmail.com

Suryawanshi, S.A., Department of Zoology, Institute of Science, 15 Madam, Canal Road, Mumbai – 400 032, Maharashtra, India.

Tripathi, Madhu, Department of Zoology, University Lucknow, Lucknow – 226 007, Uttar Pradesh, India.

Venkatessan, P., Shri Academy for Paramedical Education, Chennai – 600 011, Tamil Nadu, India.

Verma, A.K., Department of Science, Government (P.G.) College of Education, Canal Road, Jammu, J&K, India.

Verghese, Rani, Department of Marine Biology, Microbiology and Biochemistry, School of Marine Sciences. Cochin University of Science and Technology, Cochin – 682 016, Kerala, India.

Vineetha, S., Department of Marine Biology, Microbiology and Biochemistry, School of Marine Sciences, Cochin university of Science and Technology, Fine Arts Avenue, Kochi – 682 016, Kerala, India.

Wanganeo, Ashwani, Department of Environmental Science and Limnology, Barkatullah University, Bhopal – 462 026, M.P., India.

E-mail: profwanganeo@gmail.com

Wanganeo, Rajni, Department of Zoology, Govt. Banazeer College, Bhopal, M.P., India.

Section I
Fish and Limnology

Chapter 1

Review on the Reproductive Strategies of Marine Fish Species from Northeastern Brazil

Sathyabama Chellappa, Monica Rocha de Oliveira and Naithirithi T. Chellappa

ABSTRACT

Life history traits of fish in relatively stable environments suggest a model of three reproductive strategies: (1) opportunistic strategists with small body size, early first sexual maturity, and short-life span; (2) periodic or seasonal strategists of big body size with long life span and high to intermediate fecundity; and (3) equilibrium strategists of intermediate size. This work reviews the reproductive tactics and strategies of seven marine fish species from the coastal region of Northeastern Brazil. Data on the reproductive strategies of the flying fish, Hirundichythys affinis, ballyhoo half beak, Hemiramphus brasiliensis, roughneck grunt, Pomadasys corvinaeformis, maracaibo leatherjacket, Oligoplites palometa, serra Spanish mackerel, Scomberomorus brasiliensis, the lane snapper, Lutjanus synagris and white mullet, Mugil curema were verified based on body size, sex ratio, length at first sexual maturity, aspects of gonad development, fecundity, type of spawning and reproductive period. The results indicate that H. affinis, H. brasiliensis, P. corvinaeformis and O. palometa are opportunistic strategists. On the other hand S. brasiliensis, L. synagris and M. curema are considered as equilibrium strategists. This study provides information on the reproductive aspects of the fishery stocks of the coastal waters of Northeastern Brazil.

Keywords: Fish reproduction, Coastal waters, Spawning period, Sexual strategies, Tropical marine fish.

Introduction

Reproductive strategies are used by fish to maximize production and ensure the survival of offspring to adulthood. Each strategy is expressed by tactics, such as, body size, length-weight relationship, size at first sexual maturation, gonad development, fecundity, type of spawning and breeding period (Potts and Wootton, 1984), which are important information for making rational measures to regulate fishing and conservation of fish stocks (King and McFarlane, 2003; Oliveira *et al.*, 2011). These characteristics possessed by a particular population of a species could have its origin in the evolutionary past or represent fine-tuning of that population for the environmental conditions (Matthews, 1998).

Life history traits of fish in relatively stable environments suggest a model of three reproductive strategies. Winemiller and Rose (1992) used a quantitative approach to develop groupings of life history strategies by examining 16 life history traits in a large sample (216 species from 57 families) of North American freshwater and marine fish species. Based on a final selection of five life history traits for 82 freshwater species and 65 marine species, they suggested a trilateral continuum model with three endpoint strategies: (1) small, rapidly maturing, short-lived fishes (opportunistic strategists); (2) larger, highly fecund fishes with longer life spans (periodic strategists); and (3) fishes of intermediate size that often exhibit parental investment and produce fewer, larger offspring (equilibrium strategists). For this review, it was considered important that fish species, representing different reproductive strategies should be compared to assess applicability of results for fisheries management. Fish species selected for this study support commercial fisheries in the northeastern coastal waters of Brazil.

Lane snapper, *Lutjanus synagris* (Linnaeus, 1758) is distributed in the Western Atlantic, from Bermuda and North Carolina, USA to southeastern Brazil, including Gulf of Mexico and Caribbean Sea. Ballyhoo halfbeak, *Hemiramphus brasiliensis* (Linnaeus, 1758) is an inshore surface-dwelling species, which occurs in Western Atlantic: Massachusetts, USA and northern Gulf of Mexico to Brazil, including the Caribbean Sea. Roughneck grunt, *Pomadasys corvinaeformis* (Steindachner, 1868) occurs in Western Atlantic, in Mexico and the Caribbean coasts both continental and insular to the Antilles and Brazil. Serra Spanish mackerel, *Scomberomorus brasiliensis* Collette, Russo and Zavala-Camin, 1978, occurs in the Western Atlantic, along the Caribbean in Central Atlantic coasts and Southwest Atlantic from Belize to Rio Grande do Sul, Brazil. White mullet, *Mugil curema* Valenciennes, 1836, is a coastal pelagic fish which occurs in the Western Atlantic (Nova Scotia to Argentina), Eastern Atlantic (Senegal to Namibia), and in the Eastern Pacific (California to Chile). The flying fish, *Hirundichythys affinis* (Günther, 1866) occurs in the Western Atlantic and coastal waters. Maracaibo leatherjacket, *Oligoplites palometa* (Cuvier, 1832) occurs in the coastal waters of the Western Atlantic. These marine fish species are considered as important fishery resources of the Atlantic Ocean, which are of high commercial value, and form a major component of artisanal fisheries in northeastern Brazil.

Materials and Methods

Brazil is an important part of the Neotropical region which extends from Mexico to the southernmost tip of South America. The state of Rio Grande do Norte is located in Northeastern Brazil and has a coast of approximately 420 km. Fish samples for all the studies reviewed were collected from artisanal fisheries at various locations in the coastal waters of Rio Grande do Norte, Brazil, situated between latitudes 56° 44' and 5° 52' S, longitudes 35° 09' and 35° 12' W.

Literature survey was conducted on life history parameters and reproductive tactics of marine fish species from the coastal region of Rio Grande do Norte, northeastern Brazil. During the period of 1998 to 2015 a total of 28 scientific papers were published, which encompass seven marine fish species from the coastal waters of Northeastern Brazil. Reproductive aspects of the flying fish, *Hirundichythys affinis*, ballyhoo half beak, *Hemiramphus brasiliensis*, roughneck grunt, *Pomadasys corvinaeformis*, maracaibo leatherjacket, *Oligoplites palometa*, serra Spanish mackerel, *Scomberomorus brasiliensis*, the lane snapper, *Lutjanus synagris* and white mullet, *Mugil curema* (Figure 1.1) were verified considering the body size, sex ratio, length at first sexual maturity, aspects of gonad development, fecundity, type of spawning and reproductive period.

Morphometric measurements and meristic counts were used to check and confirm the taxonomical status of each fish species. All fish collected had been identified to the species level, measured (total body length to the nearest millimeter ± 1 mm) and weighed (body mass ± 1 g). Distribution analysis of total length and weight of males and females of each fish species were performed separately using the absolute frequencies (mean ± SD) in different classes of total length (Lt) and total weight (Wt). A single weight-length equation ($W=aL^b$) was fitted to estimate the value of coefficient b, using the data obtained from all individuals collected (Froese, 2006).

The macroscopic aspects and maturation stages of the gonads were observed besides determining the sex of each fish and the sex ratio was determined. Young individuals were not included as it was not possible to determine their sex by macroscopic observations. The chi-square test (χ^2) was applied at a significance level of 5 per cent. The body size at first gonadal maturity (L_{50}), where 50 per cent of the individuals exhibited maturing gonads, was estimated from the relative frequency distribution of adult males and females, using their standard length classes (mean ± SD).

Fragments of ovaries selected for histological analysis were fixed in Bouin solution for 12-24 h, washed for 24 hours in running water to remove excess fixative and were later preserved in 70 per cent Ethyl alcohol. Fragments of ovaries were embedded in paraffin, sectioned at 3-5 µm thickness, and stained with Hematoxylin-Eosin (HE) and periodic acid Schiff (PAS). The histological description of the developmental stages and classification were performed using the existing terminology (Wallace and Selman, 1981; West, 1990; Brown-Peterson *et al.*, 2011).

Fecundity was estimated using ten mature ovaries, which were removed, weighed and preserved in Gilson solution for 24 hours for complete dissociation of

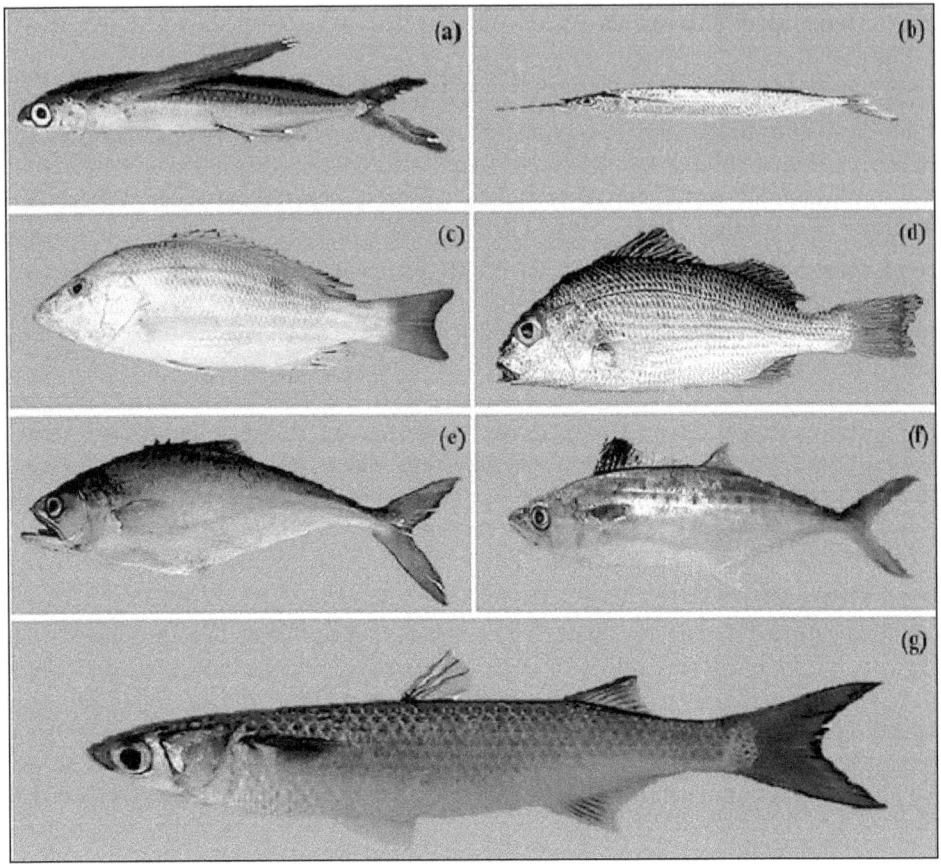

Figure 1.1: Marine Fish Species from the Coastal Waters of Rio Grande do Norte, Brazil

(a) *Hirundichythys affinis*, (b) *Hemiramphus brasiliensis*, (c) *Lutjanus synagris*, (d) *Pomadasys corvinaeformis*, (e) *Oligoplites palometa*, (f) *Scomberomorus brasiliensis* and (g) *Mugil curema*. (Illustrations provided by Dr. M. R. Oliveira, Universidade Federal do Rio Grande do Norte, UFRN/RN, Brazil).

oocytes, washed and preserved in 70 per cent ethyl alcohol. A 10 per cent sample was removed for counting the mature oocytes and the values were extrapolated to 100 per cent. The type of spawning was assessed by measuring the oocyte diameter size using a binocular microscope (x 20) and an ocular micrometer (± 1 µm). The breeding periods of the fish species were determined by the temporal relative frequency distribution of the different stages of ovarian maturation.

Results and Discussion

Body Size (Length-Weight)

The details of amplitude of total length and weight of the seven marine fish species are presented in Table 1.1. Total body length of *H. affinis* males varied from

Table 1.1: Details of the Lengths and Weights (Minimum, maximum and mean) of Seven Marine Fish Species from the Coastal Waters of Rio Grande do Norte, Brazil (M: males; F: females; SD: standard deviation)

Species	Sex	Total Length (cm)			Total Weight (g)		
		Mean ± SD	Minimum	Maximum	Mean ± SD	Minimum	Maximum
H. affinis	M	26.54± 1.40	25.25	27.71	150.24g± 31.23	126.03	199.21
	F	27.04 ± 1.22	26.30	27.72	154.75 g± 30.94	136.04	212.61
H. brasiliensis	M	23.10 ± 2.3	15.50	33.50	67.0 ± 26.9	14.04	196.1
	F	24.30 ± 2.6	19.50	33.02	73.0 ± 28.2	33.40	203.4
L. synagris	M	23.6 ± 7.10	12.20	36.04	250.9 ± 214.9	17.50	891.5
	F	28.7 ±5.08	21.50	36.50	377.2 ± 214.1	125.52	743.0
P. corvinaeformis	M	10.27± 8.20	4.90	14.70	15.95± 07.6	1.40	41.30
	F	11.80 ± 8.12	5.02	18.04	26.16± 08.4	1.42	74.02
O. palometa	M	26.10±0.66	19.80	57.50	154.3±153.3	51.60	942.0
	F	24.50±0.28	20.00	32.00	116.7±47.33	45.30	243.0
S. brasiliensis	M	31.56± 13.55	14.0	59.80	255.2± 257.6	16.00	1310
	F	33.38 ± 18.17	13.50	80.50	363.6 ± 130.9	15.00	3385
M. curema	F	24.9±4.10	15.60	34.50	160.5±82.3	35.40	382.0

25.25 to 27.71 cm (26.54 ± DP 1.40) and body mass of males varied from 126.03 to 199.21 g (150.24 ± 31.23). Total length of *H. affinis* females varied from 26.30 to 27.72 cm (27.04 ± 1.22) and their body mass varied from 13.60 to 21.26 g (15.47 ± 3.09). El-Deir (1998), obtained similar results, wherein females of *H. affinis* presented higher amplitude of body length (20.5 to 26.3 cm) than the males (18.9 to 24.3 cm).

The total body length of *H. brasiliensis* males varied from 15.5 to 33.5 cm (23.1±2.2), while body weight variation was from 14 to 196.1 g (67 ± 26.9). For the females total length varied from 19.5 to 33 cm (24.3 ± 2.6) and body weight variation was from 33.4 to 203.4 g (73±28.2). Similar results were observed for *H. brasiliensis* in Venezuela and in South of Florida (McBride and Thurman, 2003; Yelipza *et al.*, 2011; Oliveira *et al.*, 2012a; Oliveira *et al.*, 2012b). The total body length of *P. corvinaeformis* males varied from 4.9 to 14.7 cm, with a mean value of 10.27, and the body weight varied from 1.4 a 41.3 g with an average of 15.95. For the females total length varied from 5.0 to 18 cm, with a mean value of 11.8 and body weight varied from 1.4 to 74.0 g with a mean value of 26.16. However, individuals of 7.6 to 20 cm of total body length were captured in the Guaratuba bay of Paraná, Brazil (Chaves, 1998).

The total body length of *O. palometa* males varied from 19.8 to 57.5 cm (2.61±0.668) and the body weight varied from 51.6 g to 942 g (154.3±153.3). In case of the females total length varied from 20 to 32 cm (2.45±0.285), and the body weight varied from 45.3 g to 243 g (116.07±47.33). The males were longer and heavier than the females, possibly due to the capture of more males (49 males and 29 females) (Araújo *et al.*, 2012). The total body length of *S. brasiliensis* males varied from 14 cm to 59.8 cm and in females from 13.5 to 80.5 cm. The body weight of males and females of *S. brasiliensis* varied from 16 to 1,310g and 15 to 3,385 g respectively. The body length of *L. synagris* males varied from 12 to 36 cm (23.6±7.1), and the body weight varied from 17.5 to 891.5g (250.9±214.9). The body length of females of this species varied from 21.5 to 36.5 cm (28.7±5.08), and their body weight varied from 125.5 to 743 g (377.2±214.1) (Lima *et al.*, 2005; Lima *et al.*, 2007). The females of *L. synagris* were bigger and heavier than the males (Cavalcante *et al.*, 2012). Total body length of *M. curema* females varied from 15.6 to 34.5 cm (24.9±4.1) and body weight varied from 35.4 to 382g (160.5±82.3). In this study only females were captured selectively.

The females of the species *H. affinis*, *H. brasiliensis*, *L. synagris*, *P. corvinaeformis*, *S. brasiliensis* and *M. curema* were bigger and heavier than the males. This was possibly due the weight and development of ovaries in relation to testicular development (Murua *et al.*, 2003; Oliveira *et al.*, 2012a; Oliveira *et al.*, 2012b).

Length-Weight Relationship and Type of Growth

Studies on the length-weight relationship, combined with other quantitative aspects, such as, condition factor, growth, recruitment and fish mortality, provide basic information for the fishery biology and rational management of fishing stocks. The length-weight relationship functions of the flying fish, *H. affinis*, ballyhoo half beak, *H. brasiliensis*, maracaibo leatherjacket, *O. palometa*, roughneck grunt, *P. corvinaeformis*, serra Spanish mackerel, *S. brasiliensis*, the lane snapper, *L. synagris* and white mullet, *M. curema* were estimated.

It is possible to determine the type of growth of a species through the allometric coefficient (*b*), which is isometric when $b = 3$, positive allometry when $b > 3$ and negative allometry when $b < 3$. Isometric growth indicates that the body increases in all dimensions in the same proportion during growth, whereas positive allometry indicates that the body becomes more rotound as it increases in length, and negative allometry indicates a slimmer body (Jobling, 2002). Growth of males and females of *H. affinis* ($\theta = 2.208$ for males and 2.985 for females) and *O. palometa* ($\theta = 0.996$ for males and 0.913 for females) indicate negative allometric growth, where fish increases in length than in body weight, resulting in a slimmer body (Araújo *et al.*, 2011; Araújo *et al.*, 2012; Oliveira *et al.*, 2015). In the case of *H. brasiliensis* and *M. curema* ($\theta = 2.985$) there is isometric growth indicating that the body increases in both dimensions in the same proportion (Oliveira *et al.*, 2011; Oliveira *et al.*, 2012a). Growth of *L. synagris* ($\theta = 3.3647$ for males and 3.3152 for females) and *S. brasiliensis* was positively allometric indicating that the body becomes more rotound (Chellappa *et al.*, 2010; Cavalcante *et al.*, 2012).

The parameters of length-weight relationship in fish could be influenced by environmental conditions, gonadal maturity, sex, condition factor, season and variations between species (Froese, 2006).

Sex Ratio

The sex ratio of *H. affinis* was 1M:1.4F thus not differing significantly from the expected ratio of 1:1. However, there was a slight predomination of females over the males of *H. affinis*. The mature females usually migrate to the coastal waters to spawn (Araujo and Chellappa, 2002a). The sex ratio of *H. brasiliensis* was 1M:1.1F not differing from the expected sex ratio in nature (1:1). In the case of *P. corvinaeformis* the females predominated with a sex ratio of 1M:2.1F differing significantly from the expected ratio of 1:1 (Silva, 2003; Silva *et al.*, 2012). The sex ratio of *O. palometa* was 2M:1F with a significantl difference, where this population showed the occurrence of more males than females (Araújo *et al.*, 2012). A different study showed that the sex ratio of *S. brasiliensis* was 2M:1F where males predominated over females (Nobréga, 2002). A similar pattern was observed for *L. synagris* which showed a sex ratio of 4.15M:1F. The sex ratio of *M. curema* was 1M:1F not differing from the expected sex ratio. Females predominated in *H. affinis*, *H. brasiliensis* and *P. corvinaeformis*, whereas in *L. synagris*, *O. palometa* and *S. brasiliensis* the males predominated.

Usually the sex ratio is 1:1 in natural environments, but during the life cycle of fish this can vary depending on various factors that act differently on individuals of each sex. Growth and mortality are factors that can act differentially on males and females, thus determining the predominance of one sex. The sex ratio can be affected by factors related to fishing, seasons, besides the number of individuals in the feeding and spawning areas (Lasiak, 1982).

Length at First Sexual Maturity (L_{50})

The onset of sexual maturity represents a critical transition in the life history, since resource allocation is related mainly to growth before and to reproduction after sexual maturity (Potts and Wootton, 1984; Chellappa *et al.*, 1995). For rational management of fishery stocks which are subjected to exploitation, it is important to know the size at first gonadal maturation (L_{50}), since it provides information for determining the minimum size at capture and mesh dimensions of the fishing gear.

The sizes when 50 per cent of males and females of *H. affinis* were in the process of sexual maturation were 23.8 cm and 23.0 cm of total body length respectively. However, another study on the same species reported that the total length at first sexual maturity was at 27.3 cm for males and 27.1 cm for females (Araujo and Chellappa, 2002b). The males of *H. affinis* mature before the females, shown by the significant difference in size of gonadal maturation of both sexes ($t = -5.081$; df = 210; $p < 0.05$).

Total length at first sexual maturity of *H. brasiliensis* males was at 20.8 cm and of the females was at 21.5 cm. The males of *H. brasiliensis* attained first gonadal maturity at smaller body lengths than females (t =3.62, df = 408, p < 0.05) (Oliveira *et al.*, 2015). The females of *H. brasiliensis* in the coastal waters of South Florida attained maturity at 19.8 cm, however, males were not included in this study (McBride and Thurman, 2003).

Total lengths at first sexual maturity for *L. synagris* males and females were 23.5 cm and 24.5 respectively (Cavalcante *et al.*, 2012). In case of *P. corvinaeformis* the total length at first sexual maturity was 10.3 cm for males and 10.4 cm for females (Silva *et al.*, 2012). Total length at first sexual maturity of *S. brasiliensis* males was at 34.5 cm and for females it was at 28 cm. The gradual decrease in total length at first sexual maturity of *S. brasiliensis*, possibly indicates overfishing of this commercially important species (Lima *et al.*, 2007). The traditional fishing communities depend on small scale artisanal fishery, which reflects their way of making a living and sustains their lifestyle. Though it is important to preserve this traditional fishery, it is also vital to programme the sustainability and conservation of coastal fisheries resources. The predatory fishing technique of beach seine nets, in which small mesh sizes are used in order to catch marine shrimps, accounts for a large by-catch of small sized immature Serra Spanish mackerel. Total length at first sexual maturity of both sexes of *M. curema* was 24.3 cm (Oliveira *et al.*, 2014).

All three species *H. affinis*, *H. brasiliensis* and *S. brasiliensis* showed that females attained maturity earlier than the males. Conservation of fish stocks in their natural habitat are usually endangered by abusive fishing of immature fishes which have not yet completed their reproductive cycle, as recruitment via reproduction is the means by which the resource is renewed. The artisanal fishery beach-seines in Northeastern Brazil are operated in the shallow coastal waters, which captures the immature individuals of the pelagic stock. Measures should be taken to regulate this fishery in order to conserve this valuable fishery resource by increasing the size of capture.

Gonad Development

Macroscopic Characteristics of Gonads

The use of macroscopic scales of maturity of the gonads contributes to the biological knowledge in describing the reproductive cycle of fish and helps in understanding the reproductive period of the species. However, for better identification of gonad developmental stages, histological analysis is considered essential.

H. affinis had paired elongated gonads, the females with lobed ovaries and the males with flattened testes. The testicular walls were fragile when compared to the ovarian walls, and did not show much modification between the different stages of development, unlike the ovaries. The volume, coloration, thickness and blood vessels of ovaries varied according to the stage of maturation, presenting shades of light pink to dark yellow, due to the color of the mature oocytes full of yolk granules. The testes were whitish from the beginning of maturation to the mature stage. The macroscopic characteristics of the testes of *H. affinis* presented three stages of maturation, such as, immature, maturing and mature, while females showed four stages of development, immature, maturing, mature and spent (Oliveira *et al.*, 2015).

The macroscopic characteristics of *H. brasiliensis:* The ovaries and testes were paired bi-lobed structures, symmetrical, elongated and joint in the posterior part to form a short duct leading to the urogenital pore. They were located in the posterior-dorsal part of the coelomic cavity, ventral to the kidneys and swim bladder. The immature testes were small and translucent. Maturing testes were more developed and were whitish in color. The mature testes were white and spent testes were flaccid and brown in color with hemorrhagic appearance. During maturation, the ovaries were pinkish to light orange in color and developed progressively by increasing in size and vascularization. The mature ovaries were turgid and occupied 2/3 of the coelomic cavity. The mature ovaries were turgid with numerous big oocytes visible to the naked eye, and the partially spent ovaries were flaccid. The macroscopic characteristics of the gonads indicated four maturation stages: immature, maturing, mature and spent (Oliveira, 2014; Oliveira *et al.*, 2014).

The macroscopic characteristics of the gonads of *P. corvinaeformis* indicated four developmental stages for testes and ovaries: immature, maturing, mature and spent.

Immature testes and ovaries were filiform, small in size with translucent coloration. Maturing and mature ovaries appeared reddish in colour, increasing in size with visible oocytes. The testes showed varying sizes in accordance with the degree of development, become thicker and whitish in colour. The spent gonads were reduced size with a flaccid appearance (Silva *et al.*, 2012).

The macroscopic characteristics of the testes and ovaries of *O. palometa* (Figure 1.2), *S. brasiliensis, L. synagris* and *M. curema* indicated four stages of development, such as, immature, maturing, mature and spent (Chellappa *et al.*, 2010; Araújo *et al.*, 2012; Cavalcante *et al.*, 2012; Oliveira, 2010). The ovaries were bilobed, elongated, and joined posteriorly to form a short gonoduct leading to the urogenital pore. The immature testes and ovaries were small and translucid structures. During maturation the gonads occupy about one third of the coelomic cavity, while mature ovaries occupy almost two thirds of the coelomic cavity. Spent gonads have a hemorrhagic appearance and are reduced in size.

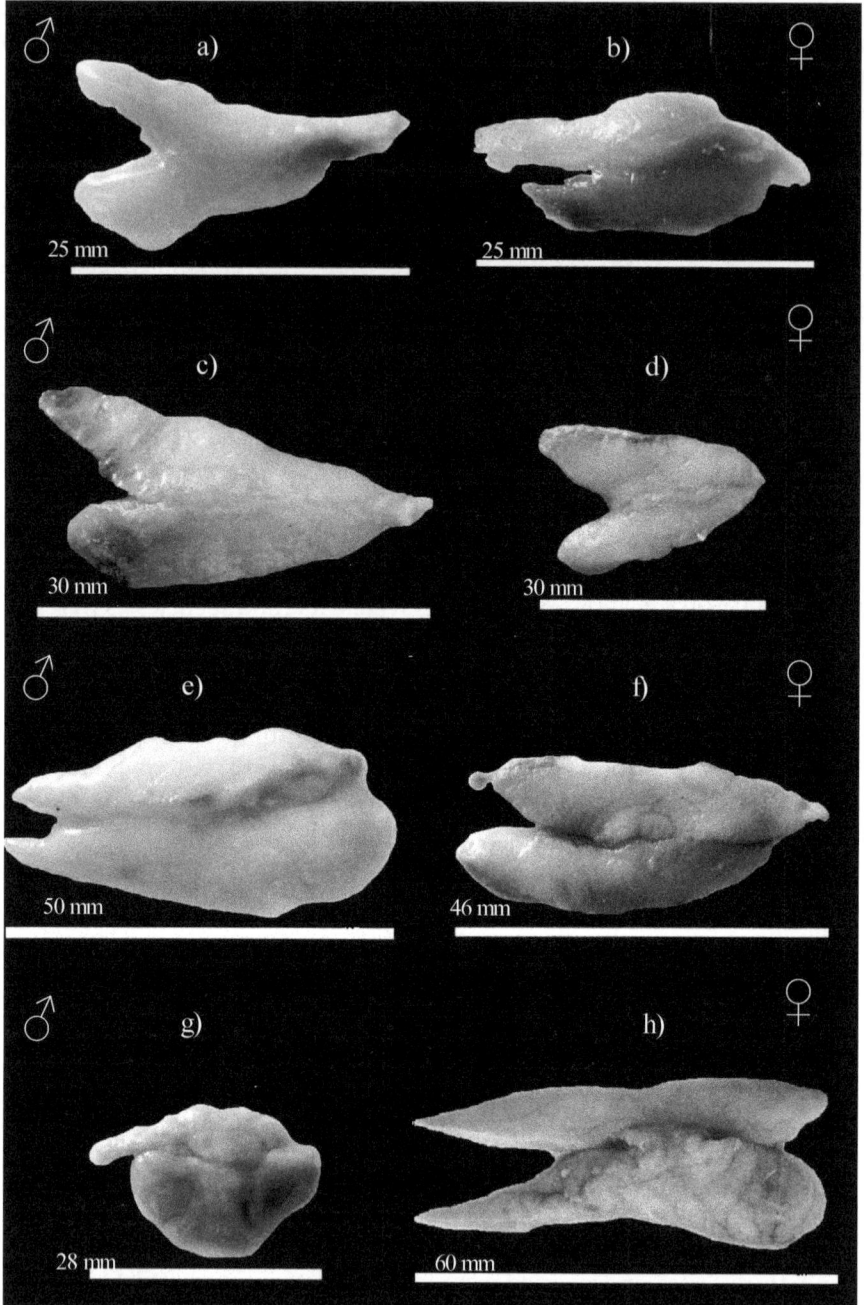

Figure 1.2: Macroscopic Stages of Development of the Gonads of *Oligoplites palometa*.

a) immature testes, b) immature ovary, c) maturing testes, d) maturing ovary, e) mature testes, f) mature ovary, g) spent testes and h) spent ovary. (Illustrations provided by G. S. Araújo M.Sc., Universidade Federal do Rio Grande do Norte, UFRN/RN, Brazil).

Microscopic Characteristics of Ovaries

Histological examinations of ovarian sections of the study species generally showed that the oocyte development was consistent along the whole length of the ovary depending on the degree of maturation. Ovaries revealed five stages of oocyte development: immature, early stage of maturing, late stage of maturing, mature and spent (Figures 1.3–1.7). Immature ovaries showed the chromatin nucleolar stage, where clusters of very small oocytes were found lying just beneath the ovigerous lamella and young germ cells compactly filled the ovaries. The ovaries in early stage of maturing showed the perinucleolar stage, oocytes with nucleoli at periphery of nucleus with thick cytoplasm. During cortical alveoli stage the oil vesicles appear and ovaries were with early yolk globule and previtellogenic stage oocytes. The ovaries in late stage of maturing revealed the yolk stage, when oocytes show the presence of yolk granules near the periphery and oil vesicles within the inner region of the cytoplasm, and cytoplasmic vesicles had a uniform distribution. Mature ovaries showed nuclear migration and hydration stages, maturation into this stage was marked by the migration of the nucleus to the periphery of the oocyte, fusion of yolk granules into yolk plates and coalescence of oil droplets. Nucleus breaks down when it reaches the periphery and hydration occurs.

The details of histological aspects of ovarian sections of the study species are presented in Figures 1.3 to 1.7.

Fecundity and Type of Spawning

The absolute fecundity of *H. affinis* varied from 7,398 to 10,021 oocytes, with an average of 9,092 (SD ±1,153.2) vitellogenic oocytes. Their diameter size varied from 100 µm to 2500 µm. The reserve stock oocytes had diameter size less than 1000 µm, and the developing oocytes were bigger than 1000 µm. This indicated that this is a total spawner, which eliminates the mature oocytes at the same time (Oliveira *et al.*, 2015).

The batch fecundity of *H. brasiliensis* varied from 862 to 1,354 with an average of 1,153 (±258.22) vitellogenic oocytes for 50 g body weight of female. The microscopic characteristics of gonad development of *H. brasiliensis* showed multiple spawning (Oliveira and Chellappa, 2014). Fecundity is a specific reproductive tactic and is adapted to the life cycle conditions of the species, varying with growth, population density, body size, food availability and mortality rate (Murua and Saborido-Rey, 2003). The absolute fecundity of *P. corvinaeformis* varied from 15,056 to 83,316 oocytes and their diameter size varied from 110µm to 390µm. The reserve stock oocytes had diameter size less than 140 µm, and the developing oocytes were bigger than that. This indicated that this is a total spawner which eliminates the mature oocytes at the same time (Silva, 2003; Silva *et al.*, 2012). The mean absolute fecundity of *S. brasiliensis* was 871,523 oocytes and is a total spawner. The reserve stock oocytes had diameter size less than 120 µm, and the developing oocytes had diameters which varied from 650 to 750 µm (Lima, 2008; Lima *et al.*, 2007). The mean absolute fecundity of *M. curema* was 245,828 oocytes and is a total spawner (Oliveira *et al.*, 2014).

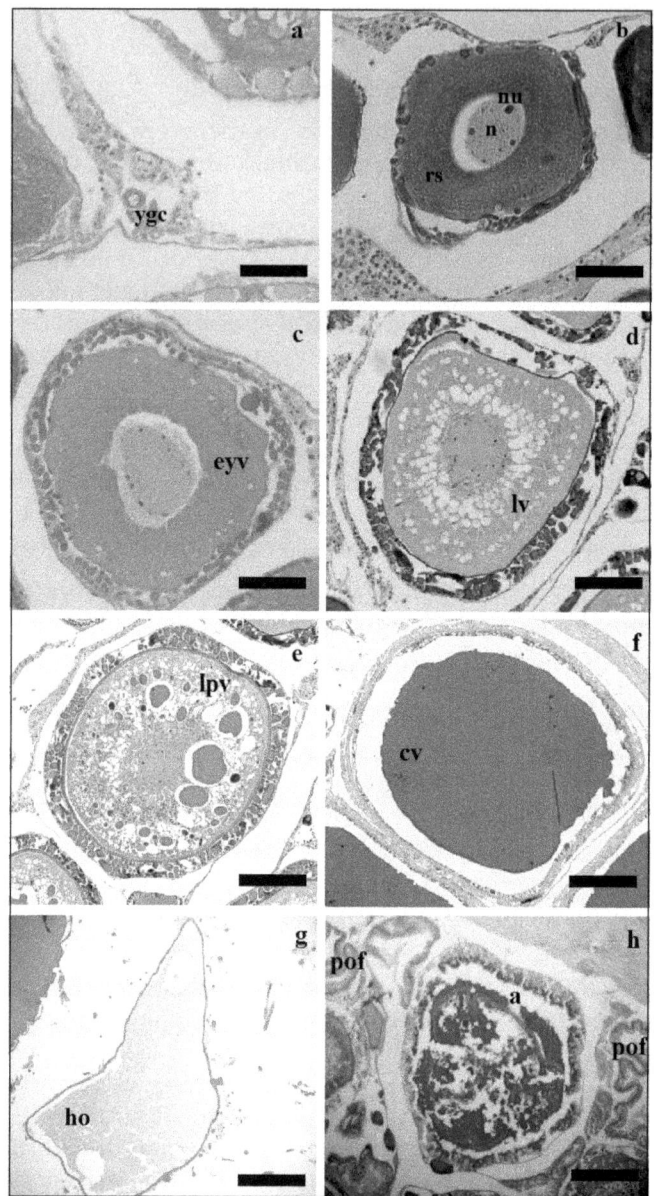

Figure 1.3: Histological Aspects of Oocyte Developmental Stages of *Hirundichthys affinis*.

(a) Young germ cell (ygc), (b) Perinucleolus stage or reserve stock (rs); (c) Early yolk vesicle (eyv); (d) Lipid vitellogenesis (lv); (e) Lipid and protein vitellogenesis (lpv) with yolk granules (yg) and lipid droplets (ld); (f) Oocytes with complete vitologenesis; (g) Oocytes in hydration (ho); (h) Oocytes in atresia (a). n,nucleus; nc, nucleolus; pof, post-ovulatory follicle. (a to d: scale bar = 200μm; e to h: scale bar = 50μm).

(Illustrations provided by Dr. M. R. Oliveira, Universidade Federal do Rio Grande do Norte, UFRN/RN, Brazil).

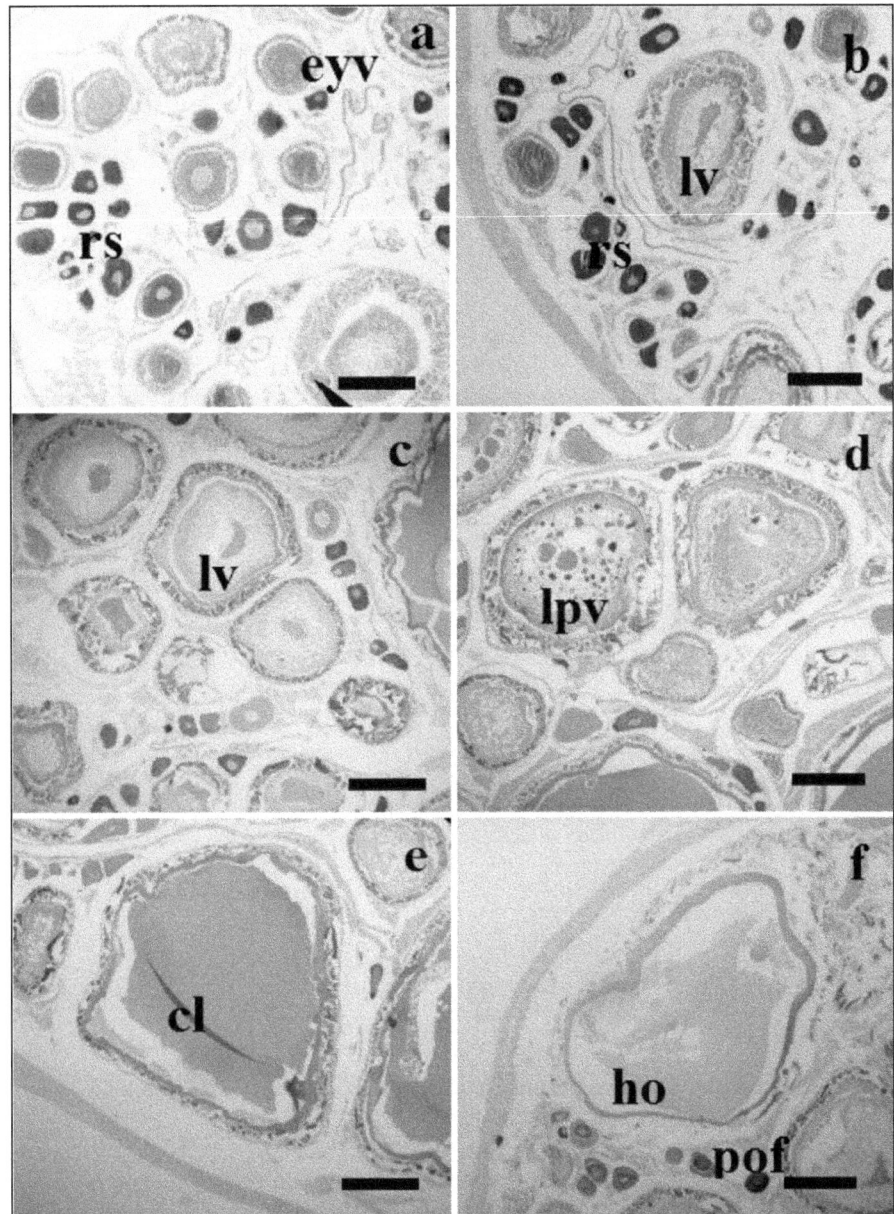

Figure 1.4: Histological Aspects of Oocyte Development Stages of *Hemiramphus brasiliensis*.

(a) Perinucleolus stage or reserve stock (rs) and early yolk vesicle (eyv); (b) Lipid vitellogenesis (lv) and rs; (c) lv (d) Lipid and protein vitellogenesis (lpv); (e) Oocytes with complete vitellogenesis (cl); (f) Oocytes in hydration (ho) and post-ovulatory follicle (pof) (scale bar = 50μm).

(Illustrations provided by Dr. M. R. Oliveira, Universidade Federal do Rio Grande do Norte, UFRN/RN, Brazil).

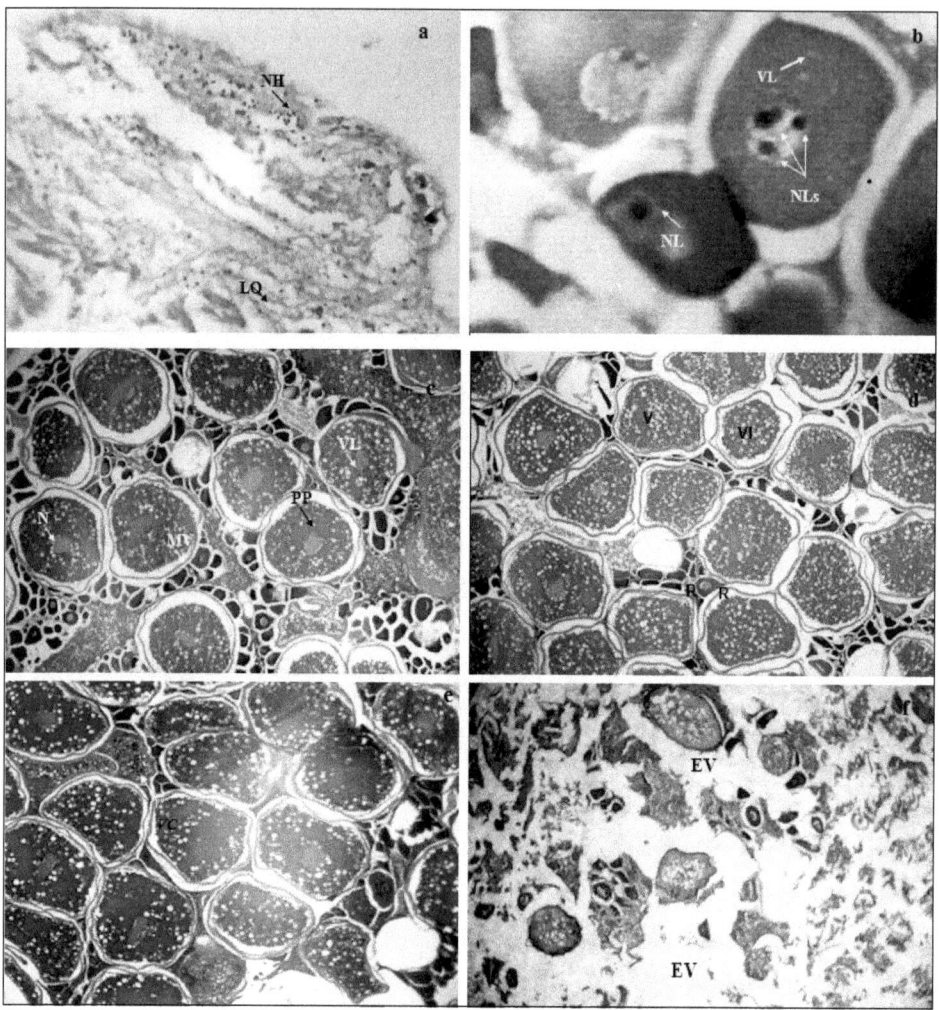

Figure 1.5: Histological Aspects of Oocyte Development Stages of *Pomadasys corvinaeformis*.

(a) Immature ovary with young germ cells (NH), ovigerous lamella (LO); (b) Initial ovarian maturation, showing reserve oocyte stock, nucleus with nucleolus (NL) and beginning of lipid vitellogenesis (VL); (c) Ovary showing oocytes with lipid vitellogenesis (VL) and protein (PP), some reserve oocyte stock (R) and central nucleus (N) *100 X*; (d) Mature ovary showing oocytes, with presence of reserve oocyte stock (R) *100 X;* (e) Ovary in final stage of maturation showing oocytes with complete vitellogenesis (VC), (f) Spent ovary, with empty space (EV)*100 X.*

(Illustrations provided by A.M. Silva, M.Sc., Universidade Federal do Rio Grande do Norte, UFRN/RN, Brazil).

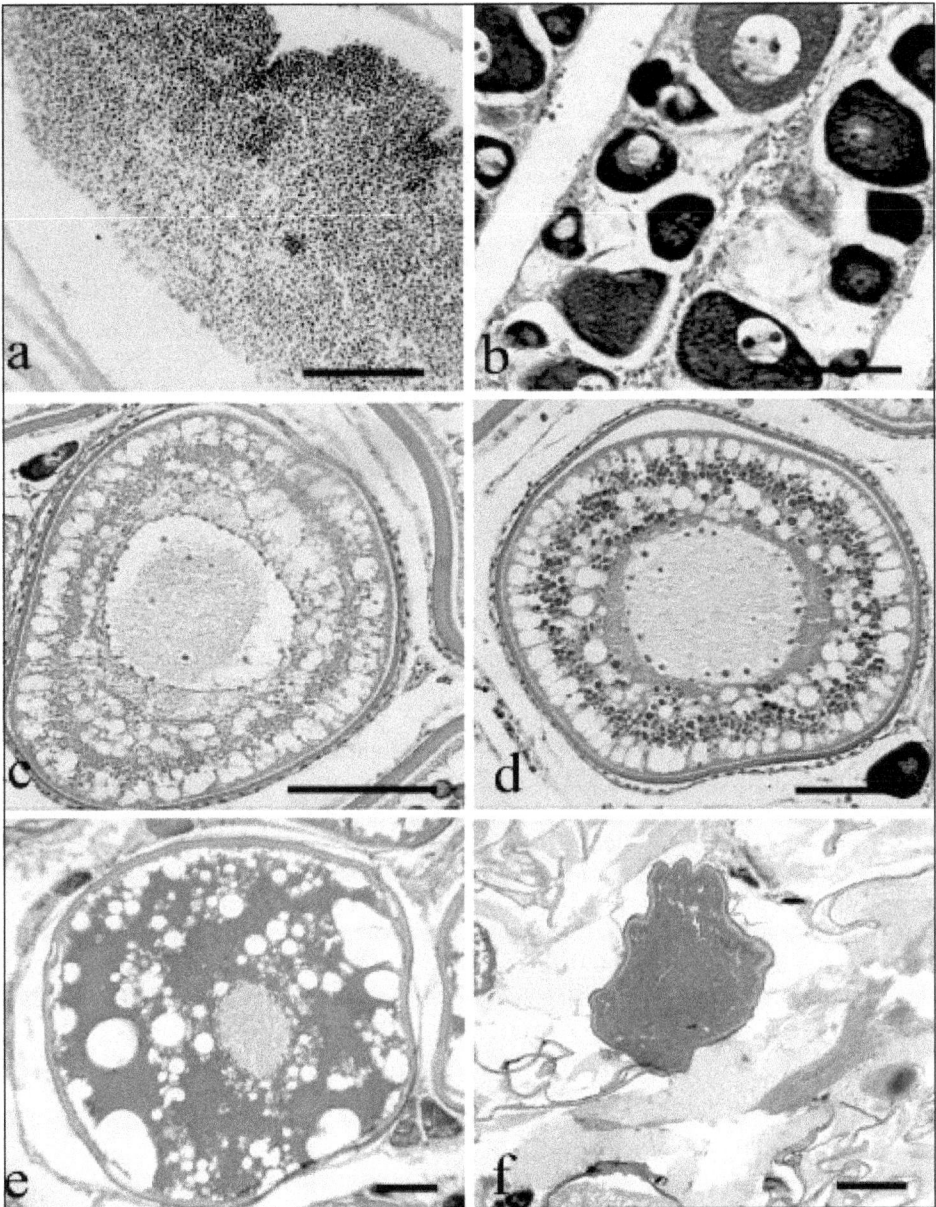

Figure 1.6: Histological Aspects of Oocyte Stages in Ovarian Development of *Scomberomorus brasiliensis*

(a) Nest of oogonia; (b) Chromatin nucleolus stage and early perinucleolar stage oocytes; (c) Oocyte in yolk vesicle stage; (d) Oocyte with yolk granules and oil vesicles stage; (e) Mature oocyte; (f) Oocyte in the process of atresia in a spent ovary (Scale bar =100 µm).

(Illustrations provided by Dr. J.T.A.X. Lima, Universidade Federal do Rio Grande do Norte, UFRN/RN, Brazil).

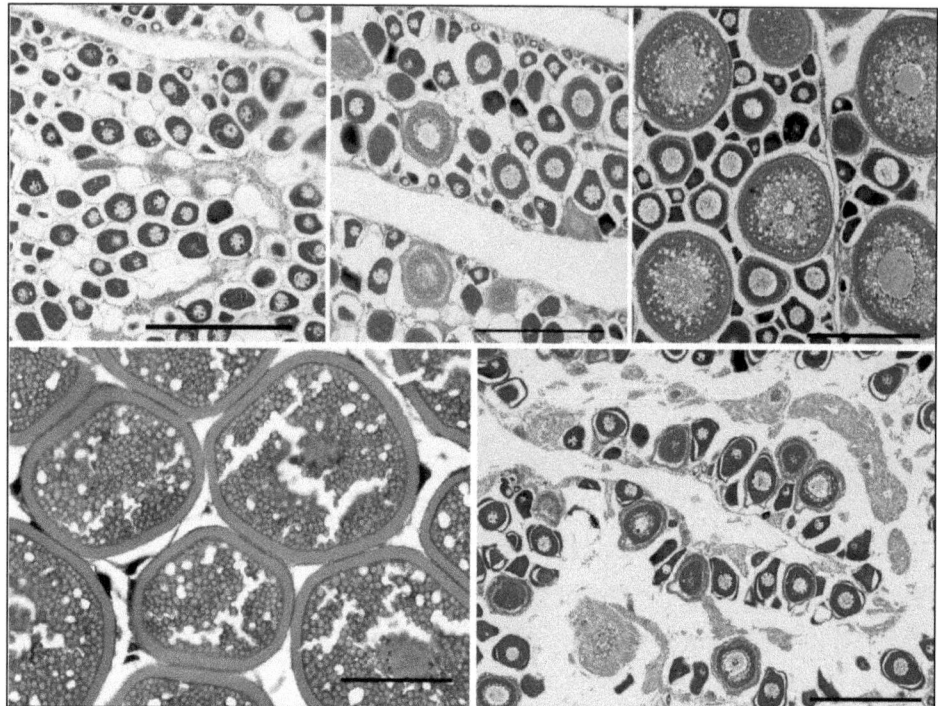

Figure 1.7: Histological Aspects of Oocytes Developmental Stages of *Mugil curema*.
(a) Immature ovary containing nest of germ cells (ngc), early perinucleolus oocytes (epo) and late perinucleolus oocytes (lpo); (b) Early maturing ovary containing lipid vitelogenic stage oocytes (lvo); (c) Maturing ovary containing lipid and protein vitelogenic stage oocytes (lpvo); (d) Mature ovary containing post-vitellogenic oocytes (pvo) with yolk granules (yg) and lipid droplets (ld) and (e) Spent ovary containing post-ovulatory follicle (pof). rs, reserve stock oocytes; es, empty spaces; n, nucleus; nc, nucleolus (scale bar = 200μm).

(Illustrations provided by Dr. M. R. Oliveira, Universidade Federal do Rio Grande do Norte, UFRN/RN, Brazil).

Gonadosomatic Index (GSI) and Reproductive Period

GSI demonstrates the functional gonadal status in relation to fish body mass, indicating the spawning period. The mean values of GSI of *H. affinis* varied from 1.25 to 17.1 for the females and from 0.1 to 8.01 for the males. There was a predominance of maturing individuals during the period of October to February and mature fish occurred during March to July. Variations in GSI and the monthly frequency of maturation stages demonstrate that *H. affinis* reproduces during the period of March to July, coinciding with the rainy season of the region (Araújo and Chellappa, 2002a; Oliveira *et al.*, 2015).

The mean monthly values of GSI of *H. brasiliensis* varied from 0.94 to 3.58 for the females and 0.18 to 0.39 for the males. 1.47 to 4.10. Frequency of monthly gonadal

Table 1.2: Reproductive Tactics and Strategies of the Marine Fish Species *Hirundichythys affinis*; *Hemiramphus brasiliensis*; *Pomadasys corvinaeformis*; *Oligoplites palometa*; *Scomberomorus brasiliensis*; *Lutjanus synagris* and *Mugil curema*

Reproductive Tatics		Estratégia reprodutiva					
		Oportunistic Strategist				Equilibrium Strategist	
		H. affinis	*H. brasiliensis*	*P. corvinaeformis*	*O. palometa*	*S. brasiliensis*	*L. synagris*	*M. curema*
Body size	Small, Females are bigger	Medium, Females are bigger	Small, Females are bigger	Medium, Males are bigger	Large, Females are bigger	Medium, Females are bigger	Medium, Females are bigger
Sex ratio	1M : 1.4F	1M : 1.1F	1M : 2.11F	2M : 1F	2M : 1F	4.15M : 1F	1M : 1.6F
Type of growth	Negative allometry	Isometric growth	Positive allometry	Negative allometry	Positive allometry	Positive allometry	Isometric growth
First sexual maturity (L_{50})	27.1cm, Females mature earlier	21.5cm, Females mature earlier	10.4cm, Males mature earlier	29.4cm	28cm, Females mature earlier	Grouped sex maturity at 25.7cm	Grouped sex maturity at 24.3 cm
Fecundity	7398 to 10021	862 to 1354	15,056 to 83,316	680,648	871,523	920,246	245,828
Type of spawning	Total	Total	Total	Total	Total	Total	Total
Spawning period	During rainy season	Reproductive peaks during rainy season	Prolonged with peaks during rainy season	During rainy season	During rainy season	During rainy season	Prolonged with peaks during rainy season

maturation stages of females indicate that mature individuals occur throughout the year. Variations of GSI and the frequency of monthly gonadal maturation stages show that *H. brasiliensis* has an active reproductive period during the months of January to June and again in October. The breeding period of this species is independent of the rainy period (Oliveira *et al.*, 2015).

The mean monthly values of GSI of *P. corvinaeformis* varied from 0.012 to 1.15 for males and 0.012 to 5.49 for females. Two peaks of GSI were observed during the months of November and March. The reproductive period extends from October to June (Silva *et al.*, 2012).

The mean monthly values of GSI of *O. palometa* varied from 0.07 to 0.15 for males and from 0.07 to 1.64 for females. During the dry period the mean monthly values of GSI of males was 0.13 ($\pm$ 0.03), and during the rainy season the mean monthly values of GSI was 0.11 ($\pm$ 0.03), without any significant difference of GSI between the seasons (Araujo *et al.*, 2012).

The mean monthly values of GSI of *S. brasiliensis* varied from 0.02 to 7.14 for females. Reproductive activity was during March to June, coinciding with the rainy season (Chellappa *et al.*, 2010; 2011).

Reproductive Strategies

The flying fish, *H. affinis*, ballyhoo half beak, *H. brasiliensis*, roughneck grunt, *P. corvinaeformis*, maracaibo leatherjacket, *O. palometa*, are considered as opportunistic stragists, with small body size, early first sexual maturity and short-life span (Table 1.2). Their spawning seasons are influenced by the rainfall of the region.

On the other hand, the three fish species Serra Spanish mackerel, *S. brasiliensis*, the lane snapper, *L. synagris* and white mullet, *M. curema* are considered as equilibrium stragists (strategy involving medium to large body and less fecund organisms). They have smaller but stable population and live in stable environments. (Table 1.2).

The semiarid region of Brazil is characterized with short spells of rain interspersed with lengthy dry season. Hence the reproductive seasonality of fish is regulated by environmental cues and rainfall seems to be the main driver which modulates the spawning period.

Conclusion

Information regarding the body size, sex ratio, length at first sexual maturity, type of growth, aspects of gonad development, fecundity, type of spawning and the spawning period are important for administration of fishery resources. The results indicate that the flying fish, *H. affinis*, ballyhoo half beak, *H. brasiliensis*, roughneck grunt, *P. corvinaeformis* and maracaibo leatherjacket, *O. palometa* are opportunistic strategists. On the other hand serra Spanish mackerel, *S. brasiliensis*, the lane snapper, *L. synagris* and white mullet, *M. curema* are considered as equilibrium strategists.

Acknowledgements

The authors wish to thank the National Council for Scientific and Technological Development (CNPq) and the Post Graduate Federal Agency of the Ministry of Education, Brazil (CAPES/MEC) for the financial support awarded during the study period. The authors thank the field team and the post-graduate scholars who participated in the relevant studies.

References

Araújo, A.S. and Chellappa, S., 2002a. Estratégia reprodutiva do peixe voador, *Hirundichthys affinis* Günther (Osteichthyes, Exocoetidae). *Revista Brasileira de Zoologia*, 19(3): 691-703.

Araújo, A.S. and Chellappa, S., 2002b. Estudo histológico das gônadas do peixe voador, *Hirundichthys affinis* Günther, 1866 (Osteichthyes: Exocoetidae) no Rio Grande do Norte, Brasil. Arquivos de Ciências do Mar, LABOMAR-Fortaleza, UFC-CE, 35: 131-134.

Araújo, A.S., Oliveira, M.R., Campos, C.E.C., Yamamoto, M.E. and Chellappa, S., 2011. Características morfométricas-merísticas, peso-comprimento e maturação gonadal do peixe voador, *Hirundichythys affinis* (Günther, 1866). *Revista Biota Amazônia*, 1(2): 33-40.

Araújo, G.S., Araújo, A.S. and Chellappa, S., 2012. Tipo de crescimento e aspectos reprodutivos do peixe marinho *Oligoplites palometa* (Osteichthyes: Carangidae), na costa do Rio Grande do Norte, Brasil. *Biota Amazônia*, 2(2): 25-30.

Brown-Peterson, N.J., Wyanski, D.M., Saborido-Rey, F., Macewicz, B.J. and Lowerre-Barbieri, S.K., 2011. A standardized terminology for describing reproductive development in fishes. *Marine and Coastal Fisheries*, 3: 52-70.

Cavalcante, L.F.M., Oliveira, M.R. and Chellappa, S., 2012. Aspectos reprodutivos do ariacó, *Lutjanus synagris* nas águas costeiras do Rio Grande do Norte. *Biota Amazônia*, 2(1): 45-50.

Chaves, P.T.C., 1998. Estrutura populacional de *Pomadasys corvinaeformis* (Steindachner) (Teleostei, Haemulidae) na Baía de Guaratuba, Paraná, Brasil. *Revista Brasileira de Zoologia*, 15(1): 203-209.

Chellappa, S., Huntingford, F.A., Strang, R.H.C. and Thomson, R.Y., 1995. Condition factor and hepatosomatic index as estimates of energy status in male three-spined stickleback. *Journal of Fish Biology*, 47(5): 775-787.

Chellappa, S., Lima, J.T.A.X., Araújo, A. and Chellappa, N.T., 2010. Ovarian development and spawning of Serra Spanish mackerel in coastal waters of Northeastern Brazil. *Brazilian Journal of Biology*, 70(2): 631-637.

Chellappa, S., Lima, J.T.A.X., Araújo, A. and Chellappa, N.T., 2011. Reproductive biology of *Scomberomorus brasiliensis* (Perciformes: Scombridae). *In*: (Org.). *Advances in Fish and Wildlife Ecology and Biology*, (Ed.) B L. Kaul. Daya Publishing House, Delhi, India, 5: 3-19.

Froese, R., 2006. Cube law, condition factor and weight-length relationships: history, metaanalysis and recommendations. *Journal of Applied Ichthyology*, 22: 241–253.

Jobling, M., 2002. Environmental factors and rates of development and growth. In: *Handbook of Fish Biology and Fisheries Vol. 1, Fish Biology*, (Eds.) P.J. Hart and J.D. Reynolds. Oxford: Blackwell Publishing Ltd. pp. 97-122.

King, Jr. and McFarlane, G.A., 2003. Marine fish life history strategies: applications to fishery management. *Fisheries Management and Ecology*, 10: 249–264.

Lasiak, A., 1982. Aspects of the reproductive biology of the Southern mullet *Liza richardsoni* from Algoa Bay, South Africa. *South African Zoology*, 18: 89-95.

Lima, J.T.A.X., 2008. Dinâmica reprodutiva e parasitária de quatro espécies de peixes das águas costeiras do Sudoeste do Oceano Atlântico, Brasil. *Doctoral Thesis*, Universidade Federal do Rio Grande do Norte, 119p.

Lima, J.T.A.X., Chellappa, S. and Thatcher, V.E., 2005. *Livoneca redmanni* leach (Isopoda, Cymothoidae) e *Rocinela signata* Schioedte and Meinert (Isopoda, Aegidae), ectoparasitos de *Scomberomus brasiliensis* collette, Russo and Zavala-Camin (Ostheichtyes, Scombridae) no Rio Grande do Norte, Brasil. *Revista Brasileira de Zoologia*, 22(4): 1104-1108.

Lima, J.T.A.X., Fonteles-Filho, A.A. and Chellappa, S., 2007. Biologia reprodutiva da Serra, *Scomberomorus brasiliensis* (Osteichthyes: Scombridae), em águas costeiras do Rio Grande do Norte. *Arquivos de Ciências do Mar*, 40(1): 24-30.

Matthews, W.J., 1998. *Patterns in Freshwater Fish Ecology*. Thomson Science, Chapman and Hall, London.

McBride, R.S. and Thurman, P.E., 2003. Reproductive biology of *Hemiramphus brasiliensis* and *H. balao* (Hemiramphidae): Maturation, Spawning Frequency, and Fecundity. *Biological Bulletin*, 204: 57-67.

Murua, H. and Saborido-Rey, F., 2003. Female reproductive strategies of marine fish species of the North Atlantic. *Journal of Northwest Atlantic Fishery Science*, 33: 23-31.

Murua, H., Kraus, G., Saborido-Rey, F., Witthames, P.R, Thorsen A. and Junquera, S., 2003. Procedures to estimate fecundity of marine fish species in relation to their reproductive strategy. *Journal of Northwest Atlantic Fishery Science*. 33: 33-54.

Nobréga, M.F., 2002. Idade, crescimento e avaliação de estoque da serra *Scombereromus brasiliensis* (Teleostei: Scombridae), na plataforma continental do Nordeste do Brasil. *Masters Dissertation*, Universidade Federal de Pernambuco.

Oliveira, M.R., 2010. Biologia reprodutiva da tainha, *Mugil curema* Valenciennes, 1836 (Osteichthyes: Mugilidae) nas águas costeiras do Rio Grande do Norte. Masters Dissertation, Universidade Federal do Rio Grande do Norte, 74p.

Oliveira, M.R. and Chellappa, S., 2014. Temporal dynamics of reproduction in *Hemiramphus brasiliensis* (Osteichthyes: Hemiramphidae). *The Scientific World Journal (Marine Biology)*, v. 2014, Article ID 837151, pp. 1-8.

Oliveira, M.R., Costa, E.F.S., Chellappa, S., 2011. Ovarian development and reproductive period of white mullet, *Mugil curema* in the coastal waters of Northeastern Brazil. *Animal Biology Journal*, 2(4): 22-237.

Oliveira, I.M.B, Oliveira, M.R., Yamamoto, M.E. and Chellappa, S., 2012.a Biologia reprodutiva de agulha-preta, *Hemiramphus brasiliensis* (Linnaeus, 1758) (Osteichthyes: Hemiramphidae) das águas costeiras do Rio Grande do Norte, Brasil. *Biota Amazônia*, 2(2): 44-53.

Oliveira, M.R., Costa, E.F.S, Araújo, A.S., Pessoa, E.K.R., Carvalho, M.M., Cavalcante, L.F.M. and Chellappa, S., 2012b. Sex ratio and length-weight relationship for five marine fish species from Brazil. *Journal of Marine Biology and Oceanography*, 1(2): 1-3.

Oliveira, M.R., 2014. Estratégia reprodutiva do peixe-voador, *Hirundichthys affinis* e do peixe agulha preta, *Hemiramphus brasiliensis* no litoral de Caiçara do Norte, Rio Grande do Norte, Brasil. *Doctoral Thesis*, Universidade Federal do Rio Grande do Norte, 169p.

Oliveira, M.R., Costa, E.F.S. and Chellappa, S., 2014. Ovarian development and reproductive period of white mullet, *Mugil curema* in the coastal waters of northeastern Brazil. *In*: (Org.). *Biology of Semiarid Tropical Fish*, (Eds.) José Rosa Gomes and Sathyabama Chellappa. Nova Scientific Publishers, New York, USA, pp. 21-34.

Oliveira, M.R., Carvalho, M.M., Silva, N.B., Yamamoto, M.E, and Chellappa, S. 2015. Reproductive aspects of the flyingfish, *Hirundichthys affinis* from the Northeastern coastal waters of Brazil. *Brazilian Journal of Biology*. 75(1): 198-207.

Potts, G.W. and Wootton, R.J., 1984. *Fish Reproduction*. Academic Press, London.

Silva, A.M., 2003. Aspectos reprodutivos do coró, *Pomadasys corvinaeformis* (Steindachner, 1868) (Osteichthyes: Haemulidae) das águas costeiras de Ponta Negra, Rio Grande do Norte. Masters Dissertation, Universidade Federal do Rio Grande do Norte, 89 p.

Silva, A.M., Oliveira, M.R. and Chellappa, S., 2012. Biologia reprodutiva do coró, *Pomadasys corvinaeformis* Steindachner (Osteichthyes: Haemulidae) das águas costeiras do Rio Grande do Norte, Brasil. *Biota Amazônia*, 2(2): 15-24.

Wallace, R.A. and Selman, K., 1981. Cellular and dynamic aspect of oocyte growth in teleosts. *Scientific Zoology*, 21: 325-343.

West, G., 1990. Methods of assessing ovarian development in fishes: a Review. *Australian Journal of Marine and Freshwater Research*, 41: 199-222.

Winemiller, K.O. and Rose, K.A., 1992. Patterns of life-history diversification in North American fishes: implications for population regulation. *Canadian Journal of Fisheries and Aquatic Sciences*, 49(10): 2196-2218.

Yelipza, L.R., Acosta, V., Parra, B. and Lista, M., 2011. Aspectos biométricos de *Hemiramphus brasiliensis* (Peces: Hemirhamphidae), Isla de Cubagua, Venezuela. *Zootecnia Tropical*, 29(4): 385-398.

Chapter 2

Biodiversity Status and Measures for Conservation of Threatened Indian Fishes: An Update

A.K. Pandey, Rehana Abidi, Madhu Tripathi and P. Das

ABSTRACT

Of the five major key issues such as water and sanitation, energy, agricultural productivity, biodiversity and human health care which were the focus of attention during the United Nations World Summit on Sustainable Development (WSSD) held during August 26-September 04, 2002 in Johannesburg (South Africa), protection of the dwindling fish stocks throughout the world constituted an important item of discussion during the special session of biodiversity conservation. Out of 29,400 taxa reported throughout the world, 2,936 species of finfishes belonging to 44 Order, 252 Family and 1,069 Genus have been recorded from different ecosystems of India. The approximate ecosystem-wise distribution of fish germplasm resources of the country are- freshwater (936; 31.78 per cent), brackishwater (113; 3.85 per cent) and marine (1,887; 64.37 per cent). Out of these, about 258 species are commercially important which include cultured, cultivable and wild taxa, 199 endemic and 275 game fishes. There is record of the introduction of 462 exotic species in Indian waters, most of which are of ornamental value. Due to anthropogenic stresses like habitat destruction, over-exploitation, indiscriminate killing of juveniles and broodfishes, excessive water abstraction, pollution, uncontrolled introductions of exotics and spread of dreaded diseases, a number of fishes are exhibiting declining trends from the conventional fishing grounds and some have also become threatened too. It appears that more than 23-29 per cent fish taxa of the world are threatened needing immediate protection. Biodiversity and community structures are being recognized as important determinants of ecosystem functioning.

As sustainable fisheries development envisages an eco-friendly, equitable mode of development that can sustain livelihoods over generations, an attempt has been made to assess the current status of fish biodiversity of India, delineate the threatened species to formulate appropriate strategies for their conservation and rehabilitation.

Keywords: *Fish germplasm resources, Threatened species, Conservation, Indian waters.*

Introduction

The Convention on Biological Diversity (CBD) which was adopted during the United Nations Conference on Environment and Development (Earth Summit) held at Rio de Janeiro (Brazil) in June 1992 reaffirms the sovereign rights of the member nations over their entire genetic resources and also envisages conservation, sustainable use and equitable sharing of the benefits arising from the biological resources (Agenda 21) (Winter and Hughes, 1997; Narain, 2000; Brahmi *et al.*, 2004). The United Nations Post-Earth Summit Assessment Report (2002) revealed that 2.4 per cent of the world forests (90 million ha, equal to the size of Venezuella) were destroyed and about 40 per cent of world population faced water shortage during this period. The need to feed rising global population of about 5 billion with exacerbated increase in food consumption from 2,100 calories to 2,700 calories/day in developing countries and from 3,000 calories to 3,400 calories/day in developed countries will put extra pressure on agricultural production which consumes about 70 per cent of the global freshwater supplies. Signs of climate change linked to global warming causing most frequent and intense droughts in parts of Asia and Africa and rising sea levels will be having impact on fish germplasm resources (Anon, 2001; Perry *et al.*, 2005; Brander, 2007. 2009; Sinha, 2009). Biodiversity and community structures are being recognized as important determinants of ecosystem functioning. As sustainable fisheries development envisages an eco-friendly, equitable mode of development that can sustain livelihoods over generations (FAO, 2005; UNCSD, 2012), an attempt has been made to assess the current status of fish biodiversity of India, delineate the threatened species to formulate appropriate strategies for their conservation and rehabilitation.

Fish Genetic Resources of India

India has vast and varied aqua-resources comprising 2.02 million square km area of Exclusive Economic Zone (EEZ) surrounding seas, more than 29,000 km length of rivers, about 1,13,000 km of canals, around 1.75 million ha of existing water-spread in the form of reservoirs, about 1 million ha in the form of tanks and ponds and nearly 0.6 million ha of stagnant, derelict, swampy water-spread areas (Jhingran, 1991). There exist reports on existence of more than 24,600 finfish species throughout the world (Groombridge, 1992), however, Nelson (1994) predicted a total of around 28,500 species representing more than half of the vertebrate diversity. Interestingly, Myers *et al.* (2000) and Myers (2003) emphasized that there could well be at least 5,000 species (more than all mammals: 4,809 species) waiting to be discovered. The recent assessment survey revealed existence of about 29,400 finfish species in different aquatic ecosyetems the world (Babin *et al.*, 2007). India has been identified as one of the mega biodiversity centres for the genetic resources in the world and the same is true in case of fishes too. Nearly 2,936 species of finfishes belonging to 44 orders, 252 families and 1,069 genera have been recorded from different ecosystems of this subcontinent. The approximate ecosystem-wise distribution of fish germplasm resources of India are- freshwater (936; 31.78 per cent), brackishwater (113; 3.85 per cent) and marine (1,887; 64.37 per cent). Out of these, about 258 species are commercially important which include cultured, cultivable and wild taxa, 199 endemic and 275 game fishes. There is record of the introduction of 462 exotic species in Indian waters, most of which are of ornamental value.

Coldwaters Fish Biodiversity

The aquatic resources above 914 m msl in Himalayas, sub-Himalayan zone and mountains of the Deccan are known as coldwaters. The Himalaya chain of mountains have been divided into (i) Greater Himalaya containing the two highest peaks in India, the Nanga Parbat (8,126 m above msl) and Nanda Devi (7,817 m above msl), (ii) Lesser Himalaya includes spurs and highly dissected uplands and (iii) the Sivaliks constituting the southern section of rocks and broken chain mountains (less than 1,200 m msl). In the Deccan plateau, Western Ghats including Sahadary, Nilgiris, Annamalai and Cardamom hills are important for coldwater fishery (Jhingran and Sehgal, 1978; Sehgal, 1992). The temparature of the upland coldwater ranges between 0-20°C with an optimal range between 10-12°C. The lakes and streams of high altitude are characterized by high transparency and dissolved oxygen (7.9-9.7 ppm) as well as sparse biodata (Sehgal *et al.*, 1988). Most of the fishes are small-sized exhibiting distribution pattern depending upon the rate of flow of water, nature of substrata and food availability. Some fishes living in turbulent streams have developed special organs for attachment. The major coldwater resources are upper stretches of Indus, Ganga, Brahmaputra rivers and their tributaries as well as coldwater lakes and reservoirs which harbour fishes belonging to more than six different families such as Cyprinidae, Cobitiidae, Salmonidae, Sisoridae, Psilorhynchidae and Homalopteridae. Though Sehgal (1992) had listed 241 species from different coldwaters of this country but NBFGR has restricted the list to 154 (Anon, 2001-2002). Some commercially important coldwater species are- *Tor tor, T. putitora, T. mosal, T. progeneius, T. khudree, T. mussullah, T. chillinoides, T. neilli, Neolissochielus hexagonolepis, Diptychus maculatus, Schizothoraichthys progastus, S. esocinus, Schizothorax richardsonii, S. plagiostomus, S. curvifrons, S. niger, S. planifrons, S. nasus, S. miropogon, S. labiatus, S. hugelli, S. longipinnis, S. kumaonensis, Barilius bendelisis, B. vagra, B. shacra, B. (Raiamas) bola, Labeo dero, L. dyocheilus, Crossocheilus diplochilus, Garra lamta, Garra gotyla gotyla, Glyptosternum pectinopterus, G. stoliczkae* etc. (Sehgal *et al.*, 1971, 1988; Sehgal, 1992; Sehgal and Shyam Sunder, 1992).

Fish Biodiversity in Warmwaters

The freshwater of inland resources below coldwater zone are known as warmwaters. Coming to the plains, the rivers become wider, the slope is slight and the current moderate to slow. The water is turbid with detritus and high temparature. In India, fourteen major river systems share about 83 per cent of the drainage and harbour 782 fish species. The important rivers being- Ganga river system with a combined length of about 8,047 km (the largest river system in India), Brahmaputra system (combined length 4.023 km), Indus river system (Beas and Sutlej), East Coast river system consisting of Mahanadi, Godavary, Krishna, Cauvery (combined length 6,437 km) and West Coast river system including Narmada and Tapti (combined length 3,380 km) (Jhingran, 1991). Though many species are common to different river systems, river Ganges harbours 382 species, Brahmaputra 126, Mahanadi 99, Cauvery 80, Narmada 95 and Tapti 57 (Jhingran, 1991). Some commercially important carps are *Catla catla, Labeo rohita, L. calbasu, L. gonius, L. bata, L. fimbriatus, L. kontius, Cirrhinus mrigala, C. cirrhosa, C. reba, Puntius dubius* and *P. carnaticus*. Catfishes are important groups contributing significantly (9.23- 28.12 per cent) to the riverine and reservoir catches. These are *Sperata aor, S. seenghala, Wallago attu, Pangasius pangasius, Silonia silondia, Bagarius bagarius, Rita rita* and *Eutropiichthys vacha* (Jhingran, 1991; Talwar and Jhingran, 1991; Yadav and Chandra, 1994; Jayaraman, 2010). Finfishes adapted to swampy areas owing to their accessory respiratory organs are known as air-breathing fishes and featherbacks. Murrels and other important species of the group are *Channa striatus, C. marulius, C. punctatus, Clarias batrachus, Heteropneustes fossilis, Anabas testudineus, Notopterus notopterus* and *Chitala chitala* (Dehadrai and Kamal, 1993).

Ichthyobiodiversity in Brackishwater

The brackishwater (estuarine) regions are consid-ered as transition zone between freshwater of the rivers and saline water of seas. The salinity of brackishwater ranges from 0.5-30 ppt. The major estuarine systems are- Hooghly-Matlah, Mahanadi, Godavari, Krishna, Cauvery, Narmada, Tapti and other estuaries of east and west coasts including large brackishwater lakes such as Chilka, Pulicat and Vembanad (Pandit *et al.*, 1994; Rao, 2001). The long coastal line of 8,129 km, large lakes and estuaries offer immense scope for expanding the coastal aquaculture (Rao, 2001). The brackishwater harbours 113 taxa including commercially important species like *Sparus sarba, Elops saurus, E. machanata, Mystus gulio, Nematolosa nasus, Pseudosciaena coibor, Gerres setifer, G. oyena, Sillago sihama, Megalops cyprinoides, Polynemus tetradactylus, P. paradiseus, Eleutheronema tetradactyla, Mugil cephalus, M. seheli, M. waigiensis, M. cunnesius, Liza macrolepis, L. tade, L. parsia, Ephinephelus tauvina, Rhinomugil corsula, Tenualosa ilisha, Chanos chanos, Etroplus suratensis, E. maculatus, Lutianus argenti-maculatus, Lates calcarifer* and *Tachysurus* spp. (Jhingran, 1991; Pandit *et al.*, 1994). The brackishwaters also harbour lucrative shellfish speices such as *Penaeus monodon, Fenneropenaeus (Penaeus) indicus, P. semisulcatus, Metapenaeus monoceros, M. dobsoni, M. affinis, M. brevicornis, Palaemon styliferus, Macrobrachium rosenbergii, M. malcolmsonii, M. rude, M. mirabilis, M. lamarrei, M. scarbiculum* and *Acetes indicus* (Jhingran, 1991; Suseelan and Nair, 1994).

Marine Fish Biodiversity

The seawater surrounding east and west coasts of the country having salinity more than 30 ppt is designated as marine water. Marine fisheries resources of the Bay of Bengal, Arabian Sea and Indian Ocean include coastal, offshore and deep sea as well as islands comprise 1.887 taxa including the commercially important species like sharks (*Carcarhinus bleekeri, C. dussumieri, C. gangeticus, C. limbatus, Scoliodon palasorrah, S. sorrakowah*), rays (*Narcine brunnea, Pristis cuspidatus, P. microdon*), Bombay duck (*Herpedon nehereus*), oil-sardine (*Sardinella longiceps, S. fimbriatus, S. gibbosa, S. albella*), Malabar sole (*Cynoglossus semifasciatus*), parrotfish (*Pseudocarus* spp.), perches (*Lethrinus* spp., *Epinephelus* spp.), whitefish (*Lactarius lactarius*), silver bellies (*Secutor muconius, S. insidiator, Leiognathus dussumieri, L. bindus, L. lineolatus*), seerfish (*Scomberomorus commersoni, S. guttatus, S. lineolattus, Acanthocybium solandri*), mackerel (*Rastralliger kanagurta*), tuna (*Auxis thazard, A. rochei, Sarda orientalis, Euthynnus affinis, Thynnus tonggol*), carangids (*Caranx caranx, Megalaspis cordyla, Decapteus russelii, D. tabl*), polynemids (*Eleutheronema tetradactylum, Polynemus indicus, P. heptadactylus*), pomfrets (*Pompus argentius, P. chinengis, Parastromateus niger*), baracuds (*Sphyraena commersoni, S. obtusata, S. acutipinnis, S. jello*), red mullets (*Upeneus sulphurus, U. vittatus, Parupeneus indicus*), ribbon fishes (*Trichurus lepturus, T. gangeticus, T. pantulli, Eupleurogrammus intermedius, E. muticus*), anchovies (*Coilia dussumieri, Anchoviella commersoni, A. indica, A. heterolobus, A. benganensis*) and catfishes (*Talchysurus thalassinus, T. tenuispinis, T. dussumieri, T. sona, T. tennuispinis, T. jella, Plotossus canius* and *P. anguillaris*) (Krishna Kartha, 1983; Talwar and Kacker, 1984) and shellfishes like *Parapenaeopsis stylifera, P. hardwickii, P. sculptilis, Penaeus merguensis, Fenneropenaeus* (*Penaeus*) *indicus, P. semisulcatus, Metapenaeus monoceros, M. dobsoni, M. affinis, M. brevicornis* and *Solanocera crassicornis* (Jhingran, 1991; Suseelan and Nair, 1994).

Exotic Fishes in Inland Ecosystems

Apart from the native species, some exotic (alien) fishes have also been introduced into the Indian waters for sport, food, vector control and ornamental purposes. A few important exotic species are *Salmo trutta fario, Oncorhynchus mykiss, Tinca tinca, Salvelinus fontinalis, Cyprinus carpio* var. *specularis, C. carpio* var. *communis, C. carpio* var. *nudus, Aristichthys nobilis, Ctenopharyngodon idella, Hypophthalmichthys molitrix, Oreochromis mossambicus, O. niloticus, Clarias gariepinus, Pangasius sutchi, Gambussia affinis, Poecilia reticulatus, P. mexicana, Betta splendens, Xiphophorus helleri, Osphronemus gourami, Carassius auratus, C. carassius, Pygocentrus nattereri* etc. (Mishra *et al.*, 1998; Gopalakrishnan and Ponniah, 1999). While the trouts (*Salmo* spp) have filled the vacant niche in upland coldwaters, grass carp (*Ctenopharyngodon idella*), silver carp (*Hypophthalmichthys molitrix*) and common carp (*Cyprinus carpio*) helped enhancing aquaculture production in the Indian subcontinent but some others like tilapia (*Oreochromis mossambicus*), *Gambussia affinis*, bighead (*Artistichthys nobilis*), African catfish (*Clarias gariepinus*), *Pangasius sutchi* and even common carp in some natural waters have created competition with the native species (Mishra *et al.*, 1998; Ramakrishniah, 1998).

Threats to Fish Genetic Diversity

Owing to anthropogenic stresses, the fish availability from natural sources has been alarmingly declining world over. This has been adversely affecting the sustainability of fisheries resources since their gene pools and genetic diversity are being eroded (Schreier *et al.*, 2012). With the rapid overall development and owing to ever-increasing demand of fish as food, the aquatic ecosystems are under constant pressure of man-induced stresses to detriment of the aquatic flora and fauna. Though the decline of individual fish species is very often related to more than one proximate factors, the various causes of imperilment of fishes in aquatic ecosystems have been identified as- (i) physical habitat loss due to construction of dams and weirs across the rivers, soil erosion due to deforestation and excessive utilization of waters, (ii) over-exploitation, (iii) indiscriminate killing of juveniles and broodfishes, (iv) chemical pollution due to industrial and municipal wastes, (iv) competition from the introduced non-indigenous species and (vi) spread of dreaded diseases (Botsford *et al.*, 1997; Pandey and Das, 2006; Atkore *et al.*, 2011; Nautiyal, 2013; Bhat and Pandit, 2016).

Habitat Modifications

This normally happens due to damming, deforestation, diversion of water for irrigation, conversion of marshy land and small water bodies for other purposes *etc*. Dams impede upstream spawning migration of fishes and displace populations from their normal spawning grounds. It may also result in separating a population into two smaller groups as in hilsa (*Tenualosa ilisha*) above and below the Farakka Barrage. Inbreeding and genetic drift are the common problems in a small population that reduce genetic variability. Habitat modifications may lead to decline of endemic species and even species extinction as observed in Madagascar (Stiassny, 1996). Due to water diversion causing habitat modification, 20 out of 24 native fish species of Aral Sea got extinct with complete collapse of the fishery (Baltz, 1991).

Siltation from the catchment areas, besides changing the ecology due to construction of dams, has destructed the feeding and breeding grounds of many fishes (Sehgal, 1994; Kirchhofer and Hefti, 1996). It is estimated that about 5,334 millin tons of soil is eroded annually from the cultivable land and forests of India. The rivers carry nearly 2,050 million tons of silt, depositing approximately 480 million tons to the reservoirs causing eutrophication and reduction in the productivity of the water bodies (Shrestha, 1990).

Habitat alterations in Himalayan waters have affected distribution and abundance of native fishes in mountain streams of India and Nepal (Shrestha, 1990; Sehgal, 1994; Joshi and Raina, 1997; Sharma, 2003; Atkore *et al.*, 2011; Gupta, 2014; Bhat and Pandit, 2016). Power dams and reservoirs have dramatically changed fish habitats and local fish communities. The migration routes of important native fishes like mahseers (*Tor putitora* and and *T. tor*) and snowtrouts (*Schizothorax richardsonii* and *S. plagiostomus*) have been blocked (Sehgal, 1994; Shrestha, 1997). The upland fast-moving habitat has been lost to reservoirs which are unfavourable for rheophylic species (Maitland, 1993: Dhanze and Dhanze, 1994). The study of Dhanze and Dhanze (1998) on the Beas river system has shown a decline in the native populations of *Tor putitora* and *Schizothorax richardsoni* and abundance of the bottom inhabiting predatory catfishes (*Sperata seenghala* and allies species) due to impoundment of the river at Talwara in the form of a big reservoir covering an area of 24,000 ha. Excessive withdrawl of water from the river courses for agriculture, domestic and industrial uses leaving inadequate water for fish is also a major factor responsible for the depletion of fish germplasm resources (Menon, 1989; Kirchhofer and Hefti, 1996).

Over-exploitation

The population size gets reduced because of disturbances in age structure and sex composition as a result of over-fishing affecting the demography of fish populations. Over-fishing affects heritable qualitative characters like growth rate and age of sexual maturity (Gjerde, 1986). Efficient gears remove large individuals, which mostly happens in the quick growing ones in the population. As a result, heterozygosity gets reduced since there is a positive correlation between heterozygosity with growth rate (Ferguson and Drahushchak, 1990). The isozyme analysis in *Hoplostethus attanticus* also showed that heterozygosity gets reduced in a population if quick growing larger individuals are removed by fishing (Smith *et al.*, 1991). Consequently, the proportions of the slow-growing ones increases and the average size of individuals in a population decreases, as observed in Pacific salmon (*Oncorhynchus* spp.) and Nile tilapia (*Oreochromis niloticus*) in Lake Georgia of Africa. Over-fishing also causes change in the genetic structure of fish populations due to loss of some alleles thus reducing genetic diversity (Sutherland, 1990).

Over-exploitation of fishery resources due to its extraordinary economic value has been a causative factor exacerbating the vulnerability of the population in different ecosystems of this country. *Tor* spp. and *Schizothorax* spp. in upland waters (Sinha, 1992; Agrawala, 1994; Dhanze and Dhanze, 1994; Nautiyal, 1994, 2013; Mahanta *et al.*, 1998; Atkore *et al.*, 2011; Gupta, 2014; Bhat and Pandit, 2016), *Notopterus chitala, Ompok pabda, Pangasius pangasius, Eutropiichthys vacha, Semiplotus semiplotus* etc. in warmwater (Menon, 1989; Rama Rao, 1992; Pandey and Awasthi, 1994; Prasad, 1994), *Mugil cephalus, Liza tade, Nematolosa nasus, Lates calcarifer* etc. in brackishwater (Algarswami, 1992; Mukhopa-dhyay, 1994; Pandit *et al.*, 1994) and *Rhiniodon typus, Polynemus indicus, P. heptadactylus, Pomadasys hassta* etc. in marine ecosystem are declining at a faster rate (James, 1994; Bensam and Menon 1994; Pravin, 2000; Menon, 2004). The IUCN has also categorized *Rhiniodon typus* under Vulnerable A2bd+3d ver 3.1 (Norman, 2005).

Wanton Destruction

Wanton killing by the use of dynamites, electric shocks and poisoning of broodfishes in spawning season and juveniles during post-monsoon periods have affected a number of food and game fishes of upland waters, especially in rivers and streams originating in Assam, Nepal, Bhutan, Garhwal, Kumaon and Himachal Pradesh (Nautiyal, 1994, 2013; Shrestha, 1997; Das and Pandey, 1998; Mahanta *et al.*, 1998; Mahapatra *et al.*, 2004; Atkore *et al.*, 2011; Gupta, 2014; Bhat and Pandit, 2016).

Aquatic Pollution

Pollution is probably the single most significant factor causing major decline in the population of many fish species (Dehadrai *et al.*, 1994; Pandey *et al.*, 1999). Industrial, sewage (municipal) and pesticides pollution have been causing deterimental environment to fish life in many water bodies (Jhingran, 1991, Kumarguru, 1995; Shyam Lal and Pandey 1995). (i) Poisonous pollutants like agrochemicals, metals, acids and phenols affect reproductve functions and even cause fish mortality in high concentration. 25 stocks of Atlantic salmon

(*Salmo salar*) in rivers of South Norway are extinct due to acid rains (Hesthagen and Hansen, 1991). In India, chemical pollution from factories and plants located in Nilgiris, Mysore and Croog have exterminated certain groups of hill-stream fishes. Certain noemacheiline loaches recorded by Sir Francis Day from Bhavani river at Mettupalayam, Coimbatore are no longer available. (ii) Suspended solids affect the respiratory processes and secretion of protective mucous making susceptable to infection. (iii) Sewage and other organic substances cause deoxygenation due to eutrophication causing mortality in fishes. (iv) Thermal pollution raises ambient temperature, reducing dissolved oxygen causing mortality of sensitive species (Padhi and Mondal, 2000). (iv) The pollutants also act as genotoxic agents causing chromosome aberrations and other cytogenetic abnormalities (Tripathi and Das, 1995). Severe reduction in population size precitipitates genetic bottleneck and drift leading to extinction of the species.

Uncontrolled Introductions of Exotics

Introduced non-native or non-indigenous species outside its natural geographical range are exotics. Exotic species are generally introduced for recreation, control of undesirable organisms, improve aquaculture or capture fishery productivity. Effects of exotic fish introduction generally are observed in the form of (i) reduction in N_e (effective population size) by the ecological and other effects of introduction and (ii) alteration/extinction of the gene pools of the species/stocks by cross-breeding/hybridization and back-crossing. Cross-breeding is interbreeding between different stocks and hybridization is interbreeding between different species. However a species may perform well in terms of objectives in one place but create problems in areas not covered by the same objectives.

Species introduction proved disastrous in many instances in abroad. Apache trout (*Oncorhynchus apache*) and Gila trout (*O. gilae*) the two native species of South West USA faced extinction primarily due to the hybridization with introduced cutthroat trout (*O. clarki*) and rainbow trout (*O. mykiss*). The exotic Nile perch (*Lates niloticus*), predator and voracious feeder, almost ousted native cichlid reducing from 99 per cent to 1 per cent in Lake Victoria which is economic as well as ecological tragedies (Arching, 1990). *Clarias gariepinus* devastated some native species in eastern Cape in South Africa, sea lamprey (*Petromyzon marinus*), rainbow smelt (*Osmerus mordax*) and common carp (*Cyprinus carpio*) affected the native fish communities of the basin of Great Lakes of North America (Crossman, 1991). Non-native *Gambusia affinis* and *Fundulus heteroclitus* threatened cyprinodont stocks and species in Europe and North America (Elvira, 1990).

Many foreign species have been introduced in Indian waters and some are now well established too. They include *Salmo salar, S. trutta fario, Oncorhynchus mykiss, O. nerka, Salveninus fontinalis, Cyprinus carpio, Carassius carassius, C. auratus, Oreochromis mossamicus, O. niloticus, Ctenopharyndon idella, Hypophthalmichthys molitrix, Tinca tinca, Osphronemus gouramy, Gambussia affinis, Lebestes reticulatus, Clarias gariepinus, Pygocentrus nattereri, Pangasianodon hypthalamus etc* (Maitland, 1993; Mishra *et al*., 1998; Ramakrishniah, 1998; Gopalakrishnan and Ponniah, 1999; Singn *et al*., 2011).).

Miller *et al.* (1989) reported that the introduced exotic species contributed to extinction of 68 per cent of the North American fish taxa during the past century. Dill and Cordone (1997) remarked that the success of introduction should be measured by its benefits to the community and the fact that it should not unduly harm existing species. Introduction of exotic fast-growing species is causing threat to the indigenous fish diversity (Menon, 1989; Pullin, 1994; Singh and Pandey, 1995; Stiassny, 1996; Mishra *et al.,* 1998). Common carp introduced into Kashmir Valley has almost exterminated the indigenous schizothoracids. Similarly, *Osteobrama belangeri*, the endemic fish to Loktak Lake (Manipur) is disappearing fast due to the introduction of common carp (Menon, 1989). In Govindsagar Dam (Himachal Pradesh), the Indian major carps, especially catla, has already been replaced by the silver carp (Kumar, 1988). Sandhu *et al.* (1994) reported that the establishment of *Cyprinus carpio* and *Hypophthalmichthys molitrix* resulted in alarming decline of the golden mahseer (*Tor putitora*) fishery in this reservoir. Sugunan (1994) has emphasized that all the three varieties of common carp, *viz., Cyprinus carpio communis, C. carpio specularis* and *C. carpio nudus* are not suitable for stocking in most of the Indian reservoirs as they are vulnerable to predators and seldom caught in passive gears. Besides, they compete with indigenous species like *Cirrhinus* spp., snowtrout and *Osteobrama belangeri*.

The tilapia, *Oreochromis mossambicus*, introduced accidently into some South Indian reservoirs like Amravathy and Vaigai has established itself firmly and completely replaced the endemic fauna (Jhingran, 1984; Sugunan, 1994). Similarly, *Gambussia* spp. practically ousted all indigenous fish fauna of Ooty Lake at one time and is also affecting the native species of Karnataka reservoirs (Ramakrishniah, 1998). Sugunan (1994) has remarked that the three exotic species like *Cirrhinus molitrorella, Molypharyn godon piceus* and *Hypophthalmichthys nobilis*, being considered for introduction in India, are unsuitable. There are apprehensions that *Clarias gariepinus* may be adversely affecting our indigenous *Clarias batrachus* (Mishra *et al.,* 1998).

The introductions of trouts in almost virgin niche at high altitude coldwater streams have, however, remained encouraging in India (Sehgal, 1992). Some exotic food fishes have also been performing excellent by enhancing production in closed culture system. Sugunan (1994) suggested that *Oreochromis niloticus* which grows to 250 g in 6 months and is prolific breeding may probably be more suitable for Indian reservoirs. However, Dehadrai (1996) has remarked that the introduction and transfer of exotic species and breeds for aquaculture purposes may change or impoverish the biodiversity and genetic resources through inter-breeding, competition for food, habitat destruction and possibly through transmission of diseases.

Diseases

Among the range of various diseases caused by bacteria, fungi, viruses *etc* the most virulent and menacing one threatening many species is the Epizootic Ulcerative Disease Syndrome (EUS) which has wiped out large populations of a number of commercial and non-commercial species in major parts of the country (Mohan and

Shankar, 1994). The erosion of genetic variability and biodiversity is serious threat from such diseases (Pandey and Das, 2002).

Categories of Threatened Animals (IUCN, 2000)

- ☆ **Extinct (EX):** A taxon is Extinct when there is no reasonable doubt that the last individual has died.
- ☆ **Extinct in the Wild (EW):** A taxon is Extinct in the Wild when it is known only to survive in cultivation, in captivity or as a naturalized population (or populations) well outside the past range. A taxon is presumed extinct in the wild when exhaustive surveys in known and/or expected habitat, at appropriate times (diurnal, seasonal, annual), throughout its historic range have failed to record an individual. Surveys should be over a time frame appropriate to the taxon's life-cycle and life-form.
- ☆ **Critically Endangered (CR):** A taxon is Critically Endangered when it is facing an extremely high risk of extinction in the wild in the immediate future.
- ☆ **Endangered (EN):** A taxon is Endangered when it is not Critically Endangered but is facing a very high risk of extinction in the wild in the near future.
- ☆ **Vulnerable (VU):** A taxon is Vulnerable when it is not Critically Endangered or Endangered but is facing a high risk of extinction in the wild in the medium-term future.
- ☆ **Rare (R):** Taxa which are not presently Endangered or Vulnerable but can become Vulnerable because of small populations ususlly located in restricted scattered over a more extensive range.
- ☆ **Lower Risk (LR):** A taxon is Lower Risk when it has been evaluated, does not satisfy the criteria for any of the categories Critically Endangered, Endangered or Vulnerable. Taxa included in the Lower Risk category can be separated into three subcategories:
    - (i) **Conservation dependent (LR-cd):** Taxa which are the focus of a continuing taxon-specific or habitat-specific conservation programme targeted towards the taxon in question, the cessation of which would result in the taxon qualifying for one of the threatened categories above within a period of five years.
    - (ii) **Near threatened (LR-nt):** Taxa which do not qualify for Conservation Dependent, but which are close to qualifying for Vulnerable.
    - (iii) **Least concern (LR-lc):** Taxa which do not qualify for Conservation Dependent or Near Threatened.
- ☆ **Data Deficient (DD):** A taxon is Data Deficient when there is inadequate information to make a direct, or indirect, assessment of its risk of extinction based on its distribution and/or population status. A taxon in this category may be well studied, and its biology well known, but appropriate data on abundance and/or distribution is lacking. Data Deficient is therefore not a

category of threat or Lower Risk. Listing of taxa in this category indicates that more information is required and acknowledges the possibility that future research will show that threatened classification is appropriate. It is important to make positive use of whatever data are available. In many cases great care should be exercised in choosing between Data Deficient and threatened status. If the range of a taxon is suspected to be relatively circumscribed, if a considerable period of time has elapsed since the last record of the taxon, threatened status may well be justified.

☆ **Not Evaluated (NE):** A taxon is Not Evaluated when it has not yet been assessed against the criteria.

☆ **Least Concern (LC):** This category was provided to differentiate species that had been evaluated, and found not to be threatened. This gives the impression that one is required to conduct a formal assessment for blatantly common (weedy) taxa. From basic observations it can easily be seen that most of these extremely common taxa would not qualify for listing even though they have not been put through a formal assessment.

Conservation Status of Fish Diversity

Minkley and Deacon (1968) were the first to highlight the threatened status of selected fishes of Southwestern part of the United States. Later, the American Fisheries Society (AFS) published a list of rare fishes of North America (Deacon *et al.*, 1979). However, the recent status assessment revealed systematic declines in the native fish distribution and abundance throughout the North America. It is estimated that about 33 per cent of the native freshwater fish taxa in the region are either endangered, threatened, or of special concern, with membership of each group exhibiting significant increase during the last decade (Williams *et al.*, 1989; Miller *et al.*, 1989; Casey and Meyers, 1998). Atleast 106 Pacific coast stocks of anadromous salmon and trouts are extinct and 214 more are at the risk of extinction or of special concerns (Nehlsen *et al.*, 1991). Similarly, out of nearly 793 freshwater species distributed throughout the heavily industrialized Europe, 101 species have been declared threatened (Lelek, 1987; Kirchhofer and Hefti, 1996). During 1996, IUCN identified 734 threatened taxa: Critically Endangered (CR) 157, Endandered (EN) 134 and Vulnerable (VU) 443 from different aquatic environments of the world (Baillie and Groombridge, 1996) while the list has been increased to 752 (CR 156, EN 144, VU 452) in 2000 (Hilton-Taylor, 2000). Out of 85 hypogean taxa evaluated, 3 are Critically Endangered, 73 Vulnerable, 4 Near Threatened, 1 Least Concern and 4 Data Deficient (Proudlove, 2001). Rosa and Menezes (1996) have tentatively identified 78 threatened (12 elasmobranchs and 66 actinopterygians) from the Brazilian waters. There exist reports that other aquatic taxa in the western world are also exhibiting even higher rates of endangerment (Williams *et al.*, 1993; Warren and Burr, 1994; Mignogno 1996; Lohoefener, 1997; Miller and Craig, 2001). Master (1990) remarked that around 36 per cent of the crayfishes and 55 per cent of mussels in North America are either extinct or imperiled. Bogan (1996) listed 35 freshwater bivalve species as Extinct, 52 taxa as Endangered, 5 taxa Threatened and 70 taxa are of Special Concens in the North America.

After exhaustive surveys of the important rivers (Gandaki, Kosi, Karnali, Mahakali, Rapti, Trisuli, Bagmati, Narayani, Sunkosi, Tamur *etc.*), Shrestha (1990) released the list of 5 Endangered (*Tor putitora, T. tor, Diptychus maculatus, Amblyceps mangois* and *Corglanis kishinouis*), 8 Threatened (*Schizothorax richardsonii, Schizothoraichthys esosinus, S. progestus, Echiloglanis hodgarti, Laguvia ribeiroi, Glyptostrnum horai, Rita rita* and *Anguilla bengalensis*) and 11 (*Acrossocheilus hexagonolepis, Garra gotyla, Nemacheilus rupicola, Botia lohachata, B. almorhae, Psilorhynchus homaloptera, Pseudecheneius sulcatus, Myersglanis blythi, Glyptosternum pectinopterum, Glyptothorax trilineatus* and *G. gracilis*) Rare taxa from Nepal. Recently, he has identified 22 catfishes (out of 52 species assessed) such as *Amblyceps mangois, Bagarius bagarius, B. yarrellii, Caraglanis kishinouyei, Erethistes pussilus, Euchilglanis hodgarti, Exostoma labiantum, Gagata cenia, G. gagata, G. sexualis, Glyptosternum maculatum, Glyptothorax conirostre, G. glacile, G. indicus, G. trilineatus, Hara hara, H. jordoni, Laguvia ribeiroi, Myerglanis blythi, Nangra viridescens, Pseudecheneis sulcatus* and *Sisor rhabdophorus* as Rare from Gandaki, Kosi, Rapti, Trisuli, Karnali, Bagmati, Narayani and Mahakali rivers of the Himalayan kingdom (Shrestha, 1998).

Menon (1989) was the first who compiled a list of 21 Vulnerable fishes - 4 Endangered (*Barilius bola, Puntius chilinoides, Semoplotus semiplotus* and *Enobarbichthys maculatus*) and 17 Threatened species (*Notopterus chitala, Acrossocheilus hexagonolepis, Cirrhinus cirrhosa, Labeo fimbriatus, Labeo potail, Labeo kontius, Puntius carnaticus, P. curmuca, P. jerdon, Tor khudree, T. putitora, T. tor, Schizothorax richardsonii, Schizothoraichthys progestus, Silonia childreni, Pangasius pangasius* and *Bagarius bagarius*) from the Indian subcontinent. However, out of 762 fishes featured in the IUCN Red Data Book of Threatened Animals throughout the world, only two species (i) *Schistira* (*Nemacheilus*) *sijuensis* (Family Bolitoridae) and (ii) *Horaglanis krishnai* (Family Clariidae) were included as Rare from the Indian waters (IUCN, 1990).

In 1992-1993, the National Bureau of Fish Genetic Resources (NBFGR) tentatively identified 4 Endangered, 21 Vulnerable, 2 Rare and 52 Indeterminate fishes from the different ecosystems of this subcontinent (Anon, 1992-93) (Table 2.1). Out of 327 warmwater taxa evaluated through Conservation Assessment and Management Plan (CAMP), 1 was categoriozed as Extinct (EX), 1 Extinct in Wild (EW), 47 Critically Endangered (CR), 98 Endangered (EN), 82 Vulnerable (VU), 67 Lower Risk-near threatened (LR-nt), 13 Lower Risk-least concern (LR-lc) and 18 Data Deficient (DD) (Molur and Walker, 1998). Interestingly, even the commonly occurring *Heteropneustes fossilis* is Vulnerable/Rare in many parts of the country whereas in certain areas it has become Extinct too ((Molur and Walker, 1998; Gadgil *et al.,* 2001; Pandey and Das, 2006). Menon (2004) has given a list of 74 threatened fishes whih includes Extinct 1, Endangered 24, Vulnerable 34 and Rare 15 species. Recently, Lakra *et al.* (2010) have released the list of 71 Endangered and 49 Vulnerable species from the freshwater ecosystem of the country. However, conservation status of these fishes requires further verification through the actual field surveys (Pandey and Das, 2006).

Table 2.1: Ecosystem-wise Conservation Status of
Fish Germplasm Resources of India

Ecosystem	Total	Endangered	Vulnerable	Rare	Indeterminate	Total Species
Coldwater	154	01	04	—	12	17
Warmwater	782 03	13	02	28	46	
Brakishwater	113	—	02	—	04	06
Marine	1,887	—	02	—	08	10
Total	2,936	04	21	02	52	79

Southern Region

After elaborate studies on the threatened fishes of Malabar region (Western Ghats) of the Peninsular India, Menon (1997) identified 8 Endangered (*Barilius canarensis, Hypselobarbus jerdoni, H. kurali, H. lithopidos, H. periarensis, H. pulchelus, H. thomsi* and *Etroplus cararensis*), 2 Vulnerable (*Labeo dussumieri* and *Pseodobagrus cryseus*) and 8 Rare species (*Osteobrama bakeri, Eechanthalakenda ophiocephalus, Puntius chalakudensis, Lepidopygopsis typus, Batasio travancoria, Horaglanis krishnai, Monopterus fossorius* and *Pristolepis fasciatus*). Easa and Shaji (1997) recorded 92 species from the Nilgiri Biosphere Reserve, 37 endemic to Western Ghats and 9 strictly to Kerala. Of these 22.83 per cent were recorded as Rare and 11.96 per cent Very Rare. Out of 27 species recorded from the Periyar Lake-Stream system of South-Western Ghats (Kerala), 14 (52 per cent) are reported threatened (Arun, 1998). Biju *et al.* (1998) reported *Sicypterus griseus* as Rare/Endangered species from Periyar river system. While evaluating the conservation status of 170 (21 endemic) freshwater species inhabiting the Western Ghats of Kerala, Kurup (2002) has delineated 18 taxa as Critically Endangered (CR), 34 Endangered (EN), 31 Vulnerable (VU), 16 Low Risk-Nearly Threatened (LR-nt) and 32 under Lower Risk-least concern (LR-lc). Thippeswamy and Hegede (1999) have identified 16 species (*Anguilla bengalensis, Barilius vagra, Danio aequipinnatus, Gonoproktopterus curmuca, Mastacembalus armatus, Neolissochilus hexagonolepis, Osteochilus nashii, O. thomassi, Puntius carnaticus, P. fasciatus, P. jerdoni, Tor khudree, T. mosal, T. putitora, T. tor* and *Xenentodon cancila*) as Threatened from the Sharavathi river (Central Western Ghats) of Karnataka. Recently, Shaji *et al.* (2000) have listed 24 fish species as Critically Endangered, 37 Endangered, 13 Vulnerable and 8 Lower Risk-Least Concern among the 287 taxa evaluated from the Western Ghats of Peninsular India.

North and North-Eastern Region

Dhanze and Dhanze (1994) recorded *Puntius chola, P. waagnei, P. chilinoides* and *Tor tor* as Endangered and *Tor putiotra, Schizothorax richardsonii, Labeo dero, Diptycus maculatus, Danio devario* and *Noemacheilus kangree* as threatened in Himachal Pradesh. Though North-Eastern states of this country has been identified as hotspot for fish biodiversity (267 species) and endemism, six taxa - *Garra litanensis, G. manipurensis, Aborichthys garoensis, Lepidocephalous goalparensis, Pangasius pangasius* and *Osteobrama belangeri* have been grouped as Critically Endangered (CR) (Ponniah and Sarkar,

2000). Preliminary surveys conducted by Mahanta *et al.* (2001) have revealed *Ompok pabda, O. pabo, Labeo dyocheilus, Semiplotus semiplotus, Balitora brucei, Barbus dukai, Olyra longicaudata, Psylorhyncus homaloptera* and *Nemacheilus elongatus* as threatened species from this region. Recently, Shyam Sunder and Joshi (2002) have released a list of 5 Endangered (*Tor mussullah, T. chelynoides, Ptychobarbus conirostris, Raiamas bola* and *Schizothorax kumaonensis*), 21 Vulnerable (*Tor khudree, T. progeneus, T. tor, T. putitora, Salmo trutta fario, Oncorhynchus mykiss, Schizothorax richradsonii, Schizothoraichthys progastus, Neolissocheilus hexagonolepis, Puntius carnaticus, P. jerdoni, P. shalynius, Ompok bimaculatus, Botia almorhae, B. lohachata, Cyprinion semiplotum, Bagarius bagarius, Aborichthys garoensis, Nemacheilus elongatus, N. reticulofasciatus* and *Chaudhria indica*) and 10 Rare fishes (*Tor mosal, Lepidopygopsis typus, Diptychus maculatus, Schizopygopsis stoliczkae, Glyptosternum reticulatum, Osteochilus brevidorsalis, O. nashii, Bhavania australis, Travancoria jonesi* and *Batasio travancoria*) from coldwater ecosystem of the country.

It is pertinent to remark that the most endangered cobitoid loach, *Enobarbichthys maculatus*, is known by a single specimen kept in the British Museum. Menon (1989) presumes that the species might have become Extinct (EX) as no specimen has so far been recorded after 1867. Sehgal (1994) has identified *Gymnocypris biswasi* as Extinct (EX) from Ladhak region. Similarly, the Gangetic shark, *Glyphis gangeticus*, has been reported as "probably Extinct" because only three museum specimens are currently known in the collections - one each in the Museum National d' Histoire Naturelle, Paris, Humboldt Museum, Berlin and Zoological Survey of India, Calcutta. All these specimens were collected during 19th century with no confirmed record after 1867 (Compagno, 1997). However, IUCN has categorized this species as Critically Endangered (CR) (Baillie and Groombridge, 1996).

West Bengal

In a preliminary aessment, Mukherjee *et al.* (2002) gave a list 39 of fish taxa including 01 Endangered, 10 Vulnerable and 16 Threatened from freshwater, 4 Vulnerable from coldwater and 2 Vulnerable and 6 Threatened from brackish and marine ecosystems. Out of 695 taxa evaluated for West Bengal, 106 species are threatened which includes 12 Critically Endangered (CR), 34 Endangered (E), 36 Vulnerable, 13 Near Vulnerable, 5 Lower Risk and 6 Remote Risk (Das *et al.*, 2004). Of the 11 taxa belonging to 9 genera, 8 familes under Order Siluriformes evaluated from north-east Sundarbans, 1 catfish was grouped under Endangered, 5 Vulnerable and 5 near Threatened catergories (Patra *et al.*, 2005).

Andaman and Nicobar Islands

Rajan *et al.* (1993) have listed *Chiloseyllium punctatum, Histrio histrio, Coelorhinchus flabellispinis, Pegasus volitans, Solenostomus cynopterus, S. paradoxus, Dactyloptera orientalis* and *Polycaulus uranoscopus* as Rare species from Andaman Islands.

Marine

Marine environment has been presumed to be resilient to human impact. However, the over-exploitation and subsequent collapse of marine fisheries has attracted the wide attention (Hutchings, 2000; Musick *et al.*, 2000a, b; Ruttenberg,

2001). Sharks, rays and skates are most vulnerable to collapse or extirpation because of slow growth rate, late maturity (dusky shark is the slowest growing marine chordates, takes 20 years to mature) and poor fecundity (the sand tiger shark produces only two youngs every alternate year) as compared to bony fishes (Brander, 1981; Casey and Meyers, 1998; Musick *et al.*, 2000c). The whale shark, *Rhiniodon typus*, occurring in the north-west coast of India has become Critically Endangered (CR) due to directed fishing off Saurashtra (Gujarat) coast and needs immediate protection (James, 1994; Baillie and Groombridge, 1996; Pravin, 2000; Menon, 2004). Other species of the Indian marine waters which have become Vulnerable due to overfishing include *Tachysurus tenuispinis, T. dussumieri, T. sona, Batrachocephalus mino, Lactarius lactarius, Polynemus indicus, P. heptadactylus, Pseudosciana diacanthus* and *Platycephalus macrolipinna* (Joseph and Menon, 2001; Menon, 2004). Fishing of gestating males during breeding season appears to be responsible for endangerment of *Tachysurus* species in the Indian marine waters (Bensam and Menon, 1994). It is pertinent to remark that several species of snappers as well as groupers (Lutjanidae, Serranidae), rockfishes (Sebastinae), some sharks (Selachii), rays (Rajidae) and sawfishes (Pristidae) seems to have become threatened due to excessive fishing as these fishes have slow growth rate, late maturity and low fecundity. Great concerns has been attributed on the by-catch (about 10 per cent of the total landings) of the trash fishes including the juveniles/sub-adults in the mechanized nets resulting in the serious decline of the marine species such as bellies, scianids, catfishes, flatfishes, lizardfishes *etc* (Bensam and Menon, 1994; Winter and Hughes, 1997; Joseph and Menon, 2001; Nair, 2001). Although the Marine Fisheries Regulation Act (MFRA) restricts the cod-end mesh size of trawlers to below 30 mm, the cod-end mesh of medium-sized commercial shrimp trawlers generally ranged between 10-15 mm (Joseph and Menon, 2001). It appears that endangerment of aquatic organisms is greater than that of terrestrial animals partly because of the social biases against small, cold-blooded and wet species and lack of information regarding the conservation status of these species (Hughes and Noss, 1992; Babbitt and Frampton, 1994; Winter and Hughes, 1997; Myers *et al.*, 2000).

Need for Ichthyodiversity Conservation

Maintenance of fish biodiversity along the other biotic resources can be viewed as prerequisite for the well being of even the human beings (Boehlert, 1996; Das and Pandey, 1998; James, 1999; Tilman, 2000; Das *et al.*, 2002). While several reasons can be ascribed to the need, there are four basic reasons for the preservation of biotic resources - (i) Diversity or variability seems aesthetically pleasing in most environments. This is not only true in general but often applies to the specific species frequently encountered by man (Simpson, 1953; Smith and Chesser, 1981; Zechendorf, 2002). (ii) There is often local pride in population or species that are characteristic of an area. People often become disturbed when some local form of animal is threatened by extinction and this concern is an important reason for conservation of atleast some species (Smith and Chesser, 1981). (iii) It is generally agreed upon by the ecologists and evolutionary biologists that species diversity and genetic variability are necessary for the long-term maintenance of stable, complex ecosystems and species (Smith and Chesser, 1981; Ryman and Utter, 1987; Tilman,

2000). Some genetic traits from the diverse germplasm resources may be useful to increase the aquaculture production through hybridization and genetic engineering (Brown et al., 1989; Das, 1996). (iv) All the living beings in an ecosystem co-evolve for their mutual benefits during the evolutionary process. Any species getting extinct upsets the ecological balance to the detriment of each species and also to the community as a whole (Frankel and Soule, 1981; Minckley and Deacon, 1991; Tilman, 2000).

Strategies for Fish Diversity Conservation

The main goal in a conservation programme is to conserve the genetic diversity. The fish genetic resources can be conserved by protecting an ecosystem which, however, is broad-based, non-specific cost-effective and relatively simplistic approach of conservation. It may aim in general or at specific species like endangered or threatened ones. It can be effected through *in situ, ex situ* including gene bank. DNA banking may include (i) genomic DNA, (ii) DNA library (genomic DNA or cDNA library), (iii) cloned DNA fragments etc (Wright, 1993; Padhi and Mondal, 2000).

Since maintenance of fish biodiversity along with other biotic resources has been viewed as prerequisite for the well-being of even human beings (Smith and Chesser, 1981; Meffe, 1986; James, 1999; Zabel et al., 2003), it is essential to prevent the further decline of the fish resources by devising all the possible measure of conservation and rehabilitation (Pavlov, 1993; Penczak, 1996; Gray, 1997; Das and Pandey, 1998, 1999; Khan and Sinha, 2000; Sakthivel, 2001; Pandey et al., 2002; Pandey and Das, 2006). The conservation policy should promote the management practices that maintain integrity of aquatic ecosystems, prevent endangerment and enhance recovery of the threatened species (Dobson et al., 1997). Allen et al. (1987) have identified five principal elements or tasks in the recovery programmes such as (i) habitat management, (ii) habitat development and maintenance, (iii) native fish stocking, (iv) non-native fish and sport-fishing and (v) research data management and monitoring. The irreparable harm caused to fish and habitats need be compensated through afforestation, eco-restoration, soil conservation, complete ban on deforestation particularly in the fragile mountains and strict implementation of Indian Fisheries Act 1897 (modified in 1956) along with the following measures would positively help in restoration of the threatened fish fauna.

In situ Conservation and Biological Reserves

In situ conservation of fish as land races and wild relatives is useful where genetic diversity exists and where wild forms are present. This is done through their maintenance within natural or man-made ecosystems in which they occur. The major advantages of *in situ* conservation are : (i) continued co-evolution wherein the wild species may continue to co-evolve with other forms providing the breeders with a dynamic source of resistance that is lost in *ex situ* conservation and (ii) national parks and biosphere reserves may provide less expensive protection for the wild relatives than *ex situ* measures (Singh and Pandey, 1995; Gray, 1997; Narain, 2000; Das et al., 2002; Ayyappan et al., 2004).

In Himachal Pradesh where the landing of prized mahseer is declining fast, the State Government has declared sanctuaries in Sidhpur and Machial (Mandi district), Renuka Lake (Sirmaur district), Baijnath and Chandra Tal (Kangra district) for protection of mahseer. Some river stretches throughout the country are also protected due to religious sentiments, as they are located vicinity to holy places and shrines (temples). Rishikesh, Har-ki-Pauri (Hardwar), Baijnath (Almora), Pushkar Lake (Ajmer), Sardamath (Shringeri) and Bichaligudda/Tirthahali (Tunga river in Karnataka) are some examples of religiously protected areas (Das, 1996). Recently, Sehgal and Malik (2002) have enumerated 29 fish sanctuaries in the Himalayan regions (Jammu and Kashmir, Himachal Pradesh and Assam). Himachal Pradesh Fishing Rule has enhanced the catchable size of golden mahseer from 300 mm to 500 mm (1.2 kg) giving the opportunity to each female to breed atleast once before being caught. Since the incorporation of this clause in the Fisheries Act during 1998, the average size of mahseer has increased from 1.2 kg to 1.7 kg in Pong Dam and 0.6 kg to 0.9 kg in Govindsagar Reservoir (Kumar, 2000). The Government of Kerala has imposed ban on trawling in marine water during monsoon season (June, July and August) since 1989. The Balakrishnan Nair Committee set up to evaluate the effect of ban on marine fisheries remarked that the ban (i) has led to an increase in fish landings, (ii) revived the stock position leading to improvement in CPUE (catch per unit effort) and (iii) resulted in real improvement in the size groups of the exploited commercial fisheries (Nair, 2001).

Procted Places

Declaration of certain Protected Areas/Biosphere Reserves for *in situ* conservation of marine resources appears to be the pragmatic approach (Huntsman, 1994; Gray, 1997). Reserves are a system approach to fishery management that allows the re-establishment of age distributions and inter- and intra-specific relationships like unaltered community (Bohnsack, 1993; Huntsman, 1994; Grimes, 1998). Establishment of Marine Parks is perhaps the best way for *in situ* conservation of marine resources (Qasim, 1980; Silas *et al.*, 1985; Huntsman, 1994; Bohnsack, 1996). In India, there exist three important National Marine Parks - (i) the Gulf of Kutch National Marine Park (established in 1980; Okha to Jodia, Gujarat coast covering 42 islands; area 400 sq km), (ii). The Gulf of Mannar National Marine Park (established in 1986; Rameswaran to Tuticorin, Tamil Nadu; area 623 ha) and (iii) the Wandoor National Marine Park (South Andaman; covering 10 islands; area 282.5 sq km) as well as two Marine Sanctuaries - (i) Bhitarkanika-Gahirmatha Sancturay (Odisha) and Malvan Marine Sanctuary (Mahrashtra). Further, there are also proposal to set up one more National Marine Park at Lakshdweep Island and extension of two existing sanctuaries (i) Konark-Balukhand Sanctuary (Odisha) and (ii) Point Calimere Sanctuary (Tamil Nadu) (Menon and Pillai, 1996). Protection of spawning habitats of has also been included in the conservation measures of the marine threatened fishes (Koenig *et al.*, 2000).

Ranching

Stock enhancement through ranching is feasible only (i) if there is incomplete colonization of available habitat by juveniles and (ii) if the trophic capacity of the habitat is under-utilized by a stock and/or its competitors *i.e.* there is available carrying capacity (Grimes, 1998; Pandey *et al.*, 2002; Pandey and Das, 2004). The successful induced spawning and larval rearing of endangered *Tenualosa* (*Hilsa*) *ilisha* (Sen *et al.*, 1990), *Tor khudree* (Kulkarni and Ogale, 1986; Nandeesha *et al.*, 1993), *Tor putitora* (Joshi, 1981; Shrestha *et al.*, 1990; Sehgal, 1991; Ogale, 1997; Pandey *et al.*, 1998a; Chaturvedi *et al.*, 2013), *Labeo dussumieri* (Kurup and Kuriakose, 1991; Kurup, 1994), *Ompok pabda* (Bhowmik *et al.*, 2000; Chaturvedi *et al.*, 2012), *Osteobrama belageri* (Reddy, 2000), *Horabragrus brachysoma* (Padmakumar *et al.*, 2011; Bindu and Padmakumar, 2014; Sahoo *et al.*, 2014), *Labeo dyocheilus* (Sarkar *et al.*, 2001),. *Chitala chitala* (Hossain, 1999; Radheyshyam and Sarani, 2005; Sarkar *et al.*, 2006; Janan *et al.*, 2015) and *Notopterus notopterus* (Srivastava *et al.*, 2010) have opened up the avenues of ranching their advanced fingerlings in the depleted water bodies for stock replenishment. In the three important lakes Bhimtal, Naukuchiatal and Sattal of the Kumaon Himalayas, there are evidences of auto-stocking due to natural breeding of mahseer (*Tor putitorta* and *T. tor*) (Pandey *et al.*, 1994). The programme for ranching of a artificially-bred golden mahseer fingerlings in Ladhiya and Sharda rivers by the National Research Centre on Coldwater Fisheries, Bhimatal has already been initiated. Ranching of the pond-reared fingerlings in Pampa river of Kerala led to the improvement in landings of the endangered *Labeo dussumieri* (Kurup, 2001).

The United States and Norway were the first to practice the concept of stock improvement through sea ranching during 1880s. The anadromous species like shad (*Alosa* spp.), herring (Clupeidae), stripped bass (*Morone saxatilis*), stugeon (Acipenseridae) as well as marine species such as cod (Gadidae), heddock (*Melanogrammus aegelfinus*) and mackerel (Scombridae) were ranched but its impact on the stock enhancement and recovery was not thoroughly studied. During recent years sea ranching of the Baramundi (*Lates calcarifer* - in Australia), red drum (*Sciaenops ocellata* - in Texas), Atlantic cod (*Gadus morhua* - in Norway), numerous species in Japan, white seabass (*Atractoscion nobilis* - in California), grey mullet (*Mugil cephalus*) and Pacific threadfin (*Polydactylus sextilis*) in Hawaii have been done successfully (Grimes, 1998; Molony *et al.*, 2003; Bailey and Zydlewski, 2013.). The success achieved in the induced maturation and breeding of the brackishwater species like grey mullet (*Mugil cephalus*), seabass (*Lates calcarifer*) and pearl spot (*Etroplus suratensis*) (Rao, 2001) as well as marine grouper (*Epinephlus tauvina*) and ornamental clownfish (*Amphiprion chrysogaster*) (Devaraj and Pillai, 2000) has paved the way for their ranching in the India waters. Recently, a feasibility study for sea ranching of mullet, pomfret, *Hilsa* and Bandit spiny lobster along the Gujarat coast has been conducted by M/s AVA Exports, London in collaboration with the Hunting Technical Services (HTS) (Anon, 2000).

Conservation Aquaculture

Though probability of inbreeding in hatchery-bred seed normally may not be ruled out (Kincaid, 1976; Eknath and Doyle, 1980; Mishra and Jain, 1993; Padhi and Mandal, 1994; Pandey *et al.*, 1998b; Mishra *et al.*, 2000; Schreier *et al.*, 2012), conservation aquaculture is gaining importance in rehabilitation programmes of endangered/threatened fishes (French *et al.*, 2004; Steffens, 2008; Schreier *et al.*, 2012). It implies aquaculture for conservation and recovery of endangered fish populations by increasing the effective population size (Ne) of the threatened species (Philippart, 1995; True *et al.*, 1996; Anders, 1998; Mukherjee *et al.*, 2002). In India too, Singh (1992), Shyam Sunder *et al.* (1995) and Sharma (2001) have successfully reared the mahseer (*Tor putitora*) fry from 0.20 gm to 105 gm in about 240 days under pond environments in Terai region of Uttar Pradesh. Interestingly, seabass (*Lates calcarifer*), a vulnerable species of the brackishwater, has been successfully cultured in West Bengal for about 6 months by stocking the hatchery-produced seed (Mondal and Mondal, 2000).

Involvement of NGOs'

Some NGOs like (i) Munnar High Range Angling Association, Kerala, (ii) Wildlife Association of South India, Bangalore, (iii) Himachal Pradesh Angling Association, Palampur and (iv) Assam (Bhoreili) Anglers Association, Tezpur have come forward to protect the coldwater fish stocks (Agarwala, 1994; Das *et al.*, 2002; Sehgal and Malik, 2002). NBFGR has also initiated an integrated project for conservation of mahseer in selected upland waters of Ladhiya and a stretch of Sharda river at Tanakpur (Uttarakhand). Population dynamics of *Tor putitora* and *T. tor* in Ladhiya river and the causes as well as magnitude of decline of mahseer fishery are under investigation. For effective implementation of conservation measures in the catchment areas, Mahseer Conservation Committees have been constituted involving fishermen, teachers, students, Yuvak Mangal Dals, Gram Sabha members and officials. A series of lectures, discussions, poster displays and photo-exhibitions are being arranged as mass awarness drives in this area from time to time for *in situ* conservation by the Bureau.

Ex situ Conservation and Gene Bank

In this measure, the threatened species are conserved outside their natural habitats. The two main pillars of *ex situ* conservation programme are (i) Live Gene Bank and (ii) Gamete/Embryo Bank.

Live Gene Bank

In a Live Gene Bank, the endangered species are reared in captivity, bred therein and genetically managed avoiding inbreeding depression, domestication and unintended selection (Franklin, 1980; Eknath and Doyle, 1990; Minckley and Deacon, 1991; Jensen, 1994; Das, 2001). The NBFGR is maintaining the wild stocks of threatened species like *Notopterus chitala, Channa marulius, Tor putitora, Labeo bata, L. dyocheilus* and *L. calbasu* in the Mini Germplasm Repository. Simultaneously, efforts are being made to establish such repository in North-Eastern Region of India.

Gamete/Embryo Bank

In Gamete/Embryo Bank, adequate representative samples of the natural genetic variations of endangered species are kept in suspended animation under extra low temperature (-196°C) in liquid nitrogen (LN_2). Establishment of Gene Bank by cryopreserved milt, eggs and embryos assures further availability of genetic materials of threatened categories and for intensive breeding programmes of economically important species (Ponniah, 1996; Basavaraja *et al.*, 1998). Long-term cryopreservation of milt of the endangered as well as commercially important fishes like *Tor putitora, T. khudree, Labeo rohita, Cyprinus carpio* var. *communis, Tenualosa (Hilsa) ilisha, Horabragrus brachysoma* and *Oncorhynchus mykiss* has been achieved by NBFGR (Ponniah, 1996; Ponniah *et al.*, 2000). Since the technique is successful only for sperm and no method for cryopreservation of eggs/embryos has yet been developed (John *et al.*, 1993; Horvath and Urbanyi, 2000; Linhart *et al.*, 2001; Tiersch, 2008), it is felt that the technique is at moment only of limited value in relation to conservation of threatened species (Maitland, 1993; Tiersch, 2008). Efforts should now be focussed on the androgenesis through which the whole genome can be constituted from the cryopreserved milt alone avoiding completely the maternal genetic contribution.

Storage of fish milt (sperms), eggs and embryos without loss of viability is of considerable value in conservation of fish genetic resources as well as in aquaculture (Tiersch, 2008). Long-term cryopreservation technique for sperms developed and standardized by the NBFGR, Lucknow has primarily helped in development of Gene Bank for conservation of endangered fishes. It has also helped in availability of male gametes all the year-round for seasonal breeders, easy transport of germplasm and genetic selection as well as hybridization programmes. While the NBFGR Mini Gene Bank holds male gametes of several species, the regular Gene Bank as the National Facility be developed

Genetic Management of Endangered Populations

When populations of a species or a group of species decline and become severely threatened or endangered in particular, some appropriate corrective measures based on genetic principles need be implemented. These should include (i) maintenance of large effective population (N_e), (ii) controlling the causative factors and (iii) maintenance of continuity of gene flow by disrupting the barriers that create discontinuity in an inter-breeding population (Kincaid, 1976; Eknath and Doyle, 1980; Soule, 1980; Stahl, 1983; Meffe, 1986a, b; Tave, 1991; Largieder *et al.*, 1996).

Captive breeding programme is a useful approach. Developmental plans that minimize genetic damage and maximize chances of long-term genetic health of small populations is important. Meffe (1986a,b) recommends action as plan as (i) Monitor genetics in field and effective populations. (ii) Maintain largest feasible genetically effective population size of captive stocks (reducing erosion of quantitative variation, loss of rare alleles and inbreeding potential). (iii) In small captive populations, avoid inbreeding through selective mating. (iv) Keep stocks in hatchery environments for as short of time as possible reducing several types of artificial selection, domestication, minimizing chances of bottleneck, drift, inbreeding and catastrophic loss of the stock. (v) Maintain separate stocks of distinctive populations to preserve among population variances (Nei *et al.*, 1975; Soule, 1980; Stahl, 1983; Dowling and Childs, 1992).

National Conservation Strategies

The Indian Fisheries Act of 1897 (modified in 1956) is a landmark in the conservation of fishes. Besides provisions to control and monitor of gears, mesh size as well as observance of fishing or closed seasons, the Act also prohibits the use of explosives or poisons to indiscriminately kill fish in any water (Menon, 1989). At present, the Ministry of Agriculture, Government of India is in the process of formulating a model Indian Fisheries Act that would incorporate all the relevant legal measures to conserve fish germplasm resources. It is proposed to incorporate the Endangered Species Act (ESA, 1973) and Sustainable Fisheries Act (SFA, 1996) in the lines of USA for protection of fish genetic heritage. Implementation of ESA has resulted in the rebound of popualtion of the greenback cutthroat trout (*Oncorhynchus clarki stomias*) in the United States (Young and Harig, 2001). Since elasmobranches (sharks, rays and skates) exhibit low intrinsic growth rates and very low resilience to fishing mortality, special attention should be accorded for their management in the Indian waters. The biomass of thsee fishes should be maintained well above the levels usually accepted to provide maximum sustainable yield (MSY) (Musick *et al.,* 2000c)/ecologically sustainable yield (ESY) (Zabel *et al.,* 2003). Legal measures for conserving fish germplasm resources may not be implemented without taking into consideration of the ground socio-economic realities (Buckworth, 1998; Dehadrai, 2001; Daw and Gray, 2005). The USA has enlisted the services of local fishermen during the lean period for collecting data from the sea under the Global Oceans Ecosystem Dynamics (GLOBEC) programme funded jointly by the National Science Foundation (NSF), National Oceanic and Atmospehreic Adminstration (NOAA) and National Marine Fisheries Service (NMFS) (Nadis, 1997). A similar type of programme may also be launched in India by involving local fishermen for collection of data in ascertaining the status of fish germplasm and in the conservation drive. However, translation and implementation of such policies even under Common Fisheries Policy (CFP) of the European Uninis is often slow and incomplete (Daw and Gray, 2005).

A baseline information in the form of a Data Bank has been made at the NBFGR. The effort of NBFGR in microlevel study of causative factors of decline in fish genetic resources needs be strengthened through involvement of all relevant agencies. Water Pollution Act need be made stricter ensuring implementation. Security of the protected water bodies including marine protected areas (MPAs) need be ensured. Declaration of more sanctuaries and establishment of more marine parks/reserves is need of the hour. Regular Gene Bank and Fish Genetic Resource Centre for genetic management of endangered species need urgently be established at NBFGR. Definite policy guidelines be framed for the introduction of exotic species in the Indian waters. Integrated Coastal Management Plans (ICCPs) are essential to protect our long coastal lines as well as mangrove forests which are the breeding and feeding grounds for a number of commercial marine species (Koenig *et al.,* 2000). Furthermore, habitat imporovement programmes for recovery of endangered populations should be initiated so that the populations rebound naturally (Angermeier and Williams, 1994; Anders, 1998; Hill and Platts, 1998). Establishment of an International Bureau of Fish Genetic Resources (IBFGR) on global basis or an Asian Bureau of Fish Genetic Resources (ABFGR) on the pattern of International Bureau of Plant Genetic Resources (IBPGR) would be useful, more particularly in view of contiguity of water bodies as well as similarities of fish germplasm between the adjoining countries and common programmes. Since India is a signatory to the

Food and Agriculture Organizations Code of Conduct for Responsible Fisheries, the provisions of conservation of fish stocks and sustainable fisheries should be sincerely implemented (FAO, 1995). Above all, a gigantic mass awareness drive for building up people opinion about the requirement of conservation of fish germplasm is an imminent requirement (Kulkarni, 1992; Singh, 1996; Das and Pandey, 1998). In a huge country like India with diverse ecosystems where enforcement of law is not an easy task, the most effective way of tackling the problem would be to arouse mass consciousness on the issues when people themselves would come forward to protect the fish genetic resources.

The fish genetic diversity conservation as a concept may sound simple but its realization and implementation is not an easy task. It is particularly so in a developing country like India with 125 crores of people, ongoing agricultural revolution, industrial modernization and where majority of the people are fish eaters and 67.3 lakhs of fishers (1992 Census) earning their livelihood from fishing and its ancillary vocations. After India became signatory to CBD, the Biological Diversity Act (BDA) was passed by the Parliament in December 2002 resulting in promulgation of Biological Diversity Rules (2004) to put administrative procedures with aim at conservation of natural biological heritage and ensure sharing of benefits of utilization of biological resources in an equitable manner. Consequenttly, the National Biodiversity Authority (NBA), State Biodiversity Boards (SBBs) and Biodiversity Management Committees at centre, state and local levels, respectively, came into the existence for effective implementation of the Rules. The recently released National Biodiversity Action Plan (NBAP, 2008) includes several plans, programmes and policies pertaining to biodiversity conservation. Fish diversity conserevation strategies should also be integrated with the various programmes associated with other policies. The conservation policy should promote management practices that maintain integrity, prevent further endangerment and enhance recovery of species and ecosystems (Dobson *et al.*, 1997). It thus calls for well-investigated scientific data for effective planning through creative approaches. However, a whole-hearted, well-conceived concerted approach by government agencies, scientists, planners, conservationists, voluntary organizations and people at large (incorpoating the Traditional Ecological Knowledge) (Drew, 2005) is required without any further loss of time for conservation, rehabilitation and sustainable utilization of the resources as envisaged in the recently concludedc United Nations Confeence on Sustainable Developent-Rio+20 (UNCSD, 2012).

References

Agrawala, H.K., 1994. Endangered sport fishes of Assam. In: *Threatened Fishes of India* (Eds.) Dehadrai, P.V., Das, P. and Verma. S.R. *Nature Conservators*, Muzaffarnagar, pp. 209-212.

Alagarswami, K., 1992. State of estuarine fisheries in India with a note on changing pattern. In: *National Seminar on Endangered Fishes of India* (April 25-26, 1992). NBFGR and Nature Conservators, Allahabad. pp. 37-38.

Allen, L., Harris, R., Martin, J., Mathews, L. and Milder, B., 1987. *Recovery Implementation Program for Endangered Fish Species in the Upper Colorado River*

Basin. United States Wildlife and Fisheries Report, U.S. Depatment of the Interior Fish and Wildlife Service, Denver, Colarodo, 94 p.

Anders, P.J., 1998. Conservation aquaculture and endangered species: can objective science prevail over risk anxiety? *Fisheries* (*Bethesda*), 23 (11): 28-31.

Angermeier, P.L. and Williams, J.E., 1994. Conservation of imperiled specis and reauthorization of the Endangered Species Act of 1973. *Fisheries (Bethesda)*, 19 (1): 26-29.

Anon, 1992-1993. *Annual Report*. National Bureau of Fish Genetic Reaources, Lucknow.

Anon, 2000. Sea ranching along Gujarat coast: feasibility study. *Fishing Chimes*, 20 (5): 47-48.

Anon, 2001. Climate and Fisheries: Interacting Paradigms, Scales, and Policy Approaches. In: *The IRI-IPRC Pacific Climate-Fisheries Workshop* (November 14-17, 2001). Columbia University, Honolulu, 63 p.

Anon, 2014-2015 *Annual Report*. National Bureau of Fish Genetic Resources, Lucknow.

Arching, A.P., 1990. The impact of the introduction of Nile perch, *Lates niloticus* (L.), on the fisheries of Lake Victoria. *J. Fish Biol.*, 37: 17-23.

Arun, L.K., 1998. Status and distribution of fishes in Periyar Lake-stream of Southern-Wastern Ghats. In: *Fish Genetics and Biodiversity Conservation* (Eds.) Ponniah, A.G., Das, P. and Verma, S.R). Nature Conservators, Muzaffarnagar, pp.77-87

Atkore, V.M., Sivakumar, K. and Johnsingh, A.J.T., 2011. Patterns of diversity and conservation status of freshwater fishes in the tributaries of rivr Ramganga in the Shiwaliks of the western Himalaya. *Curr. Sci.*, 100: 731-736.

Ayyappan, S., D.S. Malik, R. Dhanze and R.S. Chauhan (2004). *Fish Biodiversity in Protected Habitats*. Nature Conservators, Muzaffarnagar, 298 p.

Babbitt, B. and Frampton, P.G.T., 1993. The U.S. National Biological Survey: a tool to track and protect fisheries resources. *Fisheries (Bethesda)*, 18(9): 24-25.

Babin, P., Cerda, J. and Lunzens, E., 2007. *The Fish Oocyte: From Basic Studies to Biotechnological Applications*. Springer, Dordrecht, The Neetherlands.

Bailey, M.M. and Zydlewski, J.D., 2013. To stock or not to stock? Assessing the restoration potential of a remnant American shad spawning run wit hatchery supplementation. *North Am. J. Fish. Manage.*, **33**: 459-467.

Baillie, J. and Groombridge, B., 1996. *IUCN Red List of Threatened Animals*. IUCN, Gland (Switzerland).

Baltz, D.M., 1991. Introduced fishes in marine ecosystems and inland seas. *Biol. Conserv.*, 56: 151-177.

Basavaraja, N., Ahmed, I. and Hegde, S.N., 1998. Short-term and long-term preservation of fish gametes with special reference to mahseer (*Tor khudree*) spermatozoa. *Fishing Chimes*, 19 (6): 17-19.

Bensam, P. and Menon, N.G., 1994. The threatened, vulnerable and rare marine fishes of India. In: *Threatened Fishes of India* (Eds.) Dehadrai, P.V., Das, P. and Verma, S.R. Nature Conservators, Muzaffarnagar, pp. 297-305.

Bhat, J.P. and Pandit, M.K., 2016. Ebdangetred golden mahaseer, *Tor putiotora* Hamilton: a review of natural history. *Rev. Fish. Biol. Fish.*, 26: 25-38.

Bhowmik, M.L., Mondal, S.C., Chakrabarti, P.P., Das, N.K., Das, K.M., Saha, R.N. and Ayyappan, S., 2000. Captive breeding and rearing of *Ompok pabda* (Hamilton-Buchanan)-a threatened species. In: *Fish Biodiversity of North-East India* (Eds.) Ponniah, A.G. and Sarkar, U.K. NATP Pub. No.2. NBFGR, Lucknow. pp. 120-121.

Bindu, L. and Padmakumar, K.G., 2014. Spawning and early development of the endemic and threatened yellow catfish, *Horabagrus brachysoma* (Gunther, 1864) (Teleosti: Bagaridae). *J. Theat. Taxa*, 6: 5368-5374.

Biju, R., Thomas, K.R. and Ajithkumar, C.R., 1998. *Sicypterus griseus* (Day) from Periyar river, Kerala. *J. Bombay Nat. Hist. Soc.*, 95: 351-352.

Boehlert, G.W., 1996. Biodiversity and sustainability of marine fishes. *Oceanography*, 9 (1) : 28-35.

Bogan, A.E. 1996. Decline and decimation: the extirpation of the unionid freshwater bivalves of North America. *J. Shellfish Res.*, 15: 484-490.

Bohnsack, J.A., 1993. Marine reserves - they enhance fisheries, reduce conflicts and protect resources. *Oceanus*, 36: 63-72.

Bohnsack, J.A., 1996. Marine reserves, zoning, and the future of fishery management. *Fisheries (Bethesda)*, 21 (9): 14-16.

Botsford, L.W., Castilla, J.C. and Peterson, C.H., 1997. The management of fisheries and marine ecosystem. *Science*, 277: 509-515.

Brahmi, P., Dua, R.P. and Dhillon, B.S., 2004. The Biological Diversity Act in India and agro-biodiversity management. *Curr. Sci.*, 86: 659-664.

Brander, K. M., 1981. Disappearance of common skate, *Raia batis* from Irish sea. *Nature*, 290: 48-49.

Brander, K.M., 2007. Global fish production and clmate changes. *Proc. Natl. Acad. Sci. (USA)*, 104: 19709-19714.

Brander, K. M., 2009. Impact of climate change on marine ecosystems and fisheries. *J. Mar. Biol. Assoc. India*, 51: 1-13.

Brown, A.H.D., Frankel, O.H., Marshall, D.R. and Williams, T.J. (1989. *The Use of Plant Genetic Resources*. University of Cambridge Press, Cambridge.

Buckworth, R.C., 1998. World fisheries are in crisis? We must repond! In: *Reinventing Fisheries Management* (Eds.) Pitcher, T.J., Hart, P.J.B. and Pauly, D. Kluwer Acad. Pub., London, pp. 3-7.

Casey, J.M. and Meyers, R.A., 1998. Near extinction of a large, widely distributed fish. *Science*, 281: 690-692.

Chaturvedi, C.S., Singh, R.K. and Pandey, A.K., 2012. Successful induced breeding of endangered *Ompok pobda* (Hamilton-Buchanan) in Raipur, Chhattisgarh (India). *Biochem. Cell. Arch.*, 12: 321-325.

Chaturvedi, C.S., Basade, Y. and Pandey, A.K., 2013. Artificial breeding of endangered golden mahseer, *Tor putitora* (Hamilton-Buchanan). *Biochem. Cell. Arch.*, 13: 151-153.

Compagno, L.J.V., 1997. Threatened fishes of the world : *Glyphis gangeticus* (Muller and Henley 1839) (Carcharhinidae). *Environ. Biol. Fish.*, 49: 400.

Crossman, E.J., 1991. Introduced freshwater fishes: a review of North American perspective with emphasis on Canada. *Can. J. Fish. Aquat. Sci.*, 48 (Suppl. 1): 46-57.

Das, P., 1996. Endangered fishes of India and strategies for their conservation. *J. Natcon.*, 8: 191-197.

Das, P., 2001. Genetics of conservation and aquaculture. In : *International Symposium on Fish and Nutritional Security in the 21st Century (December 4-6, 2001)*. Central Institute of Fisheries Education, Mumbai. pp. 63-69.

Das, P. and Pandey, A.K., 1998. Current status of fish germplasm resources of India and strategies for conservation of endangered species. In : *Fish Genetics and Biodiversity Conservation* (Eds.)Ponniah, A.G., Das, P. and Verma, S.R. Nature Conservators, Muzaffarnagar, pp. 253-273.

Das, P. and. Pandey, A.K., 1999. Endangered fish species;measures for rehabilitation and conservation. *Fishing Chimes*, 19 (6): 31-34.

Das, P., Verma, S.R., Dhanze, J.R. and Malik, D.S., 2002. *Coldwater Fish Genetic Resources and Their Conservation*. Nature Conservators, Muzaffarnagar. 314 p.

Das, P., Bhowmick, R.M., Nandy, A.C., Pandit, P.K., Sen Gupta, R.C. and Thakurta, S.C., 2004. Diminishing trend of fish species diversity in West Bengal : field study. *Fishing Chimes*, 24 (1): 73-78.

Daw, T. and Gray, T. 2005. Fisheries Science and suatainability in internatuional policy: a study of failture in the European Union's Common Fisheries Policies. *Mar. Policy*, 29: 189-197. cy

Deacon, J.E., Kobetich, G., Williams, J.D. and Contreras-Balderas, S., 1979. Fishes of the North America : endangered, threatened, or of special concern. *Fisheries (Bethesda)*, 4 (2): 29-44.

Dehadrai, P.V., 1996. Aquaculture and environment. In : Recent Advances in Biosciences and Oceanography (Eds.) Agrawal, V.P., Prakash, R. and Abidi, S.A.H. Society of Biosciences, Muzaffarnagar, pp. 63-66.

Dehadrai, P.V., 2001. Nutritional security through fisheries and aquaculture in India. In: *International Symposium on Fish and Nutritional Security in the 21st Century* (December 4-6, 2001). Central Institute of Fisheries Education, Mumbai, pp. 6-16.

Dehadrai, P.V. and Kamal, M.Y., 1993. Role of air-breathing fish culture in rural upliftment. In: *Souvenir-Third Indian Fisheries Forum* (October 14-16, 1993).

College of Fisheries, G.B. Pant University of Agriculture and Technology, Pantnagar, pp. 28-33.

Dehadrai, P.V., Das, P. and Verma, S.R., 1994. *Threatened Fishes of India*. Nature Conservators, Muzaffarnagar, 412 p.

Devaraj, M. and Pillai, V.N., 2000. Marine fisheries research and development in India. In : *Fifty Years of Fisheries Research in India* (Eds.) Gopakumar, K., Singh, B.N. and Chitransi, V.R. Indian Council of Agricultural Research, New Delhi, pp. 124-135.

Dhanze, J.R. and Dhanze, R., 1994. An appraisal of depleting fish genetic resources of Himachal Pradesh. In: *Threatened Fishes of India* (Eds.) Dehadrai, P.V., Das, P. and Verma, S.R. Nature Conservators, Muzaffarnagar, pp. 197-204.

Dhanze, J.R. and Dhanze, R., 1998. Impact of habitat shrinkage on the indigenous fish genetic resources of Beas drainage system. In: *Fish Genetics and Biodiversity Conservation* (Eds.) Ponniah, A.G., Das, P. and Verma, S.R. Nature Conservators, Muzaffarnagar, pp. 115-126.

Dill, W.A. and Cordone, A.J., 1997. History and status of introduced fishes in California, 1871-1996 : conclusions. *Fisheries (Bethesda)*, 22 (10): 15-18.

Dobson, A.P., Bradshaw, A.D. and Baker, J.M., 1997. Hopes for the future: restoration ecology and conservation biology. *Science*, 277: 515-522.

Dowling, T.E. and Childs, M.R., 1992. Impact of hybridization on the threatened trout of the Southwest United States. *Conserv. Biol.*, 6: 355-364.

Drew, J.A., 2005. Use of Traditional Ecological Knowledge in marine conservation. *Conserv. Biol.*, 19: 1286-1293.

Easa, P.S. and Shaji, C.P., 1997. Freshwater fish diversity in Kerala part of the Nilgiri Biosphere Reserve. *Curr. Sci.*, 73: 180-182.

Eknath A.E. and Doyle, R.W., 1990. Effective population size and state of inbreeding in aquaculture of Indian major carps. *Aquaculture*, 85: 293-305.

Elvira, B., 1990) Iberian endemic freshwater fishes and their conservation status. *J. Fish Biol.*, 37 (Suppl. A): 231-232.

FAO, 1995. Code of Conduct for Responsible Fisheries. FAO, Rome, 41 p.

Ferguson, M.M. and Drahushchak, L.R., 1990. Enzyme heterogenity and disease resistance in rainbow trout. *Heredity*, 64: 413-417.

Frankel, O.H. and Soule, M.E., 1981. *Conservation Evolution*. Cambridge University Press, Cambridge and New York.

Franklin, I.R., 980. Evolutionary changes in small population. In : *Conservation Biology : An Evolutionary-Ecological Perspective* (Eds.) Soule, M.E. and Wilcox, B.A. California University Press, California, pp. 135-149.

French, T.E., Cadden, D. and Zimmerman, K., 2004. Recovery of the endangered Nechako river white sturgeon (Acipenser transmontanus) populatio. *Streamline Watershed Mange. Bull.*, 7 (4): 8-13.

Gadgil, M., Chadrasekhariah, H.N. and Bhatt, A., 2001. Freshwater fishes : out of sight, out of mind. In: *The Hindu Survey of the Environment* (Ed.) Rangarajan, S., National Press, Chennai, pp.137-142.

Gjerde, B. (1986). Growth and reproduction in fish and shellfish. *Aquaculture*, 57: 37-55.

Gopalakrishnan, A. and Ponniah, A.G., 1999. Introduction of red piranhas for aquarium purpose in Kerala. *Fishing Chimes*, 18 (12): 53-55.

Gray, J.S., 1997. Marine biodiversity: patterns, threats and conserrvation needs. *Biodiv. Conserv.*, 6: 153-157.

Grimes, C.B., 1998. Marine stock enhancement : sound management or techno-arrogance? Fisheries *(Bethesda)*, 23 (9): 18-23.

Groombridge, B., 1992. *Global Biodiversity : Status of the Earth's Living Resources.* World Conservation Monitoring Centre. Chapman and Hall, London.

Gupta, N., 2014. River conservation in the Indian Himalyan region. *Ph.D. Thesis*. University of London. London.

Hilton-Taylor, C., 2000. *IUCN Red List of Threatened Species.* The IUCN Species Survival Commission, IUCN, Gland, Switzerland and Cambridge.

Hesthagen, T. and Hansen, L., 1991. Estimates of annual loss of Atlantic salmon, *Salmo salar* L. in Norway due to acidification. *Aquacult. Fish. Manage.*, 22: 85-91.

Hill, M.T. and Platts, W.S., 1998. Ecosystem restoration : a case study in the Owens River George, California. *Fisheries (Bethedsa)*, 23 (11): 18-26.

Horvath, A. and Urbanyi, B., 2000. Cryopreservation of fish gametes. *Halaszat*, 93: 39-43.

Hughes, R.M. and Noss, R.F., 1992. Biological biodiversity and biological integrity: current concerns for lakes and streams. *Fisheries (Bethesda)*, 17(3): 11-19.

Huntsman, G.R., 1994. Endangered marine finfish : neglected resources or beasts of fiction. *Fisheries (Bethedsa)*, **19 (7)** : 8-15.

Hutchings, J.A., 2000. Collapse and recovery of marine fishes. *Nature*, 406: 882-885.

IUCN, 1990. *IUCN Red List of Threatened Animals.* World Conservation Monitoring Centre, Cambridge, United Kingdom.

IUCN, 2000. *The 2000 IUCN Red List of Threatened Species.* World Conservation Monitoring Centre, Cambridge, United Kingdom.

Janan, D.A., Rasid, J., Khan, M.M. and Mahmud, Y. 2015. Early embryonic and larval development of threated humped featherback, *Chital chitala* (Hamilton). *Tren. Fish. Res.*, 4 (2): 1-7.

James, P.S.B.R., 1994. Endangered, vulnerable and rare marine fishes and animals. In: *Threatened Fishes of India* (Eds.) Dehadrai, P.V., Das, P. and Verma, S.R. Nature Conservators. Muzaffarnagar, pp. 271-275.

James, P.S.B.R, 1999. Indian marine biodiversity heritage - an ecological and economic appraisal. *Fishing Chimes*, 19 (1): 83-86 and 19 (2): 37-38.

Jayaram, K.C., 2010. *The Freshwater Fishes of the Indian Region*. Narendra Pub. House, Delhi. 616 p.

Jenesn, B.L., 1994. Fish refugia and captive propagation : a viable aid to conservation and restoration. In: *Threatened Fishes of India* (Eds.) Dehadrai, P.V., Das, P. and Verma, S.R. Nature Conservators. Muzaffarnagar, pp. 311-320.

Jhingran, A.G., 1984. *Some Considerations on Introduction of Tilapia into Indiam Waters*. Bulletin of the Bureau of Fish Genetic Resources, CIFRI, Barrackpore, 34 p.

Jhingran, A.G., 1991. Impact of environmental stresses on freshwater fisheries resources. *J. Inland Fish. Soc. India*, 23: 20-32.

Jhingran, V.G., 1991. *Fish and Fisheries of India*. Hindustan Pub. Co., New Delhi, 727 p.

Jhingran, V.G. and Sehgal, K.L., 1978. *Coldwater Fisheries of India*. The Inland Fisheries Soiciety of India, Barrackpore, 294 p.

John, G., Ponniah, A.G., Lakra, W.S., Gopalakrishnan, A. and Barat, A., 1993. Preliminary attempts to cryopreserve enbryos of *Cyprinus carpio* (L.) and *Labeo rohita* (Ham.). *J. Inland Fish. Soc. India*, 25: 55-57.

Joseph, M.M. and Menon, N.G., 2001. Impact of coastal fisheries and other anthropogenic activities on the food security of the coastal communities of India. In: *International Symposium on Fish for Nutritional Security in the 21st Century* (December 4-6, 2001). Central Instritute of Fisheries Education, Mumbai, pp. 43-50.

Joshi C.B., 1981. Artificial breeding of golden mahseer (*Tor putitora*). *J. Inland Fish. Soc. India*, 13: 73-79

Joshi C.B. and Raina, H.S., 1997. Impact of habitat changes on mahseer fishery in Indian uplands with special reference to Kumaon Himalayas. In: *Changing Perspectives of Inland Fisheries* (Eds.) Vass, K.K. and Sinha, M. The Inland Fisheries Society of India, Barrackpore, pp. 84-88.

Khan, H.A. and Sinha, M., 2000. Status of mahseer in North and North-Eastern India with a note on their conservation. *J. Inland Fish. Soc. India*, 32: 28-36.

Kincaid, H.L., 1976. Inbreeding in rainbow trout (*Salmo gairdneri*). *J. Fish. Res. Bd. Can.*, 33: 2420-2426.

Kirchhoffer, A. and Hefti, D., 1996. *Conservation of Endangered Freshwater Fish in Europe*. Birkhauser-Verlag, Basael, Switzerland.

Koenig, C.C., Coleman, F.C., Grimes, C.B., Fitzhugh, G.R., Scanlon, K.M., Gledhill, C.T. and Grace, M., 2000. Protectio of fish spawning habitat for the conservatio of warm-temperate reef-fish fisheries of shelf-edge reefs of Florida. *Bull. Mar. Sci.*, 66: 593-616.

Krishna Kartha, K.N., 1983. *A Code List of Common Marine Living Resources of the Indian Seas*. Special Pub. No. 12. CMFRI, Cochin, 150 p.

Kulkarni, C.V., 1992. On the endangered Indian trout, *Barilius bola* (Ham.). *J. Bombay Nat. Hist. Soc.*, 89: 277-281.

Kulkarni, C.V. and Ogale, S.N., 1986. Hypophysation (induced breeding) of mahseer fish, *Tor khudree* (Sykes). *Punjab Fish. Bull.*, 10: 33-35.

Kumar, K., 1988. Govind Sagar Reservoir-a case study on the use of carp stocking for fisheries enhancement. *FAO Fish. Rep.* No. 405 (Suppl.), pp. 46-70.

Kumar, K., 2000. Conservation and development of golden mahseer (*Tor putitora*, Ham.) in Himachal waters. *Fishing Chimes*, 20 (9): 26-27.

Kumarguru, A.K., 1995. Water pollution and fisheries. *Eco. Env. Conserv.*, 1: 140-150.

Kurup, B.M., 1994. Maturation and spawning of an indigenous carp, *Labeo dussumieri* (Val) in the river Pampa. *J. Aqua. Trop.*, 9: 119-132.

Kurup, B.M. (2001). Ranching of *Labeo dussumieri* (Val.) in the river Pampa. In: *Workshop on Captive Breeding of Prioritized Cultivable and Ornamental Fishes for Commercial Utilization and Culture* (July 29-30, 2001). National Bureau of Fish Genetic Resources, Lucknow, p. 59.

Kurup, B.M., 2002. Rivers and streams of Kerala part of Western Ghats-hotspots of exceptional fish biodiversity and endemism. In: *Riverine and Reservoir Fisheries of India* (Eds.) Boopendranath, M.R., Meenakumari, B., Joseph, J., Shankar, T.V., Pravin, P. and Edwin, L. Society of Fisheries Technologists (India), Cochin, pp. 204-217.

Kurup, B.M. and Kuriakose, B., 1991. *Labeo dussumieri* (Val.): an indigenous endangered carp species of Kerala. *Fishing Chimes*, 11 (7): 39-41.

Lakra, W.S., Sarkar, U.K., Gopalakrishnan, A. and Kathirvelpandian, A., 2010. *Threatened Freshwater Fishes of India*. National Bureau of Fish Genetic Resources, Lucknow.

Largiader, C.R., Scholl, A. and Guyomard, R., 1996. The role of natural and artificial propagation on the genetic diversity of brown trout (*Salmo trutta* L.) of the river Rhone drainage. In: *Conservation of Endangered Freshwater Fish of Europe* (Eds.) Kirchhofer, A. and Hefti, D. Birkhauser-Verlag, Basel, Switzerland, pp. 181-197.

Lelek, A., 1987. The Freshwater Fishes of Europe. Vol. IX. *Threatened Fishes of Europe*. Wisbadon W. Germany, Aula-Verlag.

Linhart, O., Gela, D., Rodina, M. and Rodriguez-Gutierrez, M. (2001. Short-trem storage of ova of commomn carp and tench in extenders. *J. Fish Biol.*, 59: 616-623.

Lohoefener, R., 1997. Species and subspecies : protecting aquatic invertebrates under the Endangered Species Act 1973 as amended. *J. Shellfish Res.*, 16: 324.

Mahanta, P.C., Kapoor, D., Pandey, A.K., Srivastava, S.M., Dayal, R., Patiyal, R.S., Joshi, K.D., Singh, A.K. and Paul, S.K., 1998. Conservation and rehabilitation of mahseer in India. In : *Fish Genetics and Biodiversity Conservation* (Eds.) Ponniah, A.G., Das, P. and Verma, S.R. Nature Conservators, Muzaffarnagar, pp. 93-105.

Mahanta, P.C., Srivastava, S.M. and Paul, S.K., 2001. Preliminary assessment of fish germplasm resources of North-Eastern Region to evolve strategy for conservation. *Aquacult*, 2: 181-190.

Mahapatra, B.K., Vinod, K. and Mandal, B.K., 2004. Studies on chocolate mahseer, *Neolissocheilus hexagonolepis* (McClelland) fishery and the cause of its decline in Umiam reservoir, Meghalaya. *J. Natcon.*, 16: 199-205.

Maitland, P.S., 1993. Conservation of freshwater fish in India. In: *Advances in Fish Research* (Ed.) Singh, B.R. Narendra Pub. House, New Delhi, pp. 349-364.

Master, L.M., 1990. The imperiled status of North American aquatic animals. *Biodiversity Network News*, 3: 7-8.

Meffe, G., 1986a. Conserving fish genomes: philosophies and practices. *Environ. Biol. Fishes*, 18: 3-9.

Meffe, G., 1986b. Conservation genetics and the management of endangered fishes. *Fisheries (Bethesda)*, 11 (1): 14-23.

Menon, A.G.K, 1989. Conservation of ichthyotauna of India. In: *Conservation and Management of Inland Capture Fisheries Resources of India* (Eds.) Jhingran, A.G. and Sugunan, V.V. The Inland Fisheries Society of India, Barrackpore, pp. 25-33.

Menon, A.G.K., 1997. Rare and endangered fishes of Malabar, India. *Zoos' Print*, 12 (11): 6-19.

Menon, A.G.K., 2004. *Threatenedc Fishes of India and Their Conservation*. Zoological Survety of India, Kolkata. 170 p.

Menon, N.G. and Pillai, C.S.G., 1996. *Marine Biodiversity Conservation and Management*. Central Marine Fisheries Research Institute, Cochin.

Mignogno, D.C., 1996. Freshwater mussel conservation and the Endangered Fish Act. *J. Shellfish Res.*, 15: 486-480.

Miller, P. and Craig, J.F., 2001. Fish Biodiversity and Conservation. Annual Symposium of the Fisheries Society of the British Isles, Leicester, U.K. July 9-13, 2001. *J. Fish Biol.*, **59** (Suppl.) : 1-381.

Miller, R.R., Williams, J.D. and Williams, J.E., 1989. Extinction of North American fishes during past century. *Fisheries (Bethesda)*, 14 (6): 22-38.

Minckley, W.L. and. Deacon, J.E., 1968. Southwestern fishes and the enigma of "Endangered Species". *Science*, 159: 1424-1432.

Minckley, W.L. and Deacon, J.E., 1991. *Battle Against Extinction: Native Fish Management in the America*. Western University of Arizona Press, Tucson and London.

Mishra, A. and Jain, A.K., 1993. Annual rate of inbreeding of Indian major carps in two representative fish hatcheries of Uttar Pradesh and their growth evaluation. In : Proceedings of the Third Indian Fishries Forum, Pantnagar (Eds.) Joseph, M.M. and Mohan, C.V. College of Fisheries, Mangalore, pp. 97-100.

Mishra, A., Pandey, A.K., Singh, A.K. and Das, P., 1998) Impact of introduction of exotic and genetically-manipulated fishes on freshwater Indian conventional stock. In : *Fish Genetics and Biodiversity Conservation* (Eds.) Ponniah, A.G., Das, P. and Verma, S.R. Nature Conservators, Muzaffarnagar, pp. 275-292.

Mishra, A., Pandey, A.K. and Das, P., 2000. High incidence of body shape deformities and stunted growth in the hatchery-bred progeny of silver carp, *Hypophthalmichthys molitrix* (Valenciennes). *Proc. Zool. Soc. (Calcutta)*, 51: 55-59.

Mohan, C.V. and Shankar, K.M., 1994. Epidemiological analysis of epizootic ulcerative syndrome of fresh and brackishwater fishes of Karnataka, India. *Curr. Sci.*, **66**: 656-658.

Molur, S. and Walker, S., 1998. *Workshop Report on the Conservation Assessment and Management Plan for Freshwater Fishes of India*. Zoo Outreach Organization/ CBSG, Coimbatore, 158 p.

Mondal, B.K. and Mondal, A.K., 2000. Culture of seabass (*Lates calcarifer*, Bloch) in West Bengal - a case study. *Fishing Chimes*, 19 (12): 13-15.

Molony, B.W., Lenanton, W.R., Jackson, G and Norriss, J., 2003. Stock enhancement as a fisheries management tool. *Rev. Fish Biol. Fish.*, 13; 409-432.

Mukherjee, M., Prahraj, A. and Das, S., 2002. Conservation of endangered fish stocks through artificial propagation and larval rearing technique in West Bengal, India. *Aqua. Asia*, 7 (2): 8-11.

Mukhopadhyay, M.K., 1994. Some threatened estuarine fishes of India. In: *Threatened Fishes of India* (Eds.) Dehadrai, P.V., Das, P. and Verma, S.R. Nature Conservators, Muzaffarnagar, pp. 229-236.

Musick, J.A., Berkelet, S.A., Cailliet, G.M., Camhi, M., Huntsman, G., Nammack, M. and Warren, M.L. Jr., 2000a. Protection of marine fish stocks at risk of extinction. *Fisheries (Bethesda)*, 25 (3): 6-8.

Musick, J.A., Harbin, M.M., Berkeley, S.A., Burgees, G.H., Eklund, A.M., Findley, L., Gilmore, R.G., Golden, J.T., Ha, D.S., Huntsman, G.R., McGovern, J.C., Parker, S.J., Poss, S.G., Sala, E., Schmidt, T.W. Sedberry, G.R., Weeks, H. and Wright, S.G., 2000b. Marine, estuarine, and diadromous fish stocks at risk of extinction in North America (exclusive of Pacific salmonids). *Fisheries (Bethesda)*, 25(11): 6-29.

Musick, J.A., Burgess, G., Calliet, G., Camhi, M. and Fordham, S., 2000c. Management of sharks and their relatives (Elasmobrachii). *Fisheries (Bethesda)*, 25(3): 9-13.

Myers, N., 2003. Biodiversity hotspots revisited. *BioScience*, 53: 796-797.

Myers, N., Mittermeler, R.A., Mittermeler, C.G., da Fonseca. G.A.B. and Kent, J., 2000. Biodiversity hotspots for conservation priorities. *Nature*, 403: 853-858.

Nadis, S., 1997. Researchers bring fishermen on board. *Nature*, 386: 108.

Nair, N.B., 2001. The ban on monsoon trawling: significance and consequences. In: *National Symposium on Fishery Technologies and Their* Commercialization (January 11-12, 2001). Central Institute of Fisheries Education, Mumbai, pp. 4-12.

Nandeesha, M.C., Bhadraswamy, G., Patil, T.G., Varghese, T.J., Sarma, K. and Keshavanath, P., 1993. Preliminary results on induced spawning of pond-raised mahseer, *Tor khudree. J. Aquacult. Trop.*, 8: 55- 60.

Narain, P., 2000. Genetic diversity-conservation and assessment. *Curr. Sci.*, 79: 170-175.

Nautiyal, P., 1994. *Mahseer-the Game Fish : Natural History, Status and Conservation Practices in India and Nepal*. Rachna Publication, Srinagar-Garhwal.

Nautiyal, P., 2013. Review of art and sxcience of Indian mahseer (game fish) fron niteenth to twentieth centuary: road to extinction and conservation. *Proc. Nat. Acad. Sci. India*.

Nehlsen, W., Williams, J.E. and Lichatowich, J.A., 1991. Pacific salmon at the cross roads : stocks at risk from California, Oregon, Idaho and Washington. *Fisheries (Bethesda)*, 16 (2): 4-21.

Nei, M., Maruyama, T. and Chakraborty, R., 1975. The bottleneck effect and genetic variability in populations. *Evolution*, 29: 1-10.

Nelson, J.S., 1994. *Fishes of the World*. 3rd Edn. John Wiley and Sons, New York.

Norman, B., 2005. *Rhincodon typus. The IUCN List of Threatened Species*. Gland (Switzerland).

Norman, Ogale, S.N., 1997. Induced spawning and hatching of golden mahseer, *Tor putitora* (Hamilton) at Lonvla, Pune Dist. (Maharashtra) in Western Ghats. *Fishing Chimes*, 17 (1): 27-29.

Padhi, B.K. and Mandal, R.K., 1994. Improper fish breeding practices and their impact on aquaculture and fish biodiversity. *Curr. Sci.*, 66: 624-626.

Padhi, B.K. and Mandal, R.K., 2000. *Applied Fish Genetics*. Fishing Chimes House, Visakhapatnam. 190 p.

Padmakumar, K.G., Bindu, L., Sreerekha, P.S., Gopalkrishnan, A., Baseer, V.S., Joseph, N., Manu, P.S. and Krishnan, A., 2011. Breeding of endemic catfish, *Horabagrus brachysoma*, in captive conditions. *Curr. Sci.*, 100: 1232-1236.

Pandey, A.C., Pandey, A.K. and Das, P., 1999. Fish and fisheries in relation to aquatic pollution. In: *Environmental Issues and Resource Management* (Eds.) Das, P., Verma, S.R., Gupta, A.K and Sharma, R.K.), *Nature Conservators*, Muzaffarnagar, pp. 67-112.

Pandey, A.K. and Das, P., 2002. Fish biodiversity and strategies for conservation of threatened species. In: *Biological and Biotechnological Resources* (Eds.) Tripathi, G and Tripathi, Y.C. Campus Books International, New Delhi, pp. 42-72.

Pandey, A.K. and Das, P., 2004. Artificial fecundation and ranching for conservation of endangered fishes. In: *Fish Biodiversity in Protected Habitats* (Eds.) Ayyappan, S., Malik, D.S., Dhanze, R. and Chauhan, R.S. *Nature Conservators*, Muzaffarnagar, pp. 151-162.

Pandey, A.K. and Das, P., 2006. Current status of fish germplasm resources of India and strategies for conservation of endangered species. In: *Proceedings of Recent Advances in Applied Zoology* (Eds.) Singh, H.S., Chaubey, A.K. and. Bhardwaj, S.K Ch. Charan Singh University, Meerut, pp. 1-39.

Pandey, A.K., Patiyal, R.S., Upadhyay, J.C., Tyagi, M. and Mahanta, P.C., 1998a. Induced spawning of endangered mahseer (*Tor putitora*) with ovaprim at State Fish Farm near Dehradun. *Indian J. Fish.*, 45: 457-459.

Pandey, A.K., Mishra, A. and Das, P., 1998b. Skeletal deformities and stunted body growth in the hatchery-bred progeny of *Cirrhinus mrigala* (Hamilton-Buchanan). *J. Natcon.*, 10: 231-237.

Pandey, A.K., Mahanta, P.C., Singh, B.N. and Das, P., 2002. Fish genetic resources of India with emphasis on coldwater biodiversity conservation. In: *Coldwater Fish Genetic Resources and Their Conservation* (Eds.) Das, P., Verma, S.R., Dhanze, J.R. and Malik, D.S. *Nature Conservators*, Muzaffarnagar, pp. 301-314.

Pandey, K.D. and Awasthi, S.K. (1994). Endangered, threatened and rare fishes of Uttar Pradesh. In: *Threatened Fishes of India* (Dehadrai, P.V., P. Das and S.R. Verma, eds.). *Nature Conservators*, Muzaffarnagar. pp. 13-15.

Pandey, K.D., H.A. Khan and S.K. Awasthi, 1994. Present status of *Tor putitora* in Bhimtal Lake, U.P. In: *Threatened Fishes of India* (Eds.) Dehadrai, P.V., Das, P. and Verma, S. R. *Nature Conservators*, Muzaffarnagar, pp. 143-147.

Pandit, P.K., Bahumik, U. and Chatterjee, J.G. (1994). Threatened fishes of Sunderbans, West Bengal. In: *Threatened Fishes of India* (Eds.) Dehadrai, P.V., Das, P. and Verma, S. R. *Nature Conservators*, Muzaffarnagar, pp. 253-260.

Pavolov, D.S., 1993. Strategies for preserving rare and endangered fishes. *J. Ichthyol.*, 33(1): 109-127.

Penczak, T., 1996. Natural regeneration of endangered fish populations in the Pilica drainage basin after reducing human impact. In: *Conservation of Endangered Freshwater Fish of Europe* (Eds.) Kirchofer, A. and Hefti, D. Birkhauser-Verlag, Basel, pp.121-133.

Perry, A.L., Low, P.J., Ellis, J.R. and Reynolds, J.D., 2005. Climate change and distribution shifts in marine fishes. *Science*, 308: 1912-1915.

Philippart, J.C., 1995. Is captive breeding an effective solution for the preservation of endemic species ? *Biol. Conserv.*, 72: 281-295.

Ponniah, A.G., 1996. Fish germplasm gene banking. *J. Natcon.*, 8: 135-139.

Ponniah, A.G. and Sarkar, U.K., 2000. *Fish Biodiversity of North-East India*. NATP Pub. No. 2. NBFGR, Lucknow, 228 p.

Ponniah, A.G., Gopalakrishnan, A., Baseer, V.S., Munner, P.M.A., Paul, B., Padmakumar, K.G. and Krishnan, A., 2000. Captive breeding and gene banking of endangered endemic yellow catfish, *Horabragrus brachysoma*. In: *National Seminar on Sustainable Fisheries and Aquaculture for Nutritional Security* (November 29-December 02, 2000). National Academy of Agricultural Sciences and Madurai Kamaraj University, Chennai, p. 99.

Prasad, P.S., 1994. Status paper on endangered, vulnerable and rare fishes of Bihar. In: *Threatened Fishes of India* (Eds.) Dehadrai, P.V., Das, P. and Verma, S. R. Nature Conservators, Muzaffarnagar, pp. 24-29.

Pravin, P., 2000. Whale shark in the Indian coast - need for conservation. *Curr. Sci.*, 79: 310-315.

Proudlove, G.S., 2001. The conservation status of hypogean fishes. *Environ. Biol. Fish.*, 62: 201-213.

Pullin, R.S.V., 1994. Exotic species and genetically-modified organisms in aquacultre and enhanced fisheries : ICLARM position. *Naga (ICLARM)*, 17: 19-24.

Qasim, S.Z., 1980. *Proposal for the Development of Marine Park at Malvan (Maharashtra)*. National Institute of Oceanography, Dona Paula, Goa.

Rajan, P.T., Devi, K. and Dey, S., 1993. New records of rare fishes from Andaman Islands. *J. Andaman Sci. Assoc.*, 9: 103-106.

Ramakrishniah, M., 1998. Some aspects of biology and abundance of *Gambusia affinis* (Baird and Girard) in an artificial impoundment and its probable effects on indigenous fish fauna. *J. Inland Fish. Soc. India*, 30: 24-28.

Rama Rao, Y., 1992. Endangered, vulnerable and rare fishes of the Ganga river system and warmwaters of Indus system. In : *National Seminar on Endangered Fishes of India (April 25-26, 1992)*. NBFGR and Nature Conservators, Allahabad, pp. 9-10.

Rao, G.R.M., 2001. Brackishwater aquaculture : prospects and potentials. In: *National Symposium on Fishery Technologies and Their Commercialization* (January 11-12, 2001). Central Institute of Fisheries Education, Mumbai. pp. 51-56.

Reddy, P.V.G.K., 2000. *Captive breeding of Osteobrama belangeri* (Val.)-a threatened food fish. In : *Fish Biodiversity of North-East India* (Eds.) Ponniah, A.G. and Sarkar, U.K. NATP Pub. No.2. NBFGR, Lucknow, pp. 122-123.

Rosa, R.S. and Menezes, N.A., 1996. Preliminary list of endangered fish species (Pisces; Elasmobranchii, Actinopterigiii) in Brazil. *Rev. Bras. Zool.*, 13: 647-667.

Ruttenberg, B.I., 2001. Effects of artisanal fishing on marine communities in the Galapagos Islands. *Conerv. Biol.*, 15: 1691-1699.

Ryman, N. and Utter, F.M., 1987. *Population Genetics and Fishery Management*. Washington University Press, Seattle.

Sahoo, S.K., Giri, S.S., Paramanik, M. and Ferosekhan, S., 2014. Preliminary observation on the induced breeding and hatchery rearing of an endangered catfish, *Horabagrus brachysoma* (Gunther). *Intern. J. Fish. Aquat. Sci.*, 1 (5): 130-133.

Sakthivel, M., 2001. Need for legal framework for sustainable aquaculture. In: *Sustainable Indian Fisheries* (Ed.) Pandian, T.J. National Academy of Agricultural Sciences, New Delhi. pp. 225-231

Sandhu, G.S., Tandon, K.K. and Johal, M.S., 1994. Growth studies of an endangered fish, *Tor putitora* (Hamilton) from Govindsagar (H.P.), India. In: *Threatened Fishes of India* (Eds.) Dehadrai, P.V., Das, P. and Verma, S. R. Nature Conservators, Muzaffarnagar, pp. 137-142.

Sarkar, U.K., Patiyal, R.S. Srivastava, and S.M., 2001. Successful induced spawning and hatching of hill-stream carp, *Labeo dyocheilus* (McClelland) in Kosi river. *Indian J. Fish.*, 48: 413-416.

Sarkar, U.K., Deepak, P.K., Negi, R.S., Singh, S.P. and Kapoor, D., 2006. Captive breeding of endangered fish, *Chitala chitala* (Hamilton-Buchanan), for species conservation and sustainable utilization. *Biodiv. Conserv.*, 15: 3579-3589.

Schreier, A.D., Rodzen, J., Ireland, S. and May, B., 2012. Genetic techniques inform conservation aquaculture of the endangered Kootenai river white stugeon, *Acipenser transmontanus*. *Endang. Speces Res.*, 16: 65-75.

Sehgal, K.L., 1991. *Artificial Propagation of the Golden Mahseer, Tor putitora (Hamilton) in the Himalayas*. Special Pub. No. 2. National Research Centre on Coldwater Fisheries, Haldwani, 12 p.

Sehgal, K.L., 1992. *Review and Status of Coldwater Fishery Research in India*. Special Pub. No.3. National Research Centre on Coldwater Fisheries, Haldwani, 60 p.

Sehgal, K.L., 1994. State-of-art of endangered, vulnerable and rare coldwater fishes of India. In: *Threatened Fishes of India* (Eds.) Dehadrai, P.V., Das, P. and Verma, S. R. Nature Conservators, Muzaffarnagar, pp. 127-136.

Sehgal, K.L. and Shyam Sunder, 1992. *Coldwater Fisheries Technologies : Estimation of Biological Productivity of the Mountain Streams*. Manual No. 4. National Research Centre on Coldwater Fisheries, Haldwani, 29 p.

Sehgal, K.L. and Malik, D.S., 2002. Fish sanctuaries in Indian Himalayas -present status vis-à-vis conservation of coldwater fish stocks. In : *Coldwater Fish Genetic Resources and Their Conservation* (Eds.) Das, P., Verma, S.R., Dhanze, J.R. and Malik, D.S. *Nature Conservators*, Muzaffarnagar, pp. 19-25.

Sehgal, K.L., Shukla, J.P. and Shah, K.L., 1971. Observations on fishery of Kangra Valley and adjcent areas with special reference to mahseer and other indigenous fishes. *J. Inland Fish. Soc. India*, 3: 63-71.

Sehgal, K.L., Sunder, S., Raina, H.S. and Tyagi, B.C., 1988. *Twenty-five Years of Research on Coldwater Fisheries in India*. Special Scientific Report No. 1. National Research Centre on Coldwater Fisheries, Haldwani, 60 p.

Sen, P.R., De, D.K. and Nath, D., 1990. Experiments on artificial propagation of hilsa, *Tenualosa ilisha* (Hamilton). *Indian J. Fish.*, 37: 159-162.

Shaji, C.P., Easa, P.S. and Gopalkrishnan, A., 2000. Freshwater fish diversity of Wastern Ghats. In: *Endemic Fish Diversity of Western Ghats* (Eds.) Ponniah, A.G. and Gopalakrishnan, A. National Bureau of Fish Genetic Resources, Lucknow, pp. 33-55.

Sharma, R.C., 2001. Rearing of mahseer fry (*Tor putitora*) fed with different diets in Tarai region of Uttaranchal. In: *National Seminar on Indian Fisheries and Prospects in Relation to Environment Dynamics* (March 3-5, 2001). University of Jammu, Jammu (J&K). p. 108-109.

Sharma, R.C., 2003. Protection of an endangered fish, *Tor tor* and *Tor putitora* population impacted by transportation network in the area of Tehri Dam Project, Garhwal Himalaya, India. In: *Proceedings of the International Conference on Ecology and Transportation* (Eds.) Irwin, C.L, Garrett, P. and McDermott, K.P.). North Carolina University, Raleigh, pp. 83-90.

Shrestha, B.C., Rai, A.K., Gurung, T.B. and Mori, K., 1990. Successful artificial induced breeding of Himalayan mahseer (*Tor putitora* Hamilton) in Pokhara Valley, Nepal. In: *Proceedings of the Second Asian Fisheries Forum* (Eds.) Hirano, R. and Hanyu, I. Asian Fisheries Society, Manila, pp. 573-575.

Shrestha, T.K., 1990. Rare fishes of Himalyan waters of Nepal. *J. Fish Biol.*, 37A: 213-216.

Shrestha, T.K. (1997). *The Mahseer in the Rivers of Nepal Disrupted by Dams and Ranching Strategies*. R.K. Printers, Kathmandu.

Shrestha, T.K., 1998. Catfish diversity of Himalayan waters of Nepal, their conservation and management. In : *Fish Genetics and Biodiversity Conservation* (Eds.) Ponniah, A.G., Das, P. and Verma, S.R. *Nature Conservators*, Muzaffarnagar, pp. 177-187.

Shyam Lal and Pandey, A.K., 1995. Ecotoxicological problems in freshwater bodies with particular reference to the fertilizer factory effluents in Chilwa Lake, Gorakhpur. In: *Environmental Toxicology* (Eds.) Dwivedi, B. K. and Pandey, G. Bioved Research Society, Allahabad, pp.121-144

Shyam Sunder and Joshi, C.B., 2002. State-of-art of the threatened fishes of Himalayan uplands and strategies for their conservation. In : *Coldwater Fish Genetic Resources and Their Conservation* (Eds.) Das, P., Verma, S.R., Dhanze, J.R. and Malik, D.S. *Nature Conservators*, Muzaffarnagar, pp. 101-111.

Shyam Sunder, Raina, H.S., Mohan, M. and Joshi, C.B., 1995. Cultural possibilities of golden mahseer, *Tor putitota* (Ham.), in Himalayan uplands. *Uttar Pradesh J. Zool.*, 15: 177-181.

Silas, *e.g.*, Mahadevan, S. and Nagappan Nayyar, K., 1985. Existing and proposed marine parks and reserves in India. In: *Proceedings of the Symposium on Endangered Marine Animals and Marne Parks*. The Marine Biological Association of India, Cochin. pp. 414-428.

Simpson G.G., 1953. *The Major Features of Evolution*. Columbia University Press, New York.

Singh, A.K. and Pandey, A.K., 1995. Genetic constraints in management of endangered fishes. *J. Natcon.*, 7 : 99-105.

Singh, A.K., Sarkar, U.K., Abidi, R. and Srivastava, S.M. (2011). Fish biodiversityand invasion risks of alien species in some aquatic bodies uner forest areas of Uttar Pradesh. In: *Souvenior-National Conference on Forest Biodiversity: Earth's Living Treasure* (22 May 2011) Uttar Pradesh State Biodiversity Board, Lucknow, pp. 50-55.

Singh, C.S., 1992. Golden mahseer, *Tor putitora* (Ham.) holds promise in Terai waters of Uttar Pradesh. *Punjab Fish. Bull.*, 16 (1): 17-20.

Singh, N., 1996. Status and protection of snowtrout germplasm in the Alaknanda Valley of Garhwal Hills. In : *Proceedings of the Third Indian Fisheries Forum* (Eds.) Joseph, M.M. and Mohan, C.V. College of Fisheries, Mangalore, pp. 105-108.

Sinha, M., 1992. Mahseer fishery in the North-Eastern states. *Punjab Fish. Bull.*, 14 (1): 66-69.

Sinha, V.R.P., 2009. Fish propagation and extreme climatic events. In : *Recent Advances in Hormonal Physiology of Fish and Shellfish Reproduction* (Eds.) Singh, B.N. and Pandey, A.K. Narendra Pub. House, Delhi. pp. 105-108.

Smith, A.H. and Chesser, R.K., 1981. Rational for conserving genetic variation of fish gene pools. *Ecol. Bull. (Stockholm)*, 34: 13-90.

Smith, P.J., Francis, R. and McVeagh, M., 1991. Loss of genetic diversity due to fishing pressure. *Fisheries Res.*, 10: 309-316.

Soule, M.E., 1980. Thresholds of survival: maintaining fitness and evolutionary potential. In: *Conservation Biology: An Evolutionary-Ecological Perspective* (Eds.) Soule, M.E. and Wilcox, B.A. California University Press, California, pp. 151-169.

Srivastava, S.M., Srivastava, P.P., Dayal, R., Pandey, A.K. and Singh, S.P. (2010. Induced spawning of captive stock of threatened bronze featherback, *Notopterus notopterus*, for stock improvement and conservation. *J. Appl. Biosci.*, 36 (2): 144-147.

Stahl, G., 1983. Differences in the amount and distribution of genetic variation between natural populations and hatchery stocks of Atlantic salmon. *Aquaculture*, 33: 23-32.

Steffens, W., 2008. Significance of aquaculture for the conservation and restoration of sturgeon population. *Bulgarian J. Agric. Sci.*, 14 (2): 1255-164.

Stiassny, M.L.J., 1996. An overview of freshwater biodiversity : with some lessions from African fishes. *Fisheries (Bethesda)*, **21 (9)** : 7-13.

Sugunan, V.V., 1994. Exotic fishes and their role in reservoir fisheries of India. In: *National Seminar on Challenging Frontiers of Inland Fisheries* (December 2-3, 1994). The Inland Fisheries Society of India, Barrackpore, p. 17.

Suseelan, C. and Nair, K.P., 1994. Endangered, vulnerable and rare estuarine shellfishes of India. In: *Threatened Fishes of India* (Eds.) Dehadrai, P.V., Das, P. and Verma, S. R. Nature Conservators, Muzaffarnagar, pp. 237-252.

Sutherland, W.J., 1990. Evolution and fisheries. *Nature*, 344: 814-815.

Talwar, P.K. and A.G. Jhingran (1991). *Inland Fishes of India and Adjacent Countries. Vol.1 and 2.* Oxford and IBH Pub. Co., New Delhi. 1158 p.

Talwar, P.K. and Kacker, R.K., 1984. *Commercial Sea Fishes of India.* Zoological Survey of India, Calcutta. 997 p.

Tave, D., 1991. Effective breeding number and genetic drift. *Aquacult. Magaz.* (Sept-Oct.), 109-112.

Thippeswamy, S. and Hegde, S.N., 1999. Ichthyofaunal diversity of the Sharavathi Tail Race Hydroelectric Power Project area in the Western Ghats of India. In: *Proceedings of the Fourth Indian Fisheries Forum* (Eds.) Joseph, M.M., Menon, N.R. and Nair, N.U. College of Fisheries, Mangalore, pp. 53-56.

Tiersch, T.R., 2008. Strategies for commercialization of cryopreseved fish semen. *Rev. Bras. Zootec.*, 37: 15-19.

Tilman, D., 2000. Causes, consequences and ethics of biodiversity. *Nature*, 405: 208-211.

Tripathi, A.P. and Das, R.K., 1995. River water contaminated with paper mill effluents induces micronuclei in peripheral erythrocytes of fish. *Curr. Sci.*, 68: 708-710.

True, C.D., Silva-Lora, A. and Castro-Castro, M., 1996. Is aquaculture the answar for the endangered totoaba ? *World Aquacult.*, 27 (4): 38-43.

UNCSD, 2012. *United Nations Conference on Sustainable Development: Rio+20* (June 20-22, 2012). Rio de Janeiro, Brazil.

Warren, M.L. and Burr, B.M., 1994. Status of freshwater fishes of the United States : overview of an imperiled fauna. *Fisheries (Bethesda)*, 19 (1): 6-18.

Williams, J.D., Johnson, J.E., Hendrickson, D.A., Connteras, S., Williams, J.D., Havarro-Mendoza, M., MacAllister, D.E. and Deacon, J.E., 1989. Fishes of North America : endangered, threatened, or of special concern 1989. *Fisheries (Bethesda)*, 14 (6): 2-20.

Williams, J.D., Warren, M.L., Cummings, K.S., Harris, J.L. and Neves, R.J., 1993. Conservation status of freshwater mussels of United States and Canada. Fisheries *(Bethesda)*, 18 (9): 6-22.

Winter, B.D. and Hughes, R.M., 1997. Biodiversity. *Fisheries (Bethesda)*, 22 (1): 22-29.

Wright, J.M., 1993. DNA fingerprinting of fishes. In: *Biochemistry and Molecular Biology of Fishes. Vol. 2* (Eds.) Hochachka, P.W. and Mommsen, T.P. Elsevier Science Pub., Amsterdam, pp. 57-91.

Yadav, Y.S. and Chandra, R., 1994. Some threatened carps and catfishes of Brahmaputra river system. In: *Threatened Fishes of India* (Eds.) Dehadrai, P.V., Das, P. and Verma, S. R. Nature Conservators, Muzaffarnagar, pp. 45-66.

Young, M.K. and Harig, A.L. (2001. A critique of the recovery of greenback cutthrout trout. *Conserv. Biol.*, 15: 1575-1584.

Zabel, R.W., Harvey, C.J., Katz, S.L., Good, T.P. and Levin, P.S., 2003. Ecologically sustainable yield. *Amer. Sci.*, 91 (2); 150-157.

Zechendorf, B., 2002. Biodiversity : a critical review on major aspects. In : *Biological and Biotechnological Resources* (Eds.) Tripathi, G. and Tripathi, Y.C. Campus Books International, New Delhi, pp. 1-41.

Chapter 3

The Effect of Endocrine Disruptors on Aquatic Animals: A Short Review

**B.K. González-Pérez, S.S.S. Sarma*,
M. E. Castellanos-Páez and S. Nandini**

ABSTRACT

Endocrine disruptors are chemical substances that affect the hormonal and homeostatic systems of organisms. The importance of these systems is to communicate and respond to the changes in the environment. Among the different endocrine disruptors, bisphenol a, nonylphenol, triclosan, pharmaceuticals, estrogens and pesticides are generally considered as highly effective in influencing the homeostasis. In this work a general view of endocrine disruptors is presented with emphasis on classification, mode of action and legislation. In many countries, chemicals with potential as endocrine disruptors are directly discharged into waterbodies, sometimes without prior treatment. The non-target aquatic organisms especially the plankton in such waterbodies may be adversely affected on long term exposure. For this reason, it is necessary to conduct more studies involving the effects of these substances on aquatic organisms through the food chain.

Keywords: Zooplankton, Fish, Sublethal effects, Pesticides, Pharmaceuticals.

Introduction

Endocrine disruptors are the chemical agents that interfere with the synthesis, secretion, transport, binding or elimination of natural hormones of organisms

---

* Corresponding Author.

(USA EPA, 2002). Thus, from a physiological perspective, an endocrine disruptor is a natural or synthetic compound that can alter the hormonal and homeostatic systems which facilitate organisms to communicate and respond to the environment (Diamanti-Kandarakis *et al.,* 2009). There are also other views on the endocrine disruptors for defining them on a broad sense. For example, the International Programme on Chemical Safety (Anon., 2002) considers that endocrine disruptors are mainly the exogenous substances or their mixtures which alter the functioning of the endocrine systems, eventually affecting the health of an organism or its offspring. A well-known example is the pesticide DDT which was responsible for the reduction in the bird populations as they fail to reproduce due its adverse effects on reproductive systems (Rathore and Nollet, 2012). In addition to birds, all other groups of vertebrates (fish, amphibians, reptiles and mammals) can also be adversely affected by the presence of synthetic endocrine disruptors in the environment (Andrade-Ribeiro *et al.,* 2006).

The endocrine disruptors can be divided into two principal groups based on their natural availability: natural endocrine disruptors and synthetic endocrine disruptors. The first category of chemical substances is naturally present in the food for human and animals, which the second category of chemicals are synthesized to act up on certain target groups including humans (Diamanti-Kandarakis *et al.,* 2009). Synthetic endocrine disruptors are the chemicals used as solvents, lubricants or byproducts of industry, plastics, plasticizers, pesticides, fungicides and some drugs. Endocrine disrupters are also found in products which are widely used by the human population, from household (*e.g.,* electronic items) to personal care articles (*e.g.,* certain cosmetics). One of the most common endocrine disruptors in materials which are in contact directly with food is Bisphenol A (BPA). This material is in a variety of food containers in the form of plastics and canned foods (Gore *et al.,* 2014). The synthetic endocrine disruptors are very often found in aquatic environments in detectable concentrations.

Synthetic pesticides are the chemicals designed to be highly toxic to the reproductive and neural systems of pests. However, they are also toxic to aquatic organisms, from lower invertebrates to aquatic birds. In addition, these chemicals can affect the human body (Gore *et al.,* 2014). Effects of synthetic pesticides and herbicides on the reproductive performance of different species of aquatic organisms, mainly rotifers and crustaceans such as cladocerans have received considerable attention because of the ease with which such tests can be carried out (Kammenga and Laskowski, 2000). Some of the pesticides in extremely low concentrations affect the reproductive output of zooplankton. For example, Rao and Sarma (1986) have shown that DDT at concentrations as low as 30 µg L^{-1} had an adverse effect, not only on the survival but also on the reproduction of rotifers. For certain endocrine disruptors, in addition to a reduction in reproductive performance, the offspring may become intersexes (Schwarzenbach *et al.,* 2010). In a study on the effect of carbendazim, one of the most common fungicides, Miracle *et al.* (2011) have shown that the cladoceran *Moina micrura,* when exposed to this endocrine disruptor at a concentration of 0.04 mg L^{-1}, produced some abnormal males with short and female-like antennules.

The different endocrine disruptors have different mode of action depending on the nature of the chemical and the species to which it is exposed. In addition, some metabolic products of the endocrine disruptors can be even much more toxic than the parent compounds (Kidd et al., 2012).

Mode of Action

The mode of action of the endocrine disrupters may be in two ways: during embryonic development or throughout the life cycle (Guillete et al., 1995). However, these compounds share some common mechanisms of action and biological effects. These include mimicking and/or antagonistic effects of hormones, altered patterns in synthesis and metabolism of hormones, and modification of levels of hormone receptors (Burkhardt-Holm, 2010). There is some evidence that endocrine disruptors are also able to bioaccumulate in tissues such as adipose in different organisms (Schwarzenbach et al., 2010).

Common Endocrine Disruptors

Thousands of chemicals which are widely used by the human population (some are banned and others are still in use) have been classified as endocrine disruptors (Frye et al., 2011). Nonylphenol occurs in a mixture of ortho-, para- and meta isomers, of which the most prevalent is the para-NP (4-nonylphenol). Structurally nonylphenol resembles to some extent the endogenous hormone (Figure 3.1a) and because of this reason it has the ability to mimic or even in some cases prevent the effects of the hormone (Jie et al., 2013). It has a short half-life (10 to 15 h) in surface water but can last up to 60 years in sediments (Soares et al., 2008). This endocrine disruptor has attracted the attention of researchers because of its effects on the central nervous system especially during early stages of development of different organisms (Jie et al., 2013).

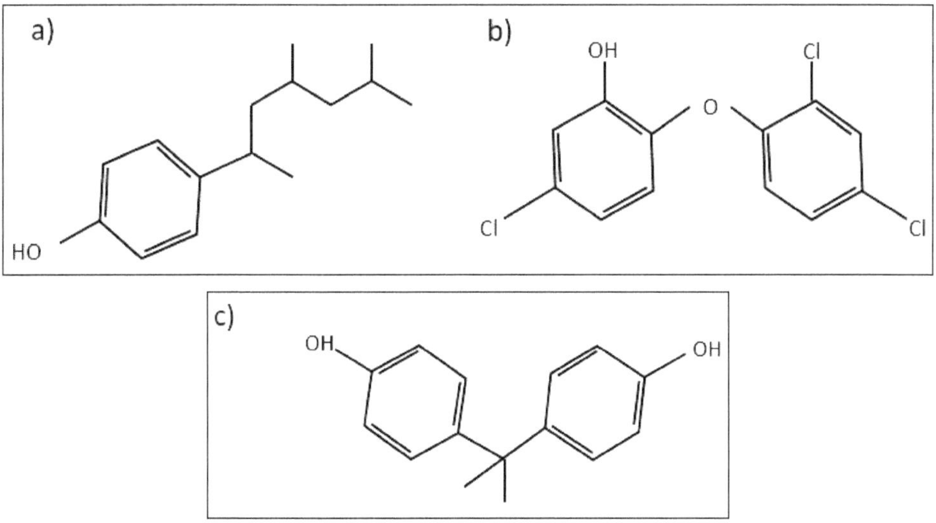

**Figure 3.1: Shemical structure of
(a) 4-nonylphenol, (b) Triclosan and (c) Bisphenol a.**

Triclosan is a halogenated phenol (Figure 3.1b) and is widely used in North America, Europe and Asia as an antimicrobial in oral cavity and skin (Russell, 2004). It is also commonly added in antibacterial soaps, deodorants, body creams and toothpaste (McAvoy et al., 2002; Sabaliunas et al., 2003). Since this is mostly used in personal care products, it is not consumed by human population and hence not metabolized (Orvos et al., 2002). However, through the food web, it can be transformed into other products which can bioaccumulate (Latch et al., 2003). Triclosan can potentially mimic as estrogen, androgen or even thyroid hormone (Dann and Hontela, 2011). Triclosan has a half-life of less than one hour in surface waters (Lyndall et al., 2010). Bisphenol A (Figure 3.1c) is yet another chemical compound that is used mainly in the production of polycarbonate plastics. This is a monomer that exhibits estrogen-like properties by altering the female reproductive systems (Vom Saal et al., 1998).

The production and consumption of chemicals have also increased in the past few decades. It is known that after the World War II several chemicals with properties similar to endocrine disrupting effects were released into the atmosphere (Colborn et al., 1993). Possibly nearly all humans are daily exposed to endocrine disruptors though in low concentrations but through hundreds of products. However, very little is known about their effects to the aquatic animals in environment (Burkhardt-Holm, 2010). According to an estimation, there are about 85,000 chemicals from which thousands could be considered as endocrine disruptors. Many of the chemicals with endocrine disrupting properties possibly come from the plastic industry, which has grown dramatically during last 40 years (Gore et al., 2014).

The Source of Endocrine Disruptors

The source of endocrine disruptors is diverse. The main sources are the industries, hospitals, agriculture, aquaculture and veterinary sectors. Though domestic wastes do contain significant quantities of these substances, it is difficult to establish their origin: whether from excretions of consumed products or dumping of unused pharmaceuticals. Though majority of compounds such as nonylphenol, chlorpyrifos with endocrine disrupting properties is regulated or even banned, some them still find their way into the aquatic systems through diverse activities (Bergman et al., 2012). For example, it has been shown that in municipal waters of different countries including Brazil, Canada, China and Germany, estrogen is present and this is primarily due to the excretion of the human population (Adler et al., 2001). Figure 3.2 shows some of the possible routes through which the endocrine disruptors reach waterbodies.

The endocrine disruptors have undesirable effects on wildlife too (Damstra et al., 2002). Most experimentally obtained data come from the test species exposed to a single chemical compound at a time. However, in nature organisms are in contact with a mixture of xenobiotics including endocrine disruptors. Some of these components are not persistent in the aquatic environment (Kidd et al., 2012) but since most of them are daily consumed and excreted, their continuous presence in the waterbodies ensures nearly a constant exposure of aquatic organisms to such chemicals.

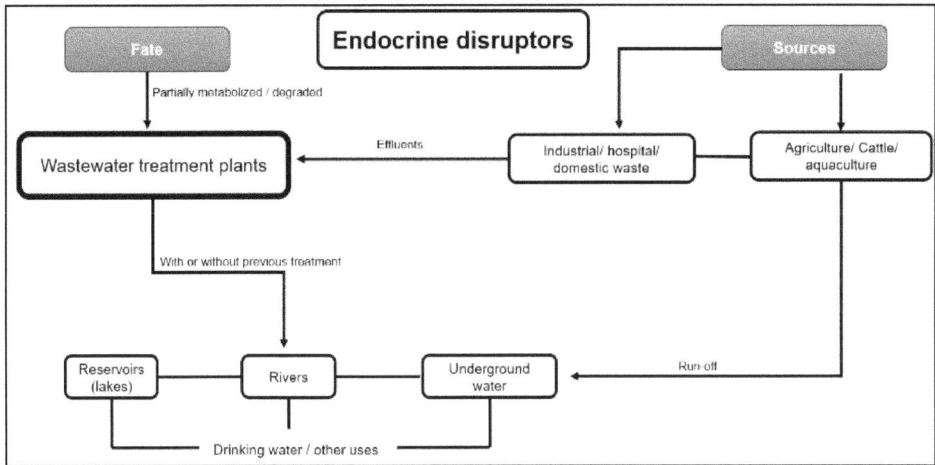

Figure 3.2: Sources and Destinations of Endocrine Disruptors in the Aquatic Environment.

Table 3.1: Concentrations of some Common Endocrine Disruptors in the Aquatic Environment from different Countries

Compound	Concentration in Waterbodies	Reference
Acetophenone	0.15 mg L^{-1} rivers in USA	Kolpin et al., 2002
Ciprofloxacin	0.2 µg L^{-1} rivers in USA	Kolpin et al., 2002
Norfloxacin	Up to 2.42 mg Kg^{-1} sediments	Golet et al., 2002
Tetracycline	0.11 µg L^{-1} rivers in USA	Kolping et al., 2002
Diclofenac	0.47 µg L^{-1} drinking water	Ternes et al., 2002
Ibuprofen	0.1 µg L^{-1} effluents in Berlin	Heberer, 2002
	2.83 ng mL^{-1} effluents in Mexico City	Peña-Álvarez and Castillo-Alanís, 2015
Paracetamol	0.11 µg L^{-1} rivers of USA	Kolpin et al., 2002
Carbamazepine	1075 ng L^{-1} effluents from Germany	Heberer et al., 2001
Estradiol	1.72 ng L^{-1} Xochimilco Lake, Mexico	Díaz-Torres et al., 2013
Testosterone	0.116 µg L^{-1} rivers in USA	Kolpin et al., 2002
Estrone	10.38 ng L^{-1} Xochimilco Lake, Mexico	Díaz-Torres et al., 2013
Triclosan	52.74-71.93 ng µl^{-1} Xochimilco Lake, Mexico	Díaz-Torres et al., 2013
	10.09 ng mL^{-1} wastewaters in Mexico City	Peña-Álvarez and Castillo-Alanís, 2015
Bisphenol A	140.33 ng µL^{-1} Xochimilco Lake, Mexico City	Díaz-Torres et al., 2013
	4.27 ng mL^{-1} wastewaters, Mexico City	Peña-Álvarez and Castillo-Alanís, 2015
Benzyl butyl phthalate	17.38 ng µL^{-1} Xochimilco Lake, Mexico	Díaz-Torres et al., 2013
Pentachlorophenol	2.49-3.89 ng µL^{-1} Xochimilco Lake, Mexico	Díaz-Torres et al., 2013
Nonylphenol	330 µg L^{-1} wastewaters, England	Blackburn and Waldock, 1995

The measurement of actual concentrations of endocrine disruptors in the aquatic environment is often not practical because of the methodological difficulties such as precision of the equipment used or even the availability of appropriate techniques for quantitative studies. Table 3.1 presents the estimated concentrations of some available endocrine disruptors in the aquatic environment.

Effects of Endocrine Disruptors in the Food Chain

Wastes from various sources (hospital, industrial, domestic, *etc.*) are usually discharged directly into the waterbodies such as reservoirs and sometimes, there is not a prior treatment before discharging (Daughton and Ternes, 1999). In addition, water treatment plants are not fully equipped to eliminate some of the endocrine disruptors. Because of this, the quality of water that has been released from the wastewater treatment plants has become questionable in many countries including Mexico (Liu *et al.*, 2009; Nandini *et al.*, 2016). In addition, some of the endocrine disruptors are found in extremely concentrations in effluent waters (ng L^{-1} to mg L^{-1}), almost undetectable levels and yet may have adverse effects on the reproductive systems of non-target organisms (Table 3.2). Further, it is known that some of the non-target species (such as zooplankton) can be much more sensitive to these xenobiotic substances than is extrapolated based on vertebrate models (Daughton and Ternes, 1999).

Table 3.2: Effects of Endocrine Disruptors on Vertebrates and Invertebrates

Compound	Species	Tests and Effects	Reference
Antiandrogen	Poecilia reticulata	Acute toxicity test, spermatogenesis	Bayley *et al.*, 2002
Androgen	Danio rerio	Acute toxicity test, sex reversal	Holbech *et al.*, 2006
Progesterone	Pimephales promelas	Acute toxicity test, effects on reproduction	Zeilinger *et al.*, 2009
Diethylstilbestrol	Coturnix japonica	Behaviour, stimulatory effects on female receptor behaviour	Adkins-Regan and Garcia, 1986
Estradiol	Pimephales promelas	Acute toxicity test, feminization	Länge *et al.*, 2009
Estrogen	Oncorhynchus mykiss	Acute toxicity test, induction of vitellogenin	Tyler *et al.*, 2009
17 β-Estradiol	Acartia tonsa (Copepoda)	Reproduction, stimulatory offspring production	Andersen *et al.*, 1999.
17 β-Estradiol	Hyella azteca	Reproduction, adverse effects on spematogenesis	Vandenbergh *et al.*, 2003
17 β-Estradiol	Oryzias latipes	Acute toxicity test, effects on fertility	Hutchinson *et al.*, 2003
17 β-Estradiol	Oreochromis sp.	Acute toxicity test, feminization	Shved *et al.*, 2008
17 β-Estradiol	Diaphanosoma celebensis	Multigenerational study, changes in reproductive rates	Marcial and Hagiwara, 2007

Contd...

Table 3.2–Contd...

Compound	Species	Tests and Effects	Reference
Bisphenol A	Acartia tonsa	Acute toxicity test, morality	Andersen et al., 1999
Diclofenac	Plationus patulus, Moina macrocopa	Population growth, effects on reproduction	Sarma et al., 2014
Carbamazepine	Ceriodaphnia dubia	Acute toxicity test, effects on reproduction	Ferrari et al., 2003
Diazepam	Hydra vulgaris	Acute toxicity test, effects on regeneration	Pascoe et al., 2003
Parathion	Daphnia magna	Acute toxicity test, morality	Guilhermino et al., 2000
Fipronil	Amphiascus tenuireis	Life cycle, inhibition of reproduction	Cary et al., 2004
Dimethoate	Brachionus calyciflorus	Multigenerational tests, reduced population growth rates	Ruixin et al., 2012
Chlorpyrifos	Daphnia carinata	Sub-acute toxicity tests, effects on survival	Zalizniak and Nugegoda, 2006
Clotrimazole	Algae	Acute toxicity test, growth inhibition	Porsbring et al., 2009
Polychlorinated biphenyls	Echnigammarus marinus	Reduced sperm count	Yang et al., 2008
Levofloxacin	Duckweed	Acute toxicity test, growth inhibition	Brain et al., 2004
Sulfamethoxazole	Cyanobacteria	Acute toxicity test, growth inhibition	Ferrari et al., 2003; 2004
Triclosan	Xenopus laevis	Acute toxicity tests, modified embryonic development	Veldhoen et al., 2006
Triclosan	Phytoplankton communities	Transgenerational tests, increase in biovolume	Pomati and Nizzetto, 2013

Laboratory tests assessing the effects of hormones on the reproduction of zooplankton are few and in some cases, the effect is not always evident nor is adverse. For example, García-García et al. (2014) have studied the effect of mixed hormones, levonorgestrel (LEV) and ethinylestradiol (ETE) on the reproductive rates of two rotifers, *Anuraeopsis fissa* and *Brachionus calyciflorus*. Their results indicate the stimulated offspring production of *A. fissa* (at 31.25/6.25 µg L^{-1} of LEV/ETE) and *B. calyciflorus* (at the concentrations of 250/50 µg L^{-1} of LEV/ETE). Such trends indicate the mode of action of these endocrine disruptors differ completely from human systems to invertebrate species. For example, the mode of action LEV and ETE is inhibition of ovulation in human females, while it causes an elevated offspring production in rotifers.

Legislation

The US Environmental Protection Agency during 1996 developed a strategy for continuous monitoring of substances that may have a potential for endocrine disruption and to develop appropriate techniques and methods (EDSTAC, 1996).

There are some guidelines available from a few other nations too such as Canada and the European Union. Many of these guidelines contain data on the permissible levels of different substances including endocrine disruptors in order to protect aquatic systems (ISTS, 2002). In addition to the governmental agencies, some individuals or groups of scientists also offer suggestions for the improvement of water quality (Gore et al., 2014).

In Mexico such vigorous laws do not exist, especially with reference to the maximum permissible limits of different substances in aquatic systems. However, the Mexican Official Standards (NOM-001-SEMARNAT-1996) establishes maximum permissible limits of pollutants and other toxic substances in wastewaters before they are discharged into rivers and other waterbodies. Industries producing pharmaceuticals, plastics, soaps and detergents are continuously or periodically monitored for the levels of pollutants in the effluents.

Conclusions

Although the modes of action of many endocrine disruptors to the target organisms, mainly the mammalian systems are known, there is a lack of information regarding their ecotoxicological effects on the aquatic organisms, which are generally the non-target species. In addition, much of the available information on the effects of these endocrine disruptors is focused on aquatic vertebrates such as fish. Many of these substances are directly discharged into waterbodies, sometimes without prior treatment and thus the non-target aquatic organisms especially the plankton may be adversely affected on long term exposure. For this reason, it is necessary to conduct more studies involving the effects of these substances through the aquatic food chain.

For quick and quantitative results, most governmental agencies use the acute toxicity testing where the endocrine disruptors are employed in concentrations much higher than even those found in effluents. Therefore, such studies do not always take into account of the sublethal effects on the whole life cycle of a given species. It is long known that under stressful conditions, reproduction-related variables are more sensitive than survival-related parameters (Rao and Sarma, 1986; Kammenga and Laskowski, 2000). Moreover, it is also important to consider that the effects of different endocrine disruptors may also vary from the parental generation to the next generation, even when the exposure concentration remains the same. It is not fully known if the different generations suffer more due to adverse effects from the endocrine disruptors than the parental generation. Also, it is possible that the antagonistic and synergistic effects of different endocrine disruptors can differ from results of the laboratory tests involving one chemical substance at a time. These point out the need for conducting further studies on the effects of endocrine disruptors to the non-target aquatic organisms, especially the invertebrates.

Acknowledgements

We thank Araceli Patricia Peña Álvarez (UNAM) for a useful discussion during the preparation of this manuscript. BKGP is grateful to the CONACyT for a doctoral scholarship (No. 492489).

References

Adkins-Regan, E. and Garcia, M., 1986. Effect of flutamide (an antiandrogen) and diethylstilbestrol on the reproductive behaviour of Japanese quail. *Physiology and Behaviour*, 36, 419–425.

Adler, P., Steger-Hartmann, T. and Kalbfus, W., 2001. Distribution of natural and synthetic estrogen steroid hormones in water samples from southern and middle Germany, *Acta hydrochimica hydrobiologica*, 29, 227–241.

Andersen, H. R., Halling-Sorensen, B. and Kusk, K. O., 1999. A parameter for detecting estrogenic exposure in the copepod *Acartia tonsa*. *Ecotoxicology and Environmental Safety*, 44, 56–61.

Andrade-Ribeiro, A., Pacheco-Ferreira, A., Nóbrega da Cunha, C., and Mendes-Kling, A., 2006. Disruptores endocrinos: potencial problema para la salud pública y medio ambiente. *Revista de Biomedicina*, 17, 146–150.

Anon., 2002. International Programme on Chemical Safety. Global Assessment of the State-of-the-Science of Endocrine Disruptors. Geneva: World Health Organization.

Bayley, M., Junge, M., and Baatrup, E., 2002. Exposure of juvenile guppies to three antiandrogens causes demasculinization and a reduced sperm count in adult males. *Aquatic Toxicology*, 56, 227–239.

Bergman Å, Heindel J.J, Jobling, S., Kidd, K.A., Zoeller, R.T., and Jobling, S.K. eds., 2012. State of the Science of Endocrine Disrupting Chemicals—2012. Geneva: United Nations Environment Programme and World Health Organization. Available: http://apps.who.int/iris/handle/10665/78101.

Blackburn, M.A., and Waldock, M.J., 1995. Concentrations of alkylphenols in rivers and estuaries in England and Wales. *Water Resources*, 29, 1623–1629

Brain, R.A., Johnson, D.J., Richards, S.M., Sanderson, H., Sibley, P.K., and Solomon, K.R., 2004. Effects of 25 pharmaceutical compounds to *Lemna gibba* using a seven-day Static renewal test. *Environmental Toxicology and Chemistry*, 23, 371–382.

Burkhardt-Holm., 2010. Endocrine disruptors and water quality: A-state of-the-art review. *International Journal of Water Resources Development*, 26, 477–493.

Cary, T.L., Chandler, G.T., Volz, D.C., Walse, S.S., and Ferry, J.L., 2004. Phenylpyrazole insecticide fipronil induces male infertility in the estuarine meiobenthic crustacean *Amphiascus tenuiremis*. *Environmental Science and Technology*, 38, 522–528.

Colborn, T., Vom Saal, F. S., and Soto, A. M., 1993. Developmental effects of endocrine-disrupting chemicals in wildlife and humans. *Environmental and Health Perspective*, 1, 378–384.

Dann, A. and Hontela, A., 2011. Triclosan: environmental exposure, toxicity, and mechanisms of action. *Journal of Applied Toxicology*, 31, 285–311.

Damstra, T., Barlow, S., Bergman, A., Kavlock, R., and Van der Kraak, G., 2002. Global assessment of the state-of-the-science of endocrine disruptors. WHO/PCS/EDC/02.2. World Health Organization, Geneva, Switzerland.

Daughton, C.G. and Ternes, T.A., 1999. Pharmaceuticals and personal care products in the environment: agents of subtle change? *Environ. Health Perspect.* 107, 907–938.

Diamanti-Kandarakis, E., Bourguignon, J.P., Guidice, L. C., Hausser, R., Prins, G.S., Soto, A. M., Zoeller, T., and Gore, A. C., 2009. Endocrine-disrupting chemicals: An endocrine society scientific statement. *Endocrine Reviews*, 30(4), 293–342.

Díaz-Torres, E., Gibson, R., González-Farías F., Zarco-Arista, A. E., and Mazari-Hiriar, M., 2013. Endocrine disruptors in the Xochimilco Wetland, Mexico City. *Water air soil pollution*, 224, 1–11.

Endocrine Disruptor Screening and Testing Advisory Committee (EDSTAC)., 1996. Final report, EPA/743/R-96/00.

Ferrari, B., Mons, R., Vollat, B., Fraysse, B., Paxeus, N., Lo Giudice, R., Pollio, A., and Garric. J., 2004. Environmental risk assessment of six human pharmaceuticals: are the current environmental risk assessment procedures sufficient for the protection of the aquatic environment? *Environmental Toxicology and Chemistry*, 23, 1344–1354.

Ferrari, B., Paxéus, N., Giudice, R.L., and Pollio, A., 2003. Ecotoxicological impact of pharmaceuticals found in treated wastewaters; study of carbamazepine, clofibric acid, and diclofenac. *Ecotoxicology and Environmental Safety*, 55, 359–370.

Frey, C. A., Bo, E., Calamandreis, G., Calza, L., Dessi-Fulgheri, F., Fernández, M., Fussani, L., Kahss, O., Kajta, M., Le Pagess, Y., Patisaul, H. B., Venerosi, A., Wojtowicz, A. K. and Panzica, G. C., 2011. Endocrine disrupters: A review of some sources, effects, and mechanisms of actions on behavior and neuroendocrine systems. *Journal of neuroendocrinology*, 24, 144–159.

García-García, G., Sarma, S.S.S., Nuñez-Ortíz, A.R. and Nandini, S., 2014. Effects of the mixture of two endocrine disruptors (ethinylestradiol and levonorgestrel) on selected ecological endpoints of *Anuraeopsis fissa* and *Brachionus calyciflorus* (Rotifera). *International Review of Hydrobiology*, 99,166–172.

Golet, E. M., Strehler, A., Alder, A.C., and Giger, W., 2002. Determination of fluoroquinolone antibacterial agents in sewage sludge and sludge-treated soil using accelerated solvent extraction followed by solid-phase extraction. *Anal. Chem.* 74, 5455–5462.

Gore, A.C., Crews, D., Doan, L. L., La Merril, M., Patisaul, H. and Zota, A., 2014. Introduction to Endocrine Disrupting Chemicals (EDCs) – A Guide for Public Interest Organizations and Policy-makers. Endocrine Society, Available at: http://www.endocrine.org//media/endosociety/Files/Advocacy per cent 20and per cent 20Outreach/Important per cent 20Documents/Introduction per cent 20to per cent 20Endocrine per cent 20Disrupting per cent 20Chemicals.pdf

Guilhermino, L., Diamantino, T., Carolina Silva, M., and Soares, M.V.M., 2000. Acute toxicity test with *Daphnia magna*: an alternative to mammals in the prescreening of chemical toxicity? *Ecotoxicology and Environmental Safety*, 46, 357–362.

Guillette, L. J., Jr., Crain, D. A., Rooney, A. A. and Pickford, D. B., 1995. Organization versus activation: the role of endocrine-disrupting contaminants (EDCs) during embryonic development in wildlife. *Environmental Health Perspectives*, 1037, 157– 164.

Heberer, T., Verstraeten, I.M. Meyer, M. T., Mechlinski, A., and Reddersen, K., 2001. Occurrece and fate of pharmaceuticals during bank filtration-preliminary results from investigations in Germany and the United States. *Water Resources Update*, 120, 4–17.

Heberer, T., 2002. Tracking persistent pharmaceutical residues from municipal sewage to drinking water. *J. Hydrol*. 266, 175–189.

Holbech, H., Kinnberg, K., Petersen, G.I, Jackson, P., Hylland, K., Norrgren, L., and Bjerregaard, P., 2006. Detection of endocrine disrupters: evaluation of a fish sexual development test (FSDT). *Comp. Biochem. Physiol. C Toxicol. Pharmacol*. 144, 57–66.

Hutchinson, T.H., Yokota, H., Hagino, S., and Ozato, K., 2003. Development of fish tests for endocrine disruptors. *Pure Appl. Chem*. 75, 2343–2353.

Instituto Sindical de Trabajo, Ambiente y Salud (ISTAS)., 2002. Curso de Introducción a los Disruptores Endocrinos: 1--34. Barcelona.

Jie, X., JianMei, L., Zheng, F., Lei, G., Biao, Z., and Jie, Y., 2013. Neurotoxic effects of nonylphenol: a review. *Wien Klien Wochenschr*, 125: 61–70

Kammenga J. and R. Laskowski (Eds)., 2000. Demography in Ecotoxicology. Wiley, New York, 318 pp.

Kidd, K.A., Becher, G., Bergman, A., Muir, D.C.G., and Woodruff, T.J., 2012. Human and wildlife exposures to EDC's. Chapter 3. State of the science of endocrine disrupting chemicals. UNEP, 189–250, Retrieved from: http: //www.unep.org/hazardoussubstances/Portals/9/EDC/SOS per cent 202012/EDC per cent 20report per cent 20Ch3.pdf

Kolpin, D.W., Furlong, E.T., Meyer, M.T., Thurman, E.M., Zaugg, S.D., Barber, L.B., and Buxton, H.T., 2002. Pharmaceuticals, hormones, and other organic wastewater contaminants in US streams, 1999–2000: a national reconnaissance. *Environ. Sci. Technol*. 36, 1202–1211.

Länge, A., Paull, G.C., Coe, T.S., Katsu, Y., Urushitani, H., Iguchi, T., and Tyler, C.R., 2009. Sexual reprogramming and estrogenic sensitization in wild fish exposed to ethinylestradiol. *Environ. Sci. Technol*. 43, 1219–1225.

Latch D, Packer J, Arnold W, and McNeill K., 2003. Photochemical conversion of triclosan to 2, 8-dichlorodibenzo-*p*-dioxin in aqueous solution. *J Photochem Photobiol A*. 158, 63–66.

Liu, Z. H. Kanjo, Y. and Mizutani, S., 2009. Removal mecanisms for endocrine disrupting compounds (EDCs) in wastewater treatment- physical means, biodegradation and chemical advance oxidation: A review. *Science of the Total Environment*, 407, 731–748.

Lyndall, J., Fuschman, P., Bock, M., Barber, T., Lauren, D., Leigh, K., Perruchon, E. and Capdevielle, M., 2010. Probabilistic risk evaluation for triclosan in surface water, sediments and aquatic biota tissues. *Integr. Environ. Assess. Manag.* 6, 419–440.

Marcial, H. S. and Hagiwara, A., 2007. Multigenerational effects of 17b-estradiol and nonylphenol on euryhaline cladoceran *Diaphanosoma celebensis*. *Fisheries Science*. 73, 324–330.

McAvoy D.C, Schatowitz, B., Jacob, M., Hauk, A., and Eckhoff, W.S., 2002. Measurement of triclosan in wastewater treatment systems. *Environmental Toxicology and Chemistry*. 21, 1323–1329.

Miracle, M. R., Nandini, S., Sarma, S. S. S., and Vicente, E., 2011. Endocrine disrupting effects, at different temperatures on *Moina micrura* (Cladocera: Crustacea) induced by carbendazim, a fungicide. *Hydrobiologia*, 668, 155–170.

Nandini, S., Ramírez-García, P. and Sarma, S.S.S., 2016. Water quality indicators in Lake Xochimilco, Mexico: zooplankton and *Vibrio cholera*. *Journal of Limnology*, 75(1), 91–100.

Orvos, R. D., Donald, J. V., Inauen, J., Capdevielle, M., Rothensein, A., and Cunningham, V., 2002. Aquatic toxicity of triclosan. *Environmental Toxicology and Chemistry*, 21, 1338–1349.

Pascoe, D., Karntanut, W., and Muller, C.T., 2003. Do pharmaceuticals affect freshwater invertebrates? A study with the cnidarian *Hydra vulgaris*. *Chemosphere*, 51, 521–528.

Peña-Álvarez, A., and Castillo-Alanís, A., 2015. Identificación y cuantificación de contaminantes emergentes en aguas residuales por microextracción en fase sólida cromatografía de gases espectrometría de masas (MEFS-CG-EM). *Revista Especializada en Ciencias Químico-Biológicas*, 18 (1), 29–42.

Pomati, F. and Nizzetto, L., 2013. Assessing triclosan-induced ecological and trans-generational effects in natural phytoplankton communities: a trait-based field method. *Ecotoxicology*, 22, 779–794.

Porsbring, T., Blanck, H., Tjellström, H., and Backhaus, T., 2009. Toxicity of the pharmaceutical clotrimazole to marine microalgal communities. *Aquatic Toxicology*, 91, 203–211.

Ramakrishna Rao, T. and Sarma, S. S. S., 1986. Demographic parameters of *Brachionus patulus* Muller (Rotifera) exposed to sublethal DDT concentrations at low and high food levels. *Hydrobiologia*, 139, 193–200.

Rathore, H.S. and Nollet, L.M.L. (Eds)., 2012. Pesticides: Evaluation of Environmental Pollution. CRC Press, Florida.

Ruixin, G., Xinkun, R. and Hongqiang, R., 2012. Effects of dimethoate on rotifer *Brachionus calyciflorus* using multigeneration toxicity tests. *Journal of Environmental Science and Health, Part B*, 47, 883–890.

Russell A. D., 2004. Whither triclosan? *J. Antimicrob. Chemother.* 53(5), 693–695.

Sabaliunas, D., Webb, S.F., Hauk, A., Jacob, M. and Eckhoff., W.S., 2003. Environmental fate of triclosan in the River Aire Basin, UK. *Water Res.* 37(13), 3145–3154.

Sarma, S.S.S., González-Pérez, B.K., Moreno-Gutiérrez, R.M. and Nandini, S., 2014. Effect of paracetamol and diclofenac on population growth of *Plationus patulus* and *Moina macrocopa*. *Journal of Environmental Biology.* 35(1),119–126.

Schwarzenbach, R. P., Egli, T., Hofstetter, T. B., Von-Gunten, U., and Wehrli, B., 2010. Annual review of environment and resources. *Global Water Pollution and Human Health*, 35, 109–136.

Semarnat. Secretaría de Medio Ambiente y Recursos Naturales., 1996. Norma Oficial Mexicana NOM-001- SEMARNAT-1996. Diario Oficial de la Federación (DOF), April, 2006.

Shved, N., Berishvili, G., Baroiller, J.F., Segner, H., and Reinecke, M., 2008. Environmentally relevant concentrations of 17alpha-ethinylestradiol (EE2) interfere with the growth hormone (GH)/insulin-like growth factor (IGF)-I system in developing bony fish. *Toxicol. Sci.* 106, 93–102.

Soares, A., Guieysse, B., Jefferson, B., Cartmell, E., and Lester, J.N., 2008: Nonylphenol in the environment: A critical review on occurrence, fate, toxicity and treatment in wastewaters. *Environment International*, 34, 1033–1049.

Ternes, T.A., Meisenheimer, M., McDowell, D., Sacher, F., Brauch, H.-J., Haist-Gulde, B., Preuss, G., Wilme, U., and Zulei-Seibert, N., 2002. Removal of pharmaceuticals during drinking water treatment. *Environ. Sci. Technol.* 36, 3855–3863.

Tyler, C.R., Filby, A.L., Bickley, L.K., Cumming, R.I., Gibson, R., Labadie, P., Katsu, Y., Liney, K.E., Shears, J.A., Silva-Castro, V., Urushitani, H., Lange, A., Winter, M.J., Iguchi, T., and Hill, E.M., 2009. Environmental health impacts of equine estrogens derived from hormone replacement therapy. *Environ. Sci. Technol.* 43, 3897–3904.

U.S. Environmental Protection Agency (US EPA)., 2002. Endocrine Disruptor Screening Program Report to Congress. (http://www.epa.gov/).

Vandenbergh, G.F., Adriaens, D., Verslycke, T. and Janssen, C.R., 2003. Effects of 17 alpha-ethinylestradiol on sexual development of the amphipod *Hyalella azteca*. *Ecotoxicology and Environmental Safety.* 54, 216–222.

Veldhoen, N., Skirrow, R.C., Osachoff, H., Wigmore, H., Clapson, D.J., Gunderson, M.P., Van Aggelen, G., and Helbing, C.C., 2006. The bactericidal agent triclosan modulates thyroid hormone38 associated gene expression and disrupts postembryonic anuran development. *Aquatic Toxicology*, 80, 217–227.

Vom Saal, F.S., Cooke, P.S., Buchanan, D.L., Palanza, P., Thayer, K.A., Nagel, S.C., Parmigiani, S., and Welshons, W.V., 1998. A physiologically based approach to the study of bisphenol A and other estrogenic chemicals on the size of reproductive organs, daily sperm production, and behavior. *Toxicol. Ind. Health*, 14, 239–260.

Yang, G., Kille, P., and Ford, A.T., 2008. Infertility in a marine crustacean: have we been ignoring pollution impacts on male invertebrates? *Aquatic Toxicology.* 88, 81–87.

Zalizniak, L. and Nugegoda D., 2006. Effect of sublethal concentrations of chlorpyrifos on three successive generations of *Daphnia carinata*. *Ecotoxicology and Environmental Safety*, 64, 207–214.

Zeilinger, J., Steger-Hartmann, T., Maser, E., Goller, S., Vonk, R., and Länge, R., 2009. Effects of synthetic gestagenes on fish reproduction. *Environmental Toxicology and Chemistry*, 28, 2663–2670.

Chapter 4

Management of Catchment Area of River Ganga to Conserve it for Posterity

Ashwani Wanganeo and Rajni Wanganeo

ABSTRACT

A number of cleanup projects/management plans carried out during the past three decades on river Ganges have not yielded desired results. Even the diversion of the upper and lower Ganges canals of the river-flow in Uttar Pradesh has reduced the capacity of the river to assimilate pollution due to inadequate flow which has affected the process of dilution.

The associated problems still remain to be addressed properly in River Ganges, even though it forms the lifeline of Indian ethos along with its tributaries, flows a distance of more than 2700 km from its origin at Gaumukh (N 30°55´, E 79°7´) up to its mouth in Bay of Bengal. It drains 29 cities of eleven states and still receives large proportion of the solid and liquid wastes besides unburnt dead bodies. Insufficient number and capacities of the sewage treatment plants along with their poor operation and maintenance have failed to treat the entire sewage entering into the Ganges (Kaushal, 2008).

During the pilgrimage season e.g. Kumbh Melas etc. though millions of liters of untreated sewage (Singh and Singh, 2007) run directly into river and millions of pilgrims visit Ganges, yet it does not interfere much with its general health as the gap between the two Kumbh is nearly six/twelve years which is sufficient time for its recuperation as, the spotting of Mahasheer and Trout at many places is testimony to this fact. A dolphin conservation action plan has also been voiced which speaks about its still better conditions, however, the pollution load bearing capacity has been jeopardized to a great extent and invites attention so that emphasis is laid on improving/upgrading the entire catchment area of the River Ganges instead of face-lifting of the riparian area to improve its quality and conserve it for posterity. As it is important to work out the impact of forcing functions (external variables/loading) whether Controllable (in and out flow of nutrients and toxic substances) or Non Controllable (Precipitation, Wind) depending on local meteorological conditions to assess the load *bearing capacity of an ecosystem. In absence of such a study/understanding all management strategies employed yield unfruitful results as is the case with present river wherein, colossal amount of money, time and manpower has been spend.*

Emphasis has been focused in the present paper on adopting realistic Land water interactions with special reference to its catchment area management from the point of view of its social, economic and scientific aspect.

Keywords: *River Ganga, Catchment areas, Tributaries, Gandak, Koshi, Kumbh mela.*

Introduction

The gradual ecological characteristics of an aquatic ecosystem are no doubt influenced by the natural processes but the rapid changes are bought about by anthropogenic factors.The forcing functions (external variables) whether controllable (in an outflow of nutrients and toxic substances, with water) or non-controllable (precipitation, wind, solar radiation *etc.*,) play a vital role in controlling/governing the internal variables. As such, development activities should be assessed in the light of protection/conservation priorities and the requirements of the resource, so that the level of water pollution does not exceed the purification capacity. Intense human activities (primary impacts) in the catchment area of water body results in multiplication of secondary impacts, difficult to control. Thus, in order to understand the wide issues of watershed area of water body both short and long term, it becomes necessary to identify specific area and the variables of the concern. In order to manage such systems attention is invited to prioritize pollution sources according to the magnitude of their contribution. Anthropogenic activities in the watershed alone allow focusing on all the effects of downhill runoff in a given area and also suggest planning conservation strategy accordingly. Development of suitable strategies for identified issues needs to prioritize the problem sources in sub catchment area, so that proper strategies are developed at the source. Since Water bodies are integral part of the entire watershed. As such, in these types of open systems the watershed influences the water body and the water body influences the watershed. It is therefore, hardly possible to manage the (water body) as a system, separated from the watershed and its environment. Further, watershed management implies the wise use of soil and water resources within a given geographical area so as to enhance the perpetual usage of the resource. The overall objective of all watershed management programmes include, increase in infiltration of soil, control of damaging excess run off (Wanganeo, 1998). The Ganga and its tributaries function as a perpetual source of water for both agricultural and fisheries in India and Bangladesh besides attracting tourists. A number of dams have comeup on the tributaries of Ganges facilitating generation of hydroelectricity. In the present paper management of catchment area of river Ganga, a transboundary river between India and Bangladesh has been Considered.This River is a lifeline to millions of dwellers in both countries. It originates from the western Himalayas (Uttarakhand) and flows south and east through the Gangetic Plain of North India into Bangladesh, where it empties into the Bay of Bengal, thus travelling a total distance of 2,510 km.

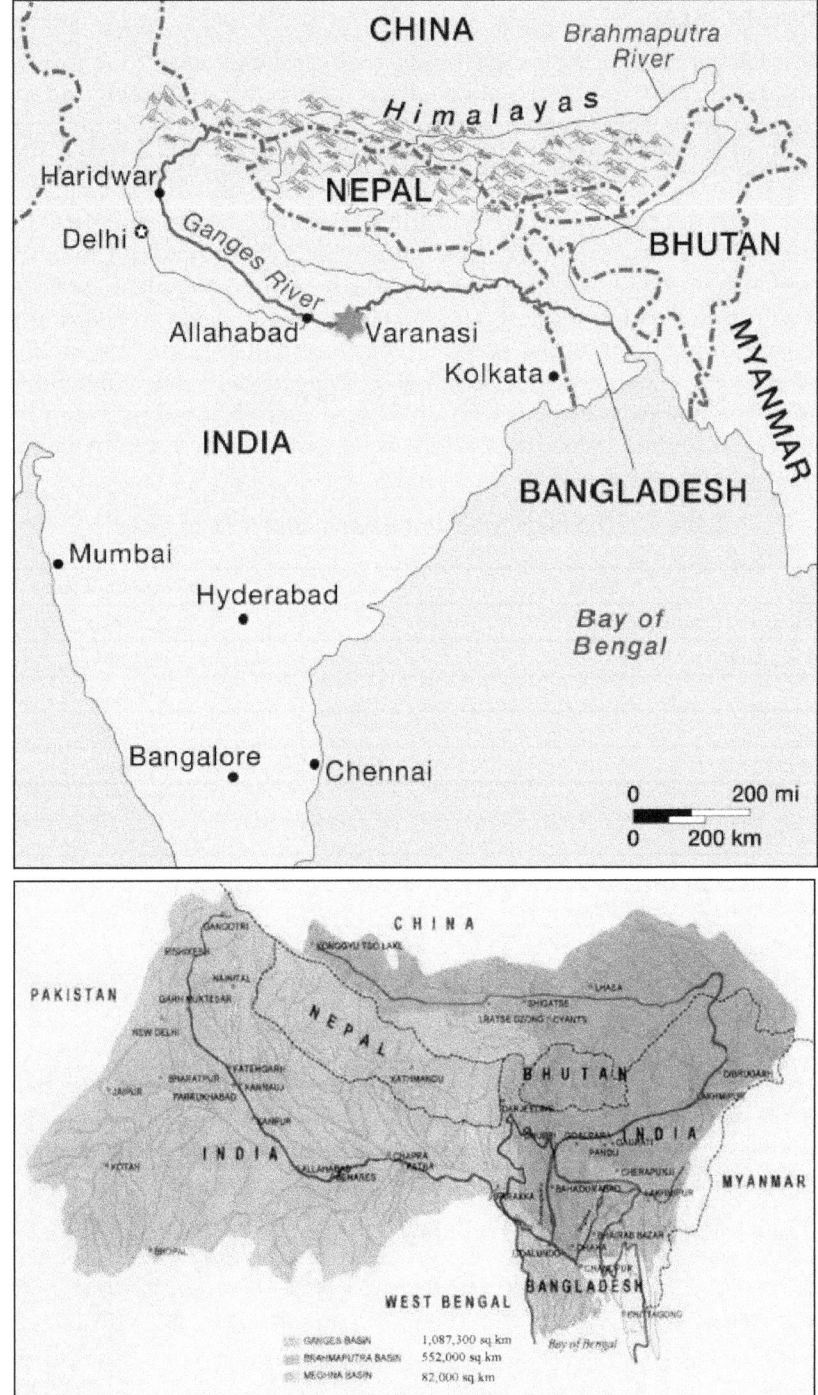

Figure 4.1: Catchment Area of River Ganga
(*Source*: Google website).

Catchment Area

The catchment area of the Ganga lies between east longitudes 73°30' to 89°0' and north latitudes 22° 30' to 31° 30' which falls in four countries, namely **India** (with a catchment area of 862,769 sq.km) - (It passes through the states of Uttarakhand, Uttar Pradesh, Bihar, Jharkhand, and West Bengal. During its course it is flanked by following tributaries *viz.*, Karnali, Mahakhali, Gandak, Koshi (Kosi), Ghaghra, and Damodar on its left side while, its right tributaries are Yamuna, Son, Mahananda and Chambal). – **Nepal** (with a catchment area of 147480 sq.km), China (with a catchment area of 33520 sq.km), and **Bangladesh** (with a catchment area of 46300 sq.Km) with major part in India (Figure 4.1) Negi (2010). Table 4.1 summarise the drainage area that falls in India. The soil types found in the basin are sand, loam, clay and their combinations, such as sandy loam, loam, siltyclay loam and loamy sand soils. The whole catchment area of Ganges basin has been divided into 23 watershed areas for understanding the impact of each wate shed area on the ecology of the river under consideration.

Table 4.1: Drainage Area of the Basin that Falls in India

State	Drainage Area (km^2)
Uttarakhand and Uttar Pradesh	294, 410
Madhya Pradesh	199,385
Bihar	143, 803
Rajasthan	112, 490
West Bengal	72, 618
Haryana	34, 271
Himachal Pradesh	4, 312
Delhi	1, 480
Total Drainage Area of Ganga Basin (km^2)	862,769

Source: www.google.com.

Since water bodies form the culmination point of all events that are being carried out in its catchment area as such, the vulnerability of these systems canot be ignored. This is the reason why water bodies of any region are designated as black boxes. Looking to these aspects of water bodies it is pertinent to understand the present land use land cover of river Ganges in order to formulate a meaningful management strategy.

Being largest river draining most part of the Northern India besides part of Madhya Pradesh and other states as mentioned in Table 4.1, river Ganges faces multiple pressures from its respective watershed areas of various tribuaries (Figure 4.2 and Tables 4.2a,b) :

The forest cover in the catchment area of whole Ganga basin is under severe stress on account of over exploitation. On an average only 16.6 per cent of forest cover has been reported in the Ganga basin.

Table 4.2a: Main Tributaries of River Ganges

Yamuna River	Uttar Pradesh
Tons river	Uttar Pradesh
Ramganga	Uttar Pradesh
Gomati	Uttar Pradesh
Ghaghara	Uttar Pradesh
Gandak	Bihar
Burhi Gandak	Bihar
Ghugri	Bihar
Kosi	Bihar
Son	Bihar
Mahananda River	West Bengal
Brahmaputra river	Bangladesh

Source: www.google.com.

Table 2b: Tributaries with Drainage Area

Name of the Tributary	Area (sq. km)
Ganga including Karmnasa Baya, Bagmari-Pagla	113,163
Yamuna including Chambal, Betwa and Ken	363,082
Sone	71,259
Ghaghra	57,647
Ramganga	32,493
Damodar including Khari-Gangur-Ghia	31,220
Gomti	30,435
Rupnarayan including Haldi, Rasulpur and Kangsabati	23,760
Mahananda	17,440
Tons	16,860
Kiul-Harohar	16,661
Kosi	11,070
Burhi-Gandak	10,150
Punpun	8,530
Mayurakshi-Babla	8,530
Gandak	7,620
Ajay	6,050
Jalangi	5,640
Badua-Chandan	4,840
Bagmati	3,720
Adhwara	2,600
Kamla-Balan	2,980
Tidal rivers	15,650

Source: www.google.com.

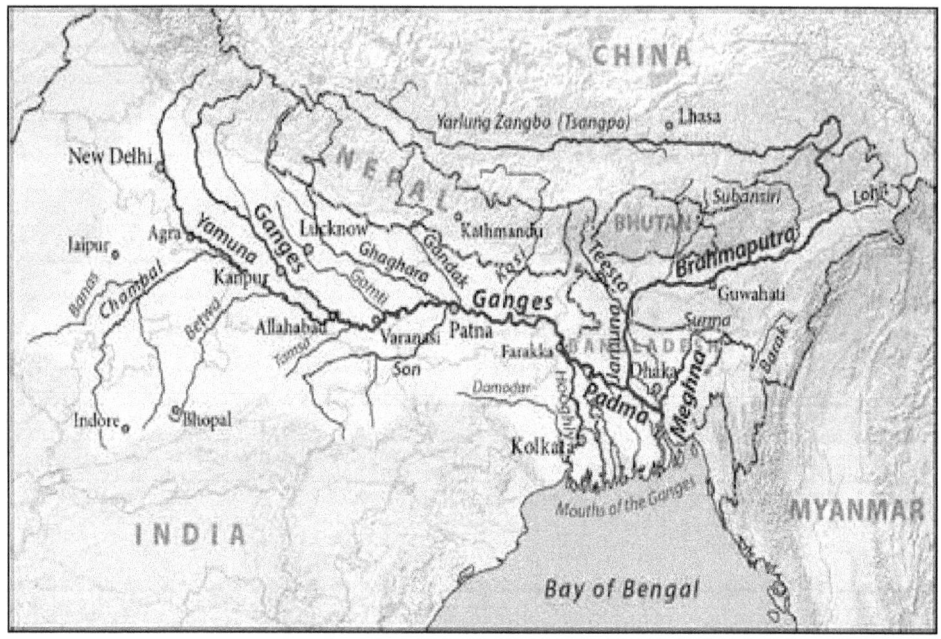

**Figure 4.2: The Tributaries of River Ganga
(Source: www.google.com).**

The River Ganges comes into existence only after Alaknanda and Bhagirathi streams unite at Devaprayag to form the main stream known as Ganga. It flows from the mountains at Rishikesh and then flows onto the plain at Haridwar. The principal tributary (Yamuna River with twelve tributaries) joins river Ganges near Allahabad – {the streatch of river Ganges from Haridwar up to Allahabad experiences lot of hydraulic variations (both ebb and spate) round the year. This streatch experiences maximum pollution load on account of its low flow and poor dilution as most of the tributaries join the River Ganges after Allahabad region. Attention is invited to link one of the rivers in this streatch so that dilution takes care of the pollution load.}-. The Tons River, which flows north from the Vindhya Range in Madhya Pradesh, joins the Ganges just below Allahabad. The other major tributaries along with their sub tributaries joining river Ganges are Chambal, Kosi, Damodar, Betwa, Gomti River, Ghaghara (Karnali), Son River, Ramganga River and Mahananda River.

This open net work of small and big tributaries upholding high density of population besides mushrooming of industries contributes towards the pollution load of Ganges River. Thus, river Ganges can easily be called as a system of rivers with individual wter sheds experiencing varying land use and land cover which pose various types of pressures on the entire system.

Pressures Faced from its Sub Catchment Areas

1. In puts from highly productive agricultural land in the form of biologically active nutrients, pesticide and heavey metals *etc.*

2. Wastewater generated from various types of industries which have been allowed a mushroom growth all along its course.
3. Increasing demand of water for both potable and other purposes. It has been observed that the domestic water demand in high-population density urban areas scattered throughout the Ganga basin has outstripped the supplies.
4. Rapid industrialization has generated large demands for water and hydropower. Though generation of hydroelectricity is need of the hour - (A few smaller projects have been already completed and several larger ones are planned in the Himalayan catchments of the northern tributaries to the Ganga and Brahmaputra) - however, its mismanagement during the construction of large dams on the perennial rivers leaves an indelible mark on the overall well being of the system nurturing it. The major fauna remains at the mercy of dams thus jeopardizing their existence.

The basic problem in utilizing water resources in the Ganga basin is that in relation to the relatively large annual flow in the basin, the storage capacity of existing and foreseeable reservoirs in India is not large enough to permit conservation of flows during high flow season. The live storage capacity of all reservoirs in the Ganga basin is less than one-sixth of the annual flow, which does not permit a significant degree of flow regulation. Lean season flows in the basin without an adequate storage backup are not sufficient to meet the requirements for various demands while monsoon flows are so high that the Ganga and its tributaries remain in spate almost every year.

Many of the diversion works are not backed by any large upstream storage. Therefore the supply of water for irrigation is limited by the flow of the rivers. Only a few tributaries, namely Chambal, Betwa and Sone have large reservoirs and are relatively better developed.

Since no catchment study has been conducted before hand that is why most of the proposed projects have not seen the light of day, *e.g.,* Ken (Ken river, a major tributary of Yamuna) multipurpose project has been under consideration for several decades; the North Koel dam project remains to be completed for several decades; two other reservoirs, namely Kishau and Renuka are in pipeline for more than 10 years; the Lakhwar Vyasi Dam is under construction for over 3 decades. In the Ganga basin, the flooding problem is mainly confined to the middle and terminal reaches. In general, the severity of the problem increases from west to east and from south to north. The worst flood affected states in theGanga basin are Uttar Pradesh, Bihar, and West Bengal. In Uttar Pradesh, flooding is largely confined to the eastern districts where the rivers that cause flooding include the Sarada, the Ghagra, the Rapti, and the Gandak. The major causes of flooding here are drainage congestion and bank erosion.

Four major tributaries of the Ganga — Pilee Nadhi, Barasati Nallah, Rawasan river and Kotawali river - on the Haridwar- Nijababad highway, had been mercilessly targeted by illegal miners. Most tributaries of the Ganga are drying because of construction of dams, and destruction of the ecology, in upstream areas, there is possibility of elimination of major flora and fauna especially Dolphins the in Nepal stretch of Ganges tributary.

Water Quality of Ganga Basin

In the hilly reaches up to Rishikesh, Ganga water is quite clean except for sediments. From Rishikesh onwards, disposal of sewage into Ganga begins. Besides the municipal waste of Rishikesh and Haridwar, industrial units discharge partly treated effluents into the river. Haridwar City has a population of 1.5 lakh and nearly 60,000 people visit the city every day on an average. This number rises to a few lakh on important religious days and may go up to 15 lakh on the auspicious days during Kumbha Mela (fair).

The profusely growing population together with expansion of industries in the Ganges baisn (Nepal, India, and Bangladesh) puts extreme pressure on the quality of water requirement.

A sharp decline in the quality of Ganga water is due to increasing level of pollution from urban and industrial areas. The problem has arisen largely due to the discharge of untreated urban wastes and industrial effluents from the cascade of large and medium cities located along the course of Ganga and its tributaries. Numerous cities located in the Ganga basin generate and discharge huge quantities of wastewater, a large portion of which eventually reaches the river through natural drainage system. Over the years, the Ganga and its tributaries have become the channels of transport of industrial effluents and the drains for the wastewater of the cities.

Moving downstream, the situation changes for the worse at Kanpur from the quality point of view. Sewage from the city (population 2.7 million) coupled with untreated toxic waste discharge from about 150 industrial units results in severe damage to water quality (Beg and Ali, 2008 and Praveen *et al.*, 2013).

It is estimated that some 900 million litres of sewage is dumped into the Ganga every day; three-fourths of the pollution in the Ganga is from untreated municipal sewage. In particular the middle reach of the basin between Kanpur and Buxar is the most urbanized and industrialized, as also the most polluted segment of the basin. At Allahabad with population of more than a million, municipal wastes are the major contributor to river pollution.

Yamuna whose water is highly polluted joins Ganga at Sangam. Large volume of municipal and industrial waste is dumped in the river at Varanasi, a city with approximately 1.2 million population. Entering in Bihar, a number of industries (including fertilizer and oil refining) have come up along Ganga River. At Kolkota in West Bengal, the Hooghly (Ganga) river basin is highly populated as the waster from numerous industries as well as municipal sewage is dumped in the river. Ganga finds its name in the list of the five most polluted rivers of the world. In certain areas in Ganga River the bacteria levels are more than 100 times higher than the limits set by the government.

In view of the magnitude of water quality problems in the Ganga basin, two actions plans were launched by the government of India: the Ganga action plan and the Yamuna action plan. However, at present it ranks among the top most polluted rivers of the world, threatening not only humans, but also both minor and major flora and fauna. The Ganga Action Plan, an environmental initiative to clean up the river, has been a major failure so far on account of not adopting a wholestic approach of Land water interactions.

Though The Ganga River has been declared as India's National River besides beeing revered as a holy river since time immemorial yet people do not hesitate dumping domestic and industrial waste in it.

The Ganga has been described by the World Wildlife Fund as one of the world's top ten rivers at risk. It has over 140 fish species, 90 amphibian species, and five areas which support birds found nowhere else in the world. A species of dolphins found in the Ganges river know as the Ganges Dolphin was recently declared as the National Aquatic Animal of India.Gangetic dolphins were once found in abundance in the river Ganges. But over the years a steady increase in pollution in the river has declined the population of Dolphins.

Suggestions

Being of stream nature in the first part of its downward journey Ganges experiences high rate of aeration absorbing good amount of atmospheric oxygen. However, regular inputs of Nutrients from exogenous sources play a vital role in comparison to endogenous sources in enriching the adjacent and drown stream ecosystems. The work conducted by various workers clearly signifies the drop down in its nutrient carrying capacity during the low water flux from the upper reaches of Himalayas. However, on yearly basis the nutrient influx does not raise an alarming situation due to its high flushing, this really does the trick for the Government for being pathetically unconcerned towards the brewing situation.

Therefore, in order to either retain its present carrying capacity or enhance the same, it is very important to manage the different opened up watershed areas properly by adopting/maintaining natural vegetation cover. Further, in order to reduce the increasing fallow areas in different watersheds it is important to use proper crop soil management and among the most of the problems related with aquatic ecosystem is the nutrient income from its watershed area.

Since we are basically dealing with dynamic system it is, as such important to consider a time sequence of controls, to be set into operation.

In order to achieve meaningful management following steps have been suggested.

1. Identification of catchment wide issue: short and long term. Development of suitable strategies for identified issues.
2. Prioritize sub-catchment for action: the areas where maximum benefit can be derived, using criteria such as fitness for use, development status and maintenance of integrity of potable water supply.

3. Identify specific areas and variable of concern: priority could be afforded to those areas that contain the most significant sources of pollution and those which threaten the health of communities.
4. Prioritize pollution sources: according to the magnitude of their contribution and the possibility of managing the problem, specific site inspections and meetings will be needed to prioritize these sources.

Management of watershed is important due to the following reason:

1. A well-managed watershed allows percolation of the storm water (as most of the storm water soaks into the soil, which would otherwise resulting frequent flash floods, loss of so cover, washing out of crops, polluting of downstream water bodies or fills lake with sediments resulting in changing of groundwater supplies and providing crops, pastures and trees with needed moisture.
2. A well protected forest canopy maintains the humidity of that area, besides other numerous benefits.
3. It makes better provision for the conservation, allocation, use and quality of water.
4. It promotes soil conservation and helps in preventing damage by flood and erosion.
5. It provides controls and multiple use of water and drainage of land.
6. It helps in protecting wildlife and scenic values.

Integrated Management

Following integrated management strategy has been suggested.

1. Demand and supply management: water management has traditionally focused on ensuring adequate water supplies. But meeting growing demands for water only by seaking more supplies creates environmental, social and economic problems.

   Effective water management must manage demand and supply, through conservation guidelines and regulations, economic incentives and education.
2. Land and water management: land and water management must be integrated, with decisions regarding water use made in full awareness of its effect on land and vice versa.
3. Environment, economic and social well-being: the full range of economic, social and environmental benefits and costs must be assessed in decision making.
4. Water quality management: providing better water quality protection and coordinating drinking water quality measures through:
   a) Specific inclusion of water quality criteria and objectives in the water act and public involvement.

b) Considerations of water quality in water allocation and management decisions.

c) Setting priorities under strict legislative authority keeping in mind the role of stake holders.

Integrated approach for catchment area treatment plan (CAT plan) for sustainable utilisation of the aquatic resource.

The purpose of the development of water shed is to increase the economic and social well-being of the people of chosen catchment resource.

The main objective of CAT plan

1. To suggest the treatment that will reduce the surface run-off and will improve the sub-soil moister regime, so that impact on a water body is minimized.

2. To suggest treatment that will reduce the rate of erosion and nutrient loss in the catchment and minimize the rate of siltation and eutrophication of an aquatic system.

3. To maintain the balance, preservation of biodiversity and the maintenance of gene pool in the existing area and to bring about eco development of the catchment area.

4. To suggest alternate land use besides providing gainful employment during lean period in order to stabilize the local ecological condition.

Thus by integrated water managements following goals can be achieved.

☆ Better provision for the conservation, allocation use and quality of water.

☆ Soil conservation and prevention of damage caused by erosion.

☆ Proper drainage of land can be achieved by controlling multiple uses of water.

☆ Wildlife and scenic values together with water related recreation can be met with.

References

Beg, K.R. and Ali, S., 2008. Chemical contaminants and toxicity of Ganga river sediment from up and down stream area at Kanpur. *American Journal of Environmental Sciences,* 4(4): 362–366.

Kaushal, Kishore, 208). *The Holy Ganga.* Pub. Rupa and Co., New Delhi, p. 295.

Negi, Sharad Singh, 2010. *Himalayan Rivers, Lakes and Glaciers.* Indus Pub. Co., New Delhi, p. 182.

Praveen Anjum, Rajesh Kumar, Pratima and Rajat Kumar, 2013. Physio-chemical properties of the water of river Ganga at Kanpur. *International Journal of Computational Engineering Research,* 3(4): 134-137.

Singh, Munendra and Singh, Amit K., 2007. Bibliography of environmental studies in natural characteristics and anthropogenic influences on the Ganga river. *Environmental Monitoring and Assessment*, 129(1-3): 421-432.

Wanganeo, A., 1998. Impact of Anthropogenic Activities on Bhoj Wetland with Particular Emphasis on nutrient Dynamics. Ministry of Environment and Forests, New Delhi.

www.google.com

Chapter 5

Review on Reproductive Tactics and Strategies of Marine Fish Species from the Coastal Waters of Rio Grande do Norte, Brazil

Sathyabama Chellappa and Naithirithi T. Chellappa

ABSTRACT

This work reviews the reproductive tactics and strategies of seven marine fish species from the coastal region of Rio Grande do Norte, Northeastern Brazil. Life history traits of fish in relatively stable environments suggest a model of three reproductive strategies: (1) opportunistic strategists with small body size, early first sexual maturity, and short-life span; (2) periodic or seasonal strategists of big body size with long life span and high to intermediate fecundity; and (3) equilibrium strategists of intermediate size. Studies on the reproductive strategies of the flying fish, Hirundichythys affinis, ballyhoo half beak, Hemiramphus brasiliensis, roughneck grunt, Pomadasys corvinaeformis, maracaibo leatherjacket, Oligoplites palometa, serra Spanish mackerel, Scomberomorus brasiliensis, the lane snapper, Lutjanus synagris and white mullet, Mugil curema were verified using data on body size, sex ratio, length at first sexual maturity, aspects of gonad development, fecundity, type of spawning and reproductive period. The results indicate that H. affinis, H. brasiliensis, P. corvinaeformis and O. palometa are opportunistic strategists. On the other hand S. brasiliensis, L. synagris and M. curema are considered as equilibrium strategists. This study provides information on the reproductive aspects of the fishery stocks of the coastal waters of Rio Grande do Norte, Brazil.

Keywords: Fish reproduction, Coastal waters, Spawning period, Sexual strategies, Tropical marine fish.

Introduction

The reproductive strategies are used by fish to maximize production and ensure the survival of offspring to adulthood. Each strategy is expressed by tactics, such as, body size, length-weight relationship, size at first sexual maturation, gonad development, fecundity, type of spawning and breeding period (Potts and Wootton, 1984), which are important information for making rational measures to regulate fishing and conservation of fish stocks (King and McFarlane, 2003; Oliveira *et al.*, 2011). These characteristics possessed by a particular population of a species could have its origin in the evolutionary past or represent fine-tuning of that population for the environmental conditions (Leggett and Carscadden, 1978; Matthews, 1998).

Life history traits of fish in relatively stable environments suggest a model of three reproductive strategies. Winemiller and Rose (1992) used a quantitative approach to develop groupings of life history strategies by examining 16 life history traits in a large sample (216 species from 57 families) of North American freshwater and marine fish species. Based on a final selection of five life history traits for 82 freshwater species and 65 marine species, they suggested a trilateral continuum model with three endpoint strategies: (1) small, rapidly maturing, short-lived fishes (opportunistic strategists); (2) larger, highly fecund fishes with longer life spans (periodic strategists); and (3) fishes of intermediate size that often exhibit parental investment and produce fewer, larger offspring (equilibrium strategists). For the present study, it was considered important that fish species, representing different reproductive strategies should be compared to assess applicability of results for fisheries management. Fish species selected for this study support commercial fisheries in the northeastern coastal waters of Brazil.

Lane snapper, *Lutjanus synagris* (Linnaeus, 1758) is distributed in the Western Atlantic, from Bermuda and North Carolina, USA to southeastern Brazil, including Gulf of Mexico and Caribbean Sea. Ballyhoo halfbeak, *Hemiramphus brasiliensis* (Linnaeus, 1758) is an inshore surface-dwelling species, which occurs in Western Atlantic: Massachusetts, USA and northern Gulf of Mexico to Brazil, including the Caribbean Sea. Roughneck grunt, *Pomadasys corvinaeformis* (Steindachner, 1868) occurs in Western Atlantic, in Mexico and the Caribbean coasts both continental and insular to the Antilles and Brazil. Serra Spanish mackerel, *Scomberomorus brasiliensis* Collette, Russo and Zavala-Camin, 1978, occurs in the Western Atlantic, along the Caribbean in Central Atlantic coasts and Southwest Atlantic from Belize to Rio Grande do Sul, Brazil. White mullet, *Mugil curema* Valenciennes, 1836, is a coastal pelagic fish which occurs in the Western Atlantic (Nova Scotia to Argentina), Eastern Atlantic (Senegal to Namibia), and in the Eastern Pacific (California to Chile). The flying fish, *Hirundichythys affinis* (Günther, 1866) occurs in the Western Atlantic and coastal waters. Maracaibo leatherjacket, *Oligoplites palometa* (Cuvier, 1832) occurs in the coastal waters of the Western Atlantic. These marine fish species are considered as important fishery resources of the Atlantic Ocean, which are of high commercial value, and form a major component of artisanal fisheries in northeastern Brazil.

Materials and Methods

Brazil is an important part of the Neotropical region which extends from Mexico to the southernmost tip of South America. The state of Rio Grande do Norte is located in Northeastern Brazil and has a coast of approximately 420 km. Fish samples for all the studies reviewed were collected from artisanal fisheries at various locations in the coastal waters of Rio Grande do Norte, Brazil, situated between latitudes 56° 44' and 5° 52' S, longitudes 35° 09' and 35° 12' W.

Literature survey was conducted on life history parameters and reproductive tactics of seven marine fish species from the coastal region of Rio Grande do Norte, northeastern Brazil. During the period of 1998 to 2015 a total of 28 scientific papers were published, which encompass seven marine fish species from the coastal waters of this region in Brazil (Table 1.1). Reproductive aspects of the flying fish, *Hirundichythys affinis,* ballyhoo half beak, *Hemiramphus brasiliensis,* roughneck grunt, *Pomadasys corvinaeformis,* maracaibo leatherjacket, *Oligoplites palometa,* serra Spanish mackerel, *Scomberomorus brasiliensis,* the lane snapper, *Lutjanus synagris* and white mullet, *Mugil curema* were verified considering the body size, sex ratio, length at first sexual maturity, aspects of gonad development, fecundity, type of spawning and reproductive period.

Morphometric measurements and meristic counts were used to check and confirm the taxonomical status of each fish species. All fish collected were identified to the species level, measured (total body length to the nearest millimeter ± 1 mm) and weighed (body mass ± 1 g). Distribution analysis of total length and weight of males and females of each fish species were performed separately using the absolute frequencies (mean ± SD) in different classes of total length (Lt) and total weight (Wt). A single weight-length equation ($W=aL^b$) was fitted to estimate the value of coefficient b, using the data obtained from all individuals collected (Froese, 2006).

The macroscopic aspects and maturation stages of the gonads were observed besides determining the sex of each fish and the sex ratio was determined. Young individuals were not included as it was not possible to determine their sex by macroscopic observations. The chi-square test (χ^2) was applied at a significance level of 5 per cent. The body size at first gonadal maturity (L_{50}), where 50 per cent of the individuals exhibited maturing gonads, was estimated from the relative frequency distribution of adult males and females, using their standard length classes (mean ± SD).

Fecundity was estimated using ten mature ovaries, which were removed, weighed and preserved in Gilson solution for 24 hours for complete dissociation of oocytes, washed and preserved in 70 per cent ethyl alcohol. A 10 per cent sample was removed for counting the mature oocytes and the values were extrapolated to 100 per cent. The type of spawning was assessed by measuring the oocyte diameter size using a binocular microscope (x 20) and an ocular micrometer (± 1 μm). The breeding periods of the fish species were determined by the temporal relative frequency distribution of the different stages of ovarian maturation (De Martini and Fountain 1981).

Results and Discussion

Body Size (Length-weight)

The details of amplitude of total length and weight of the seven marine fish species are presented in Table 5.2. Total body length of *H. affinis* males varied from 25.25 to 27.71 cm (26.54 ± DP 1.40) and body mass of males varied from 126.03 to 199.21 g (150.24 ± 31.23). Total length of *H. affinis* females varied from 26.30 to 27.72 cm (27.04 ± 1.22) and their body mass varied from 13.60 to 21.26 g (15.47 ± 3.09). El-Deir (1998), obtained similar results, wherein females of *H. affinis* presented higher amplitude of body length (20.5 to 26.3 cm) than the males (18.9 to 24.3 cm).

The total body length of *H. brasiliensis* males varied from 15.5 to 33.5 cm (23.1±2.2), while body weight variation was from 14 to 196.1 g (67 ± 26.9). For the females total length varied from 19.5 to 33 cm (24.3 ± 2.6) and body weight variation was from 33.4 to 203.4 g (73±28.2). Similar results were observed for *H. brasiliensis* in Venezuela and in South of Florida (McBride; Thurman, 2003; Yelipza *et al.*, 2011; Oliveira *et al.*, 2012). The total body length of *P. corvinaeformis* males varied from 4.9 to 14.7 cm, with a mean value of 10.27, and the body weight varied from 1.4 a 41.3 g with an average of 15.95. For the females total length varied from 5.0 to 18 cm, with a mean value of 11.8 and body weight varied from 1.4 to 74.0 g with a mean value of 26.16. However, individuals of 7.6 to 20 cm of total body length were captured in the Guaratuba bay of Paraná, Brazil (Chaves, 1998).

The total body length of *O. palometa* males varied from 19.8 to 57.5 cm (2.61±0.668) and the body weight varied from 51.6 g to 942 g (154.3±153.3). In case of the females total length varied from 20 to 32 cm (2.45±0.285), and the body weight varied from 45.3 g to 243 g (116.07±47.33). The males were longer and heavier than the females, possibly due to the capture of more males (49 males and 29 females) (Araújo *et al.*, 2012). The total body length of *S. brasiliensis* males varied from 14 cm to 59.8 cm and in females from 13.5 to 80.5 cm. The body weight of males and females of *S. brasiliensis* varied from 16 to 1,310g and 15 to 3,385 g respectively. The body length of *L. synagris* males varied from 12 to 36 cm (23.6±7.1), and the body weight varied from 17.5 to 891.5g (250.9±214.9). The body length of females of this species varied from 21.5 to 36.5 cm (28.7±5.08), and their body weight varied from 125.5 to 743 g (377.2±214.1). The females of *L. synagris* were bigger and heavier than the males (Cavalcante *et al.*, 2012). Total body length of *M. curema* females varied from 15.6 to 34.5 cm (24.9±4.1) and body weight varied from 35.4 to 382g (160.5±82.3). In this study only females were captured selectively.

The females of the species *H. affinis, H. brasiliensis, L. synagris, P. corvinaeformis, S. brasiliensis* and *M. curema* were bigger and heavier than the males. This was possibly due the ovarian weight and development in relation to testicular development (Murua *et al.*, 2003; Oliveira *et al.*, 2012).

Length-Weight Relationship and Type of Growth

Studies on the length-weight relationship, combined with other quantitative aspects, such as, condition factor, growth, recruitment and fish mortality, provide basic information for the fishery biology and rational management of fishing stocks.

The length-weight relationship functions of the flying fish, *H. affinis*, ballyhoo half beak, *H. brasiliensis*, maracaibo leatherjacket, *O. palometa*, serra Spanish mackerel, *S. brasiliensis*, the lane snapper, *L. synagris* and white mullet, *M. curema* were estimated. However, there is no information for the roughneck grunt, *P. corvinaeformis*.

It is possible to determine the type of growth of a species through the allometric coefficient (*b*), which is isometric when $b = 3$, positive allometry when $b > 3$ and negative allometry when $b < 3$. Isometric growth indicates that the body increases in all dimensions in the same proportion during growth, whereas positive allometry indicates that the body becomes more rotound as it increases in length, and negative allometry indicates a slimmer body (Jobling, 2002). Growth of males and females of *H. affinis* ($\theta = 2.208$ for males and 2.985 for females) and *O. palometa* ($\theta = 0.996$ for males and 0.913 for females) indicate negative allometric growth, where fish increases in length than in body weight, resulting in a slimmer body (Araújo *et al.*, 2011; Araújo *et al.*, 2012; Oliveira *et al.*, 2015). In the case of *H. brasiliensis* and *M. curema* ($\theta = 2.985$) there is isometric growth indicating that the body increases in both dimensions in the same proportion (Oliveira *et al.*, 2011; Oliveira *et al.*, 2012). Growth of *L. synagris* ($\theta = 3.3647$ for males and 3.3152 for females) and *S. brasiliensis* was positively allometric indicating that the body becomes more rotound (Chellappa *et al.*, 2010; Cavalcante *et al.*, 2012).

The parameters of length-weight relationship in fish could be influenced by environmental conditions, gonadal maturity, sex, condition factor, season and variations between species (Froese, 2006).

Sex Ratio

The sex ratio of *H. affinis* was 1M:1.4F, thus not differing significantly from the expected ratio of 1:1. However, there was a slight predomination of females over the males of *H. affinis*. The mature females usually migrate to the coastal waters to spawn (Araujo and Chellappa, 2002). The sex ratio of *H. brasiliensis* was 1M:1.1F not differing from the expected sex ratio in nature (1:1). In the case of *P. corvinaeformis* the females predominated with a sex ratio of 1M:2.1F differing significantly from the expected ratio of 1:1 (Silva *et al.*, 2012). The sex ratio of *O. palometa* was 2M:1F with a significantl difference, where this population showed the occurrence of more males than females (Araújo *et al.*, 2012). A different study showed that the sex ratio of *S. brasiliensis* was 2M:1F where males predominated over females (Nobréga, 2002). A similar pattern was observed for *L. synagris* which showed a sex ratio of 4.15M:1F. The sex ratio of *M. curema* was 1M:1F not differing from the expected sex ratio. Females predominated in *H. affinis*, *H. brasiliensis* and *P. corvinaeformis*, whereas in *L. synagris*, *O. palometa* and *S. brasiliensis* the males predominated.

Usually the sex ratio is 1:1 in natural environments, but during the life cycle of fish this can vary depending on various factors that act differently on individuals of each sex. Growth and mortality are factors that can act differentially on males and females, thus determining the predominance of one sex. The sex ratio can be affected by factors related to fishing, seasons, besides the number of individuals in the feeding and spawning areas (Lasiak, 1982).

Length at First Sexual Maturity (L_{50})

The onset of sexual maturity represents a critical transition in the life history, since resource allocation is related mainly to growth before and to reproduction after sexual maturity (Potts and Wootton, 1984; Chellappa *et al.*, 1995). For rational management of fishery stocks which are subjected to exploitation, it is important to know the size at first gonadal maturation (L_{50}), since it provides information for determining the minimum size at capture and mesh dimensions of the fishing gear.

The sizes when 50 per cent of males and females of *H. affinis* were in the process of sexual maturation were 23.8 cm and 23.0 cm of total body length respectively (Araujo and Chellappa, 2002). However, another study on the same species reported that the total length at first sexual maturity was at 27.3 cm for males and 27.1 cm for females. The males of *H. affinis* mature before the females, shown by the significant difference in size of gonadal maturation of both sexes ($t = -5.081$; df = 210; $p < 0.05$).

Total length at first sexual maturity of *H. brasiliensis* males was at 20.8 cm and of the females was at 21.5 cm. The males of *H. brasiliensis* attained first gonadal maturity at smaller body lengths than females (t =3.62, df = 408, $p < 0.05$) (Oliveira *et al.*, 2015). The females of *H. brasiliensis* in the coastal waters of South Florida attained maturity at 19.8 cm, however, males were not included in this study (McBride and Thurman, 2003).

Total lengths at first sexual maturity for *L. synagris* males and females were 23.5 cm and 24.5 respectively (Cavalcante *et al.*, 2012). In case of *P. corvinaeformis* the total length at first sexual maturity was 10.3 cm for males and 10.4 cm for females (Silva *et al.*, 2012). Total length at first sexual maturity of *S. brasiliensis* males was at 34.5 cm and for females it was at 28 cm. The gradual decrease in total length at first sexual maturity of *S. brasiliensis*, possibly indicates overfishing of this commercially important species (Lima *et al.*, 2007). The traditional fishing communities depend on small scale artisanal fishery, which reflects their way of making a living and sustains their lifestyle. Though it is important to preserve this traditional fishery, it is also vital to programme the sustainability and conservation of coastal fisheries resources. The predatory fishing technique of beach seine nets, in which small mesh sizes are used in order to catch marine shrimps, accounts for a large by-catch of small sized immature Serra Spanish mackerel. Total length at first sexual maturity of both sexes of *M. curema* was 24.3 cm (Oliveira *et al.*, 2014).

All three species *H. affinis*, *H. brasiliensis* and *S. brasiliensis* showed that females attained maturity earlier than the males. Conservation of fish stocks in their natural habitat are usually endangered by abusive fishing of immature fishes which have not yet completed their reproductive cycle, as recruitment via reproduction is the means by which the resource is renewed. The artisanal fishery beach-seines in Northeastern Brazil are operated in the shallow coastal waters, which captures the immature individuals of the pelagic stock. Measures should be taken to regulate this fishery in order to conserve this valuable fishery resource by increasing the size of capture.

Gonad Development

Macroscopic Characteristics of Gonads

The use of macroscopic scales of maturity of the gonads contributes to the biological knowledge in describing the reproductive cycle of fish and helps in understanding the reproductive period of the species. However, for better identification of gonad developmental stages, histological analysis is considered essential.

H. affinis had paired elongated gonads, the females with lobed ovaries and the males with flattened testes. The testicular walls were fragile when compared to the ovarian walls, and did not show much modification between the different stages of development, unlike the ovaries. The volume, coloration, thickness and blood vessels of ovaries varied according to the stage of maturation, presenting shades of light pink to dark yellow, due to the color of the mature oocytes full of yolk granules. The testes were whitish from the beginning of maturation to the mature stage. The macroscopic characteristics of the testes of *H. affinis* presented three stages of maturation, such as, immature, maturing and mature, while females showed four stages of development, immature, maturing, mature and spent (Oliveira *et al.*, 2015).

The macroscopic characteristics of *H. brasiliensis:* The ovaries and testes were paired bi-lobed structures, symmetrical, elongated and joint in the posterior part to form a short duct leading to the urogenital pore. They were located in the posterior-dorsal part of the coelomic cavity, ventral to the kidneys and swim bladder. The immature testes were small and translucent. Maturing testes were more developed and were whitish in color. The mature testes were white and spent testes were flaccid and brown in color with hemorrhagic appearance. During maturation, the ovaries were pinkish to light orange in color and developed progressively by increasing in size and vascularization. The mature ovaries were turgid and occupied 2/3 of the coelomic cavity. The mature ovaries were turgid with numerous big oocytes visible to the naked eye, and the partially spent ovaries were flaccid. The macroscopic characteristics of the gonads indicated four maturation stages: immature, maturing, mature and spent (Oliveira *et al.*, 2015).

The macroscopic characteristics of the gonads of *P. corvinaeformis* indicated four developmental stages for testes and ovaries: immature, maturing, mature and spent.

Immature testes and ovaries were filiform, small in size with translucent coloration. Maturing and mature ovaries appeared reddish in colour, increasing in size with visible oocytes. The testes showed varying sizes in accordance with the degree of development, become thicker and whitish in colour. The spent gonads were reduced size with a flaccid appearance (Silva *et al.*, 2012).

The macroscopic characteristics of the testes and ovaries of *O. palometa* (Figure 1.2), *S. brasiliensis, L. synagris* and *M. curema* indicated four stages of development, such as, immature, maturing, mature and spent (Chellappa *et al.*, 2010; Araújo *et al.*, 2012; Cavalcante *et al.*, 2012; Oliveira *et al.*, 2014). The ovaries were bilobed, elongated, and joined posteriorly to form a short gonoduct leading to the urogenital pore. The immature testes and ovaries were small and translucid structures. During maturation the gonads occupy about one third of the coelomic cavity, while mature ovaries occupy almost two thirds of the coelomic cavity. Spent gonads have a hemorrhagic appearance and are reduced in size.

Microscopic Characteristics of Gonads

Histological examinations of ovarian sections of the study species generally showed that the oocyte development was consistent along the whole length of the ovary depending on the degree of maturation. Ovaries revealed five stages of oocyte development: immature, early stage of maturing, late stage of maturing, mature and spent (Figures 1.3–1.7). Immature ovaries showed the chromatin nucleolar stage, where clusters of very small oocytes were found lying just beneath the ovigerous lamella and young germ cells compactly filled the ovaries. The ovaries in early stage of maturing showed the perinucleolar stage, oocytes with nucleoli at periphery of nucleus with thick cytoplasm. During cortical alveoli stage the oil vesicles appear and ovaries were with early yolk globule and previtellogenic stage oocytes. The ovaries in late stage of maturing revealed the yolk stage, when oocytes show the presence of yolk granules near the periphery and oil vesicles within the inner region of the cytoplasm, and cytoplasmic vesicles had a uniform distribution. Mature ovaries showed nuclear migration and hydration stages, maturation into this stage was marked by the migration of the nucleus to the periphery of the oocyte, fusion of yolk granules into yolk plates and coalescence of oil droplets. Nucleus breaks down when it reaches the periphery and hydration occurs.

The details of histological aspects of ovarian sections of the study species are presented in Figures 1.3–1.7.

Fecundity and Type of Spawning

The absolute fecundity of *H. affinis* varied from 7,398 to 10,021 oocytes, with an average of 9,092 (SD ±1,153.2) vitellogenic oocytes. Their diameter size varied from 100 µm to 2500 µm. The reserve stock oocytes had diameter size less than 1000 µm, and the developing oocytes were bigger than 1000 µm. This indicated that this is a total spawner, which eliminates the mature oocytes at the same time (OLIVEIRA et al., 2015a).

The batch fecundity of *H. brasiliensis* varied from 862 to 1,354 with an average of 1,153 (±258.22) vitellogenic oocytes for 50 g body weight of female. The microscopic characteristics of gonad development of *H. brasiliensis* showed multiple spawning (Oliveira et al., 2015b). Fecundity is a specific reproductive tactic and is adapted to the life cycle conditions of the species, varying with growth, population density, body size, food availability and mortality rate (Murua et al., 2003).

The absolute fecundity of *P. corvinaeformis* varied from 15,056 to 83,316 oocytes and their diameter size varied from 110µm to 390µm. The reserve stock oocytes had diameter size less than 140 µm, and the developing oocytes were bigger than that. This indicated that this is a total spawner, which eliminates the mature oocytes at the same time (Silva et al., 2012). The mean absolute fecundity of *S. brasiliensis* was 871,523 oocytes and is a total spawner. The reserve stock oocytes had diameter size less than 120 µm, and the developing oocytes had diameters which varied from 650 to 750 µm (Lima *et al.*, 2007). The mean absolute fecundity of *M. curema* was 245,828 oocytes and is a total spawner (Oliveira *et al.*, 2014).

Gonadosomatic Index (GSI) and Reproductive Period

GSI demonstrates the functional gonadal status in relation to fish body mass, indicating the spawning period. The mean values of GSI of *H. affinis* varied from 1.25 to17.1 for the females and from 0.1 to 8.01 for the males. There was a predominance of maturing individuals during the period of October to February and mature fish occurred during March to July. Variations in GSI and the monthly frequency of maturation stages demonstrate that *H. affinis* reproduces during the period of March to July, coinciding with the rainy season of the region (Araújo and Chellappa, 2002; Oliveira *et al.*, 2015).

The mean monthly values of GSI of *H. brasiliensis* varied from 0.94 to 3.58 for the females and 0.18 to 0.39 for the males. 1.47 to 4.10. Frequency of monthly gonadal maturation stages of females indicate that mature individuals occur throughout the year. Variations of GSI and the frequency of monthly gonadal maturation stages show that *H. brasiliensis* has an active reproductive period during the months of January to June and again in October. The breeding period of this species is independent of the rainy period (Oliveira *et al.*, 2015).

The mean monthly values of GSI of *P. corvinaeformis* varied from 0.012 to 1.15 for males and 0.012 to 5.49 for females. Two peaks of GSI were observed during the months of November and March. The reproductive period extends from October to June (Silva *et al.*, 2012).

The mean monthly values of GSI of *O. palometa* varied from 0.07 to 0.15 for males and from 0.07 to 1.64 for females. During the dry period the mean monthly values of GSI of males was 0.13 (± 0.03), and during the rainy season the mean monthly values of GSI was 0.11 (± 0.03), without any significant difference of GSI between the seasons (Araujo *et al.*, 2012).

The mean monthly values of GSI of *S. brasiliensis* varied from 0.02 to 7.14 for females. Reproductive activity was during March to June, coinciding with the rainy season (Chellappa *et al.*, 2010; 2011).

Reproductive Strategies

The flying fish, *H. affinis,* ballyhoo half beak, *H. brasiliensis,* roughneck grunt, *P. corvinaeformis,* maracaibo leatherjacket, *O. palometa,* are considered as opportunistic stragists, with small body size, early first sexual maturity and short-life span (Table 1.2). Their spawning seasons are influenced by the rainfall of the region.

On the other hand, the three species Serra Spanish mackerel, *S. brasiliensis*, the lane snapper, *L. synagris* and white mullet, *M. curema* are considered as equilibrium stragists. (strategy involving medium to large body and less fecund organisms) (Table 1.2). They have smaller but stable population and live in stable environments.

The semiarid region of Brazil is characterized with short spells of rain interspersed with lengthy dry season. Hence the reproductive seasonality of fish is regulated by environmental cues and rainfall seems to be the main driver which modulates the spawning period.

Conclusion

Information regarding the body size, sex ratio, length at first sexual maturity, type of growth, aspects of gonad development, fecundity, type of spawning and the spawning period are important for administration of fishery resources. The results indicate that the flying fish, *H. affinis*, ballyhoo half beak, *H. brasiliensis*, roughneck grunt, *P. corvinaeformis* and maracaibo leatherjacket, *O. palometa* are opportunistic strategists. On the other hand serra Spanish mackerel, *S. brasiliensis*, the lane snapper, *L. synagris* and white mullet, *M. curema* are considered as equilibrium strategists.

Acknowledgements

The authors wish to thank the National Council for Scientific and Technological Development (CNPq) and the Post Graduate Federal Agency of the Ministry of Education, Brazil (CAPES/MEC) for the financial support awarded during the study period.

References

Albieri, R.J., 2009. Biologia reprodutiva da tainha *Mugil liza* Valenciennes e do parati *Mugil curema* Valenciennes (Actinopterygii, Mugilidae) na Baia de Sepetiba, RJ, Brasil. *Masters Dissertation,* Universidade Federal Rural do Rio de Janeiro, 63p.

Albieri, R.J., Araújo, F.G. and Ribeiro, T.P., 2009. Gonadal development and spawning season of white mullet *Mugil curema* (Mugilidae) in a tropical bay. *Journal of Applied Ichthyology*, 26: 105-109.

Araújo, A.S., 2000. Estratégia de reprodução e produção pesqueira do peixe voador, *Hirundichthys affinis* Günther, 1866 (Osteichthyes: Exocoetidae) de Caiçara do Norte e Galinhos, RN. *Masters Dissertation*, Universidade Federal do Rio Grande do Norte, 86 p.

Araújo, A.S., Santos, G.R. and Chellappa, S., 2000. Peixe voador, *Hirundichthys affinis* Günther, 1866 (Osteichthyes: Exocoetidae) de Caiçara do Norte, RN. *Revista de Ecologia Aquática Tropical*, Editora da UFRN, Natal, RN, 10: 123-128.

Araújo, A.S., Oliveira, J.C.S. and Chellappa, S., 2001. Alguns aspectos da dinâmica populacional de *Hirudichthys affinis* Gunther, 1866 (Osteichthyes: Exocoetidae) no litoral norte do estado do RN. *Boletim Técnico Científico do CEPENE, Pernambuco*, 9(1): 181-190.

Araújo, A.S. and Chellappa, S., 2002a. Estratégia reprodutiva do peixe voador, *Hirundichthys affinis* Günther (Osteichthyes, Exocoetidae). *Revista Brasileira de Zoologia*, 19(3): 691-703.

Araújo, A.S. and Chellappa, S., 2002b. Estudo histológico das gônadas do peixe voador, *Hirundichthys affinis* Günther, 1866 (Osteichthyes: Exocoetidae) no Rio Grande do Norte, Brasil. *Arquivos de Ciências do Mar, LABOMAR-Fortaleza, UFC-CE*, 35: 131-134.

Araújo, G.S., 2008. Ecologia parasitária de isópodos e biologia reprodutiva em Tibiro, *Oligoplites* spp (Osteichthyes: Carangidae) das águas costeiras de Natal, Rio Grande do Norte. *Masters Dissertation,* Universidade Federal do Rio Grande do Norte, 97 p.

Araújo, A.S., Oliveira, M.R., Campos, C.E.C., Yamamoto, M.E. and Chellappa, S., 2011. Características morfométricas-merísticas, peso-comprimento e maturação gonadal do peixe voador, *Hirundichythys affinis* (Günther, 1866). *Revista Biota Amazônia,* 1(2): 33-40.

Araújo, G.S., Araújo, A.S. and Chellappa, S., 2012. Tipo de crescimento e aspectos reprodutivos do peixe marinho *Oligoplites palometa* (Osteichthyes: Carangidae), na costa do Rio Grande do Norte, Brasil. *Biota Amazônia*, 2(2): 25-30.

Carvalho, M.M., Morais, A.L.S., Gurgel, T.A.B., Oliveira, M.R. and Chellappa, S., 2014. Frequência de ocorrência e características morfológicas externos de peixes marinhos de Caiçara do Norte, Rio Grande do Norte, Brasil. *Biota Amazônia*, 4(2): 55-63.

Cavalcante, L.F.M., Oliveira, M.R. and Chellappa, S., 2012. Aspectos reprodutivos do ariacó, *Lutjanus synagris* nas águas costeiras do Rio Grande do Norte. *Biota Amazônia*, 2(1): 45-50.

Costa, P.S.R., Santos, M.A.M., Espínola M.F.A. and Monteiro-Neto, J., 1995. Biologia e biometria do coró, *Pomadasys corvinaeformis* (Steindachner) (Teleostei: Pomadasyidae), em Fortaleza, Ceará, Brasil. *Arquivos de Ciências do Mar,* 29(1-2): 20-27.

Chaves, P.T.C., 1998. Estrutura populacional de *Pomadasys corvinaeformis* (Steindachner) (Teleostei, Haemulidae) na Baía de Guaratuba, Paraná, Brasil. *Revista Brasileira de Zoologia*, 15(1): 203-209.

Chaves, P.T.C. and Corrêa, C.E., 2000. Temporary use of a coastal ecosystem by the fish, *Pomadasys corvinaeformis* (Perciformes: Haemulidae), at Guaratuba Bay, Brazil. *Revista Brasileira de Oceanografia*, 48(1): 1-7.

Chellappa, S., Lima, J.T.A.X., Araújo, A. and Chellappa, N.T., 2010. Ovarian development and spawning of Serra Spanish mackerel in coastal waters of Northeastern Brazil. *Brazilian Journal of Biology*, 70(2): 631-637.

Chellappa, S., Lima, J.T.A.X., Araújo, A. and Chellappa, N.T., 2011. Reproductive biology of *Scomberomorus brasiliensis* (Perciformes: Scombridae). In: (Org.). *Advances in Fish and Wildlife Ecology and Biology*, (Ed.) B. L. Kaul. Daya Publishing House, Delhi, India, 5: 3-19.

Dias, T.L.P., Rosa, R.S. and Damasceno, L.C.P., 2007. Aspectos socioeconômicos, percepção ambiental e perspectivas das mulheres marisqueiras da Reserva de Desenvolvimento Sustentável Ponta do Tubarão (Rio Grande do Norte, Brasil). *Gaia Scientia*, 1(1): 25-35.

Duque-Nivia, G., Arthuro, A. P. and Santos-Martinez, A., 1995. Aspectos reproductivos de *Oligoplites saurus* y *O. palometa* (PISCES: CARANGIDAE) en La Ciénaga Grande de Santa Marta, Caribe Colombiam. *Caribean Journal of Science*, 31(3-4): 317-326.

El-Deir, A.C.A., 1998. Reprodução e Caracterização Morfométrica e Merística do Peixe-voador *Hirundichthys affinis* (Günther,1866) em Caiçara–RN. Recife, UFRPE, *Masters Dissertation*, Universidade Federal Rural de Pernambuco, 92p.

Engelhard, G.H. and Heino, M., 2004. Maturity changes in Norwegian spring-spawning herring before, during, and after a major population collapse. *Fisheries Research*, 66: 299-310.

Fonteles-Filho, A.A., 1988. Sinopse de informações sobre a cavala, *Scomberomorus cavalla* (Cuvier) e a serra, *Scomberomorus brasiliensis* Collette, Russo and Zavala-Camin (Pisces: Scombridae), no Estado do Ceará, Brasil. *Arquivos Ciência do Mar, Fortaleza*, 27: 21-48.

Gesteira, T.C.V., 1972. Sobre a reprodução e fecundidade da serra, *Scomberomorus maculatus* (Mitchill), no Estado do Ceará. *Arquivos Ciência do Mar, Fortaleza*, 12(2): 117-122.

Gesteira, T.C.V. and Mesquita, A.L.L., 1976. Época de reprodução, tamanho e idade na primeira desova da cavala e da serra, na costa do Estado do Ceará (Brasil). *Arquivos Ciência do Mar, Fortaleza*, 16(2): 83-86.

Gondolo, G.F., 2008. Idade e crescimento de *Hemiramphus brasiliensis* (Linnaeus, 1758) no litoral de Pernambuco. *Masters Dissertation*, Universidade Federal Rural de Pernambuco, 48p.

IBAMA (Instituto Brasileiro do Meio Ambiente e dos Recursos Naturais Renováveis), 2008. Monitoramento da atividade pesqueira no litoral nordestino – Projeto ESTATPESCA – 2006. Tamandaré, PE. 384p.

Ibañez-Aguirre, A.L. and Gallardo-Cabello, M., 2004. Reproduction of *Mugil cephalus* and *M. curema* from a coastal lagoon to the northwest of the Gulf of Mexico. *Bulletin Marine Science*, 75: 37–49.

Jobling, M., 2002. Environmental factors and rates of development and growth. *In*: *Handbook of Fish Biology and Fisheries Vol. 1: Fish Biology*, (Eds.) Hart, P.J. and J.D. Reynolds. Blackwell Publishing Ltd., Oxford, p. 97-122.

King, M.G., 1997. *Fisheries Biology: Assesment and Management*. Osney Mead, Oxford, England: Fishing News Books, p. 341.

King, Jr. and McFarlane, G.A., 2003. Marine fish life history strategies: applications to fishery management. *Fisheries Management and Ecology*, 10: 249–264.

Lasiak, A., 1982. Aspects of the reproductive biology of the Southern mullet *Liza richardsoni* from Algoa Bay, South Africa. *South African Zoology*, 18: 89-95.

Lima, J.T.A.X., 2004. Biologia reprodutiva e parasitismo por isópodes do serra, *Scomberomorus brasiliensis* (Collette, Russso and Zavala-Camin, 1978) (Osteichthyes: Scombridae) no litoral do Rio Grande do Norte. *Masters Dissertation*, Universidade Federal do Rio Grande do Norte, RN. 153p.

Lima, J.T.A.X., 2008. Dinâmica reprodutiva e parasitária de quatro espécies de peixes das águas costeiras do Sudoeste do Oceano Atlântico, Brasil. *Doctoral Thesis*, Universidade Federal do Rio Grande do Norte, 119p.

Lima, J.T.A.X., Chellappa, S. and Thatcher, V.E., 2005. *Livoneca redmanni* leach (Isopoda, Cymothoidae) e *Rocinela signata* Schioedte and Meinert (Isopoda, Aegidae), ectoparasitos de *Scomberomus brasiliensis* collette, Russo and Zavala-Camin (Ostheichtyes, Scombridae) no Rio Grande do Norte, Brasil. *Revista Brasileira de Zoologia*, 22(4): 1104-1108.

Lima, J.T.A.X., Fonteles-Filho, A.A. and Chellappa, S., 2007. Biologia reprodutiva da Serra, *Scomberomorus brasiliensis* (Osteichthyes: Scombridae), em águas costeiras do Rio Grande do Norte. *Arquivos de Ciências do Mar*, 40(1): 24-30.

Lucena, F. Lessa, R. and Nóbrega, M., 2001. Presente status do estoque da serra *Scomberomorus brasiliensis* no Nordeste do Brasil. Anais do XII Congresso Brasileiro de Engenharia de Pesca, Foz do Iguaçu.

Luckhurst, B.E., Dean, J.M. and Reichert, M., 2000. Age, growth and reproduction of the lane snapper *Lutjanus synagris* (Pisces: Lutjanidae) at Bermuda. *Marine Ecology Progress Series*, 203: 255-261.

Marín, B.J., Quintero, A., Bussière, D. and Dodson, J.J., 2003. Reproduction and recruitment of white mullet (*Mugil curema*) to a tropical lagoon (Margarita Island, Venezuela) as revealed by otolith microstructure. *Fishery Bulletin*, 101: 809–821.

McBride, R.S., Foushee, L. and Mahmoudi, B., 1996. Florida's halfbeak, *Hemiramphus* spp., bait fishery. *Marine Fishery Reviews*, 58: 29-38.

McBride, R.S. and Thurman, P.E., 2003. Reproductive biology of *Hemiramphus brasiliensis* and *H. balao* (Hemiramphidae): Maturation, Spawning Frequency, and Fecundity. *Biological Bulletin*, 204: 57-67.

Monteiro, A., 2003. Biologie et pêche des Aiguilles *Hemiramphus brasiliensis* (Linnaeus, 1758) et *Hyporhamphus unifasciatus* (Ranzani, 1842) (Poissins – Téléosteens – Hemiramphidae) dans la région Nord-Est du Brésil. *Doctoral Thesis*, Universite de Bretagne Occidentale, 210 p.

Moreno, T., Castro, J.J. and Socorro, J., 2005. Reproductive biology of the sand smelt *Atherina presbyter* Cuvier, 1829 (Pisces: Atherinidae) in the central-east Atlantic. *Fisheries Research*, 72: 121–131.

Morgan, M.J., 2004. The relationship between fish condition and the probability of being mature in American plaice (*Hippoglossoides platessoides*). *ICES Journal of Marine Science*, 61: 64-70.

MPA (Ministério da Pesca e Aquicultura), 2012. Boletim Estatístico da Pesca e Aquicultura/Brasil 2010. *Brasília*, 129 p.

Murua, H. and Saborido-Rey, F., 2003. Female reproductive strategies of marine fish species of the North Atlantic. *Journal of Northwest Atlantic Fishery Science*, 33: 23-31.

Murua, H., Kraus, G., Saborido-Rey, F., Witthames, P.R, Thorsen A. and Junquera, S., 2003. Procedures to estimate fecundity of marine fish species in relation to their reproductive strategy. *Journal of Northwest Atlantic Fishery Science*, 33: 33-54.

Nobréga, M.F., 2002. Idade, crescimento e avaliação de estoque da serra *Scombereromus brasiliensis* (Teleostei: Scombridae), na plataforma continental do Nordeste do Brasil. *Masters Dissertation*, Universidade Federal de Pernambuco.

Oliveira, I.M.B., 2001. Aspectos reprodutivos e produção pesqueira do peixe agulha, *Hemiramphus brasiliensis* (Linnaeus, 1758) (Osteichthyes: Hemiramphidae) no litoral norte do RN. *Masters Dissertation*, Universidade Federal do Rio Grande do Norte, 86p.

Oliveira, M.R., 2007. Aspectos reprodutivos da tainha, *Mugil curema* Valenciennes,1836 (Osteichthyesa: Mugilidae) capturados nas águas costeiras de Ponta Negra Rio Grande do Norte. *Bachelors Degree Monograph*, Universidade Federal do Rio Grande do Norte, 73p.

Oliveira, M.R., 2010. Biologia reprodutiva da tainha, *Mugil curema* Valenciennes, 1836 (Osteichthyes: Mugilidae) nas águas costeiras do Rio Grande do Norte. *Masters Dissertation*, Universidade Federal do Rio Grande do Norte, 74p.

Oliveira, M.R. and Chellappa, S., 2014. Temporal dynamics of reproduction in *Hemiramphus brasiliensis* (Osteichthyes: Hemiramphidae). *The Scientific World Journal (Marine Biology)*, Article ID 837151, p. 1-8.

Oliveira, M.R., Costa, E.F.S. and Chellappa, S., 2011. Ovarian development and reproductive period of white mullet, *Mugil curema* in the coastal waters of Northeastern Brazil. *Animal Biology Journal*, 2(4): 22-237,.

Oliveira, I.M.B., Oliveira, M.R, Yamamoto, M.E. and Chellappa, S., 2012. Biologia reprodutiva de agulha-preta, *Hemiramphus brasiliensis* (Linnaeus, 1758) (Osteichthyes: Hemiramphidae) das águas costeiras do Rio Grande do Norte, Brasil. *Biota Amazônia*, 2(2): 44-53.

Oliveira, M.R., Costa, E.F.S., Araújo, A.S., Pessoa, E.K.R., Carvalho, M.M., Cavalcante, L.F.M. and Chellappa, S., 2012. Sex ratio and length-weight relationship for five marine fish species from Brazil. *Journal of Marine Biology and Oceanography*, 1(2): 1-3.

Oliveira, M.R., Carvalho, M.M., Souza, A.L., Molina, W.F. Yamamoto, M.E. and Chellappa, S., 2013. Caracterização da produção do peixe-voador, *Hirundichthys affinis* em Caiçara do Norte, Rio Grande do Norte, Brasil: durante 1993 a 2010. *Biota Amazônia*, 3: 23-32.

Oliveira, M.R., 2014. Estratégia reprodutiva do peixe-voador, *Hirundichthys affinis* e do peixe agulha preta, *Hemiramphus brasiliensis* no litoral de Caiçara do Norte, Rio Grande do Norte, Brasil. *Doctoral Thesis*, Universidade Federal do Rio Grande do Norte, 169p.

Oliveira, M.R., Costa, E.F.S. and Chellappa, S., 2014. Ovarian development and reproductive period of white mullet, *Mugil curema* in the coastal waters of northeastern Brazil. *In*: (Org.) *Biology of Semiarid Tropical Fish*, (Eds.) José Rosa

Gomes and Sathyabama Chellappa. Nova Scientific Publishers, New York, USA, p. 21-34.

Oliveira, M.R., Carvalho, M.M., Silva, N.B., Yamamoto, M.E. and Chellappa, S., 2015. Reproductive aspects of the flyingfish, *Hirundichthys affinis* from the Northeastern coastal waters of Brazil. *Brazilian Journal of Biology*. 75(1): 198-207.

Oscoz, J., Campos, F. and Escala, M.C., 2005. Weight length relationships of some fish species of the Iberian Penisula. *Journal Applied Ichthyology*, 21: 73-74.

Potts, G.W. and Wootton, R.J., 1984. *Fish Reproduction*. Academic Press, London.

Sampaio, J.R., 1996. Índice dos Peixes Marinhos Brasileiros. Fortaleza: Gráfica Editora VT. 124p.

Quiñonez-Velázquez, C. and López-Olmos, J.R., 2011. Juvenile growth of white mullet *Mugil curema* (Teleostei: Mugilidae) in a coastal lagoon southwest of the Gulf of California. *Latin American Journal of Aquatic Research*, 39(1): 25-32.

Santos, E.P., 1978. Dinâmica de Populações aplicadas á pesca e piscicultura. *São Paulo, HUCITEC*, 129 p.

Silva, E.I.L. and Silva, S.S., 1981. Aspects of the biology of grey mullet, *Mugil cephalus* L., adult populations of a coastal lagoon in Sri Lanka. *Journal of Fish Biology*, 19: 1-10.

Silva, A.M., 2003. Aspectos reprodutivos do coró, *Pomadasys corvinaeformis* (Steindachner, 1868) (Osteichthyes: Haemulidae) das águas costeiras de Ponta Negra, Rio Grande do Norte. *Masters Dissertation*, Universidade Federal do Rio Grande do Norte, 89 p.

Silva Junior, L.A., 2009. Pesca com covo e reprodução do ariocó *Lutjanus synagris* (Perciformes: Lutjanidae) na Costa de Pernambuco. *Masters Dissertation*, Universidade Federal de Pernambuco, 55p.

Silva, A.M., Oliveira, M.R. and Chellappa, S., 2012. Biologia reprodutiva do coró, *Pomadasys corvinaeformis* Steindachner (Osteichthyes: Haemulidae) das águas costeiras do Rio Grande do Norte, Brasil. *Biota Amazônia*, 2(2): 15-24.

Solomon, F.N. and Ramnarine, I.W., 2007. Reproductive biology of white mullet, *Mugil curema* (Valenciennes) in the Southern Caribbean. *Fisheries Research*, 88: 133–138.

Souza, L.M. and Chaves, P.T., 2007. Atividade reprodutiva de peixes (Teleostei) e o defeso da pesca de arrasto no litoral norte de Santa Catarina, Brasil. *Revista Brasileira de Zoologia*, 24(4): 1113-1121.

Stratoudakis, Y., Bernal, M., Ganias, K. and Uriarte, A., 2006. The daily egg production method: recent advances, current applications and future challenges. *Fish and Fisheries*, 7: 35-57.

Vazzoler, A.E.A.M., 1996. Biologia da Reprodução de Peixes Teleósteos: Teoria e Prática. Maringá: EDUEM, 169p.

Vieira, E.M.M. and Lima, I.M.M.R., 2003. Um novo olhar para a extensão pesqueira: gênero na prática organizativa das mulheres marisqueiras. Prorenda Rural, PE. Extensão Pesqueira: Desafios Contemporâneos. Edições Bagaço: Recife, pp. 137-152.

West, G., 1990. Methods of assessing ovarian development in fishes: a Review. *Australian Journal of Marine and Freshwater Research*, 41: 199-222.

Winemiller, K.O. and Rose, K.A., 1992. Patterns of life-history diversification in North American fishes: implications for population regulation. *Canadian Journal of Fisheries and Aquatic Sciences*, 49(10): 2196-2218.

Yelipza, L.R., Acosta, V., Parra, B. and Lista, M., 2011. Aspectos biométricos de *Hemiramphus brasiliensis* (Peces: Hemirhamphidae), Isla de Cubagua, Venezuela. *Zootecnia Tropical*, 29(4): 385-398.

Chapter 6

Macrobenthic Community Structure along the Temporal Shift from Non-paddy Cultivating (Flooded Phase) to Paddy Cultivating (Paddy Phase) Phases in Kole Paddy Fields, Vembanad Kole Wetland, India

S. Vineetha, S. Bijoy Nandan and K.P. Rakhi Gopalan

ABSTRACT

Paddy fields (man managed temporary wetlands) are characterized by their dynamic, temporary and transitional nature, the allied changes alter the environmental conditions eventually reflecting on the paddy field biota. The temporal variation in macrobenthic community, an integral component in paddy field fertility and aquatic food web, was analyzed along the shift from non-paddy cultivating (flooded phase; July-December 2010) to paddy cultivating (paddy phase; January-May 2011) phases in Maranchery Kole wetland, a part of the Ramsar site Vembanad Kole wetland. The macrobenthic fauna belonged to 4 phyla (Annelida, Arthropoda, Mollusca and Chordata) and 6 classes (Oligochaeta, Insecta, Gastropoda, Bivalvia, Crustaeca and Pisces). In flooded phase the major benthic class was Oligochaeta (69 per cent) followed by Insecta (30 per cent) and Pisces (1 per cent); in paddy phase, it was Insecta (70 per cent), Oligochaeta (25 per cent) Gastropoda (2 per cent) Bivalvia (2 per cent) and Crustaeca (1 per cent). A shift in benthic composition was apparent from flooded phase to paddy phase; oligochaetes were the major benthic group from July to December 2010 (flooded phase) and *insects from January to May 2011 (paddy phase). The fragmented nature of aqueous body in paddy phase favored insect dominance as insects were characterized by flight dispersal mode (active dispersal), thus less affected by habitat fragmentation compared to oligochaetes which were benthic crawlers (passive dispersal). The mean macrobenthic abundance was 335.79±460.37 ind./m^2 in flooded phase and 277±189.82 ind./m^2 in paddy phase (ANOVA $F_{5,27}$)=1.50 P> 0.05. Higher habitable area resulted in higher abundance in flooded phase whereas the bottom of the paddy fields were compartmented by paddy root structures providing insufficient space for macrofaunal development resulting in less abundance in paddy phase. Diversity indices (richness and diversity) of benthic families were comparable among flooded phase (d= 0.40±0.18, H'= 1.40±0.47) and paddy phase (d= 0.5± 0.22, H'= 1.68±0.35) implying that in spite of various agricultural practices that disrupt the aquatic habitat in rice fields, benthic fauna were well adapted to this ecosystem. No significant correlation emerged between benthic abundance and environmental parameters revealing that available limited habitable area for benthic fauna and agricultural practices might have determined abundance pattern unlike the usual environmental factors in wetlands.*

Keywords: *Macrobenthic fauna, Paddy fields, Temporary wetlands, Vembanad Kole wetlands.*

Introduction

The chief focus of biological conservation until the late 1980's was on undisturbed natural habitats that cover only a small proportion of the world land area. But this priority has changed when attention was called on the fact that a major part of the terrestrial environment consisted of managed ecosystems, including agricultural systems. As a huge fraction of the world's biological diversity coexists in these ecosystems (Western and Pearl, 1989) they also became the focus of research (Pimental *et al.*, 1992). Furthermore there is growing interest in the concepts of eco-agriculture (Mcneely and Scherr, 2001) where agricultural systems are managed both as a food production and biodiversity conservation system. Paddy fields that existed since the beginning of organized agriculture comprise a mosaic of rapidly changing ecotones, harboring a rich biological diversity, maintained by rapid colonization as well as by rapid reproduction and growth of organisms (Fernando 1996, Edirisinghe *et al.*, 2006). Recently researchers opined that, due to its biological diversity paddy fields could even surrogate the loss of natural wetlands (Angelini *et al.*, 2008; Nathuhara, 2013).

Benthic invertebrates, the organisms that live their entire or part of life cycle on or in the sediment of the water body is considered as key components of rice field fertility due to its significant role in nutrient translocation and organic matter decomposition (Roger *et al.*, 1987). Apart from being a critical link in food chain, they release dissolved nutrients by their feeding activities, excretion and burrowing into sediments and increase the rate of decomposition of particulate matter (Covich *et al.*, 1999; Henry and Stripari, 2005). In paddy fields, as the system transforms from an open littoral environment to a vegetated littoral system a seasonal variation of aquatic biota is expected. Rice fields are the most manipulated and frequently disturbed ecosystem resulting from various seral stages in crop cycle, agricultural operations and agrochemicals usage. The alterations in hydrology and rice field ecology have a clear impact on the organisms making it less favorable for certain organisms and more favorable for others (Bambaradeniya and Amerasinghe, 2003).

This study analyzed the variation in macrobenthic community in a seasonal paddy field during non-cultivating (submerged/flooded phase) and paddy cultivating (paddy phase) periods in Kole wetlands, a part of Vembanad kole wetlands, a Ramsar site. Kole wetlands are among the water-logged, paddy cultivating areas in Kerala such as Kuttanad, Pokkali and Kaipad; and were under rice cultivation for the past 200 years since the erstwhile Maharaja permitted to reclaim the wetland into paddy fields in the early 18th century (Anon., 1989). They are renowned for its high rice production, even the term Kole in Malayalam (the regional language in Kerala, India) means 'bumper yield of high returns in case flood does not damage the crops' (Johnkutty and Venugopal, 1993). Even though studies has been done on paddy field benthic community in different countries (Roger, 1989; Roger *et al.*, 1995; Pereira *et al.*, 2000; Al-Shami *et al.*, 2008) there is a lacuna in the literature available from benthic fauna in paddy fields even though India stands first in area under rice cultivation, second in rice production and has an agricultural based economy (Balachandran, 2007).

Materials and Methods

Study Area

The Kole lands are saucer shaped tracts, lying 0.5 to 1.5 m below the mean sea level, spread over Thrissur and Malappuram districts of Kerala extending from Northern bank of Chalakkudy river in the South to the Southern bank of Bharathappuzha river in the North. The intrusion of salt water to the paddy fields is prevented by Viyyam dam which is situated at the downstream of end of Kole lands. The Kole lands are assumed to be lagoons formed by the recession of the seas centuries back. A shallow portion of the sea along the western periphery of the main land was isolated and they were gradually silted up during rains making the lagoons shallow (Kurup and Varadachar, 1975). The farmers then bunded the fields, dewatered and raised rice in summer months. During the rains, the inflow into the basin submerges the kole areas. The cyclical nutrient recharging of the wetland during the flood season made the area as one of the most fertile soils of Kerala. Usually two rice crops, a summer crop (*punja* in December/January- April/ May) and an additional crop (*Kadumkrishi*) are raised in Kole wetlands (Raj and Azeez, 2009). The main crop is *Punja* (Summer crop).

The study area, with an area of 100 acres, is a part of the Ponnani Kole lies in between Maranchery and Veliyamkodu panchayats (a village council is called panchayat) in Malappuram district (Figure 6.1). From July to December 2010, the study area was flooded due to the South West Monsoon (June to September). By the end of the north east monsoon (October to December), water from the area was pumped out and paddy cultivation was begun by January. Dewatering was done by an indigenous centrifugal pumping device (*petti* and *para*) after protecting the paddy fields (*Padavu* or *Padashekharam*) with permanent or temporary earthen bunds (*Mattoms*). The crop was harvested by the end of May, soon after which the field gets flooded due to the South West Monsoon.

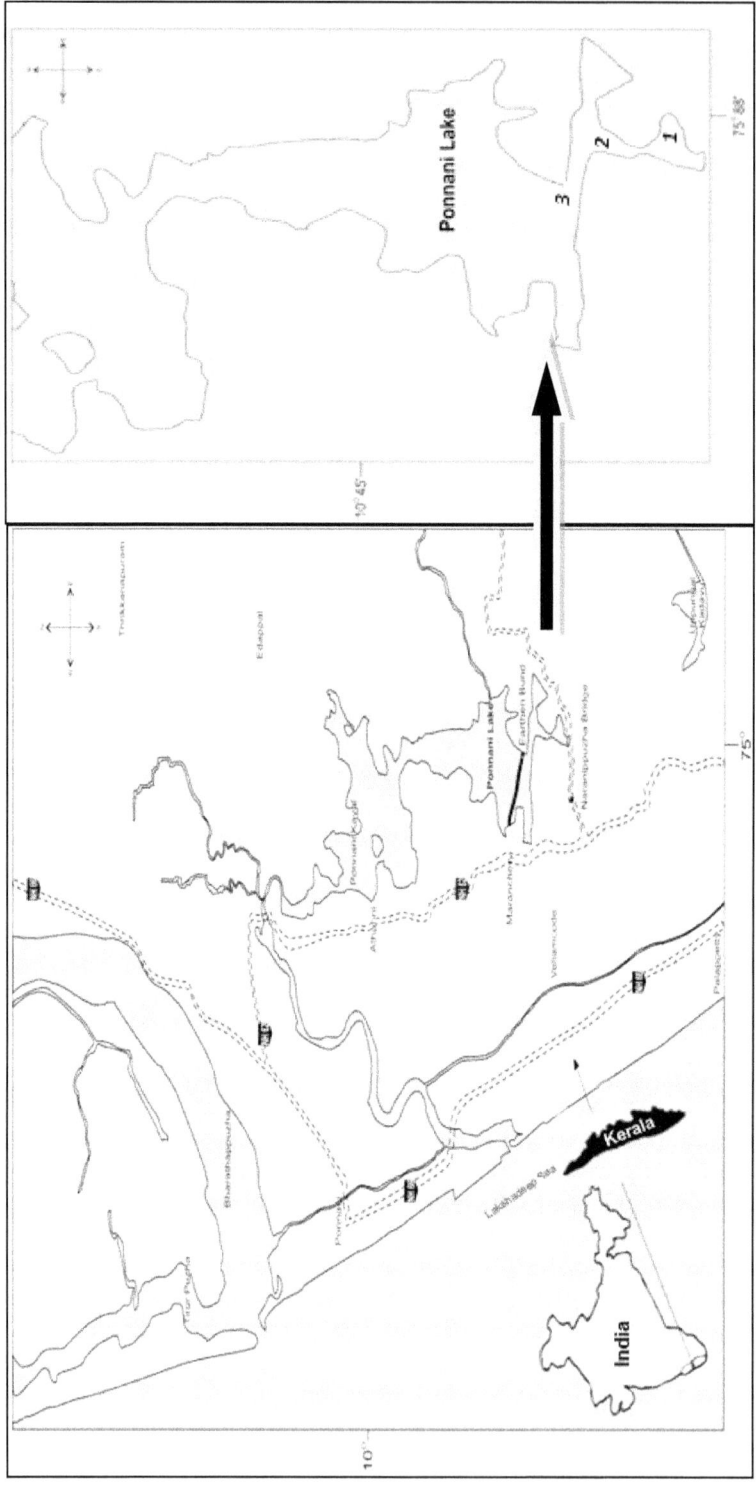

Figure 6.1: Location of Sampling Stations in Maranchery Kole Paddy Fields.

Sampling Procedure and Methods

The field sampling was carried out from three sampling stations for 11 months including a submerged/flooded season extending from July to December 2010 and a complete crop season *'punja'* extending from January to May 2011 on a monthly basis for the study of benthic fauna and environmental parameters.

During flooded phase water samples were collected using a a Niskin water sampler (Hydrobios 5 L) and during paddy phase, as the water body was shallow (average depth 0.39m), water samples were collected using a locally fabricated shallow water sampler of 1 litre capacity. The samples were stored in plastic containers and kept frozen till analysis. The sediment samples for the analysis were collected using a Van Veen grab of size $0.45m^2$. Temperature was measured in the field using a standard degree centigrade thermometer of 0°C to 50°C range and 0.1°C accuracy. pH was measured using Systronics digital pH meter model MK VI. Dissolved oxygen was analyzed by modified Winkler method (Strickland and Parsons, 1972). Organic carbon was determined by Walkley - Black method then converted to organic matter by multiplying with Van Bemmelen factor of 1.742 (Jackson, 1973). Particle size was analyzed using particle analyzer SympatrecT 100 laser diffraction granulometer, made in Germany.

Sediment samples in replicate were collected for the analysis of macrobenthos using a VanVeen grab of size $0.45m^2$. The samples were washed in the field itself through a sieve of mesh size 500 µm and those that were retained in the sieve were collected and preserved in 5 per cent formalin (Holme and McIntyre, 1971; McIntyre and Eleftheriou, 2005). The organisms were separated into different taxonomic groups. Identification of benthic families was done using standard keys (Yule and Sen, 2004; Morse *et al.*, 1994). Oligochaetes were identified up to species level by temporarily mounting the specimens using Amman's Lactophenol (Phenol, Lactic acid, Glycerol, and water in the ratio of 1:1:2:1) using taxonomic keys of Brinkhrust and Jamieson (1971) and Naidu (2005).

Statistical Analysis

One way ANOVA was used to determine the significant difference in environmental parameters and numerical abundance of benthos between the flooded phase and paddy phase using SPSS 16.0. Abundance data was square root transformed to meet the ANOVA assumptions. The data on benthic families was subjected to multivariate and univariate analysis by using the Primer software version 6.0 (Clarke and Gorley, 2006). ANOSIM was used to analyze the similarity of benthic family assemblages between flooded phase and paddy phase. Non metric multi-dimensional scaling, a hierarchical cluster analysis was used for the pictorial representation of the pattern of benthic family assemblage in the flooded and paddy phases (relative abundance). Ordinations were based on distance matrices, which were computed using Bray Curtis coefficient. The univariate indices of diversity such as species richness by Margalefs index (Margalef, 1958), species diversity by Shannon index (Shannon Wiener, 1949) and species dominance by Simpson's index (Simpson, 1949) were calculated for flooded and paddy phases.

Results

Flooded and paddy phases showed significant difference in many of the physico chemical parameters. As Flooded phase existed in monsoon and post monsoon, and paddy phase in pre monsoon seasons, a significant difference in rain fall was observed (ANOVA $F_{5,27}$)=11.37 $P < 0.01$. Rain fall in flooded phase was 290±160 mm and paddy phase was 49± 52 mm. Depth was the most variable physical parameter with a significant variation among the two phases (ANOVA $F_{5,27}$)=53.26 $P < 0.01$. Depth in flooded and paddy phases were 2.54±1.01 and 0.39±0.08 m respectively. Dissolved oxygen in flooded phase was 6.54±1.14 and paddy phase was 6.37±1.70 mg/L. Sediment temperature recorded in flooded and paddy phases were 26.91±1.01°C and 27.22±1.6 °C respectively. Sediment pH in flooded and paddy phases were 6.68±0.2 and 6.35±0.58 respectively, the difference in pH was significant (ANOVA $F_{5,27}$)=4.25 $P < 0.01$. A significant difference was observed in sediment oxidation reduction potential (Eh) between the phases (ANOVA $F_{5,27}$)=3.11 $P < 0.05$. Flooded phase was more reduced (-249.60±3.45 mV) than paddy phase (-222.41±4.01 mV). Moisture content recorded from flooded and paddy phases were 28.5 and 26.4 per cent respectively. Organic matter showed no significant variation among the two phases, the values observed in flooded and paddy phases was 3.73±1.05 and 3.95±0.78 per cent respectively. Available nitrogen in flooded phase was 0.02±.005 and paddy phase was 0.01±0.006 per cent. A significant difference in available phosphorus existed among both the phases (ANOVA $F_{5,27}$)=7.77 $P < 0.01$. Available phosphorus in flooded and paddy phases was 0.98 ±0.60 and 0.34 ±0.17 ppm respectively. Clay, silt and sand in flooded phase was 28.82±4.76, 48.74±7.47, 22.43±6.62 per cent respectively and that in paddy phase was 26.45±5.50, 46.46±10.49 and 46.42±10.49 per cent respectively.

The macrobenthic fauna in Maranchery wetlands belonged to 4 phyla (Annelida, Arthropoda, Mollusca and Chordata) and 6 classes (Oligochaeta, Insecta, Gastropoda, Bivalvia, Crustacea and Pisces). The major benthic groups were oligochaetes and insects whereas gastropod, bivalvia, crustacea and pisces showed sparse appearances. In flooded phase the major benthic class was Oligochaeta (69 per cent) followed by Insecta (30 per cent) and Pisces (1 per cent) whereas in paddy phase, the major benthic class was Insecta (70 per cent) followed by Oligochaeta (25 per cent), Gastropoda (2 per cent), Bivalvia (2 per cent) and Crustacea (1 per cent). A shift in benthic composition was apparent from flooded phase to paddy phase, oligochaetes were the major benthic group from August to December 2010 (flooded phase) and insects from January to May 2011 (paddy phase) (Figure 6.2).

The class Insecta consisted of Diptera (true flies) represented by Chironomidae, Ceratopogonidae, Tipulidae; Coleoptera (aquatic beetle) represented by Gyrinidae, Dysticidae, Hydrophilidae; Hemiptera (True bugs) represented by Aphelecherinidae; Odonata (Dragon fly and Damsel fly nymph) represented by Chlorocyphidae, Calopterygidae; Megaloptera (Alderflies) represented by Corydalidae. The class Oligochaeta consisted of Tubificidae and Naididae. The class Gastropoda was represented by Bithinidae and Lymnaeidae. The class Crustacea and Pisces consisted of Palaemonidae and Mysticidae respectively. About macrobenthic families present in flooded phase, Tubificidae (53 per cent) formed the major group followed by

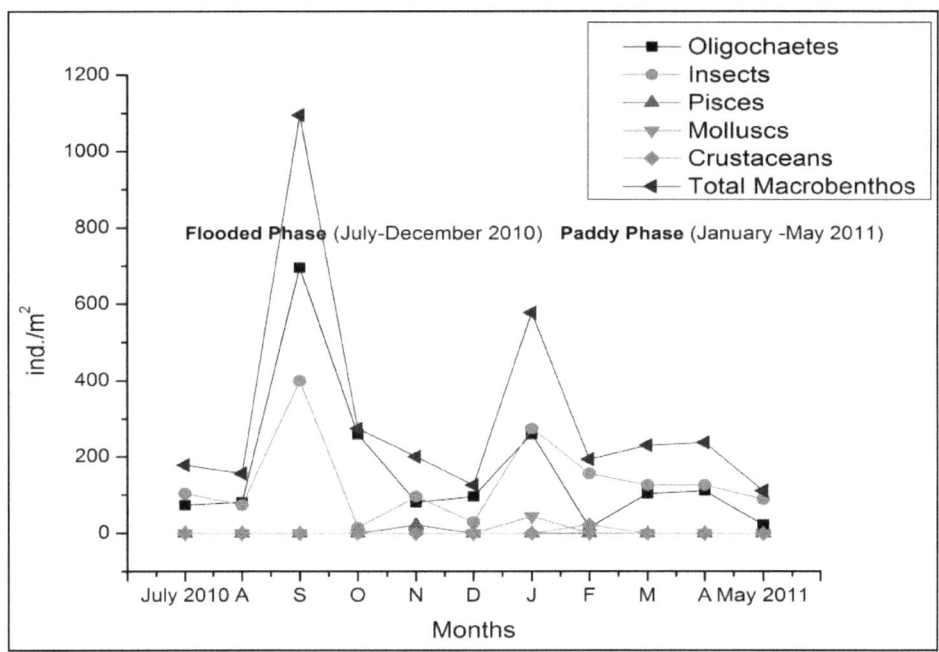

Figure 6.2: Temporal Variation in Numerical Abundance of Macrobenthic Fauna in Maranchery Kole Paddy Fields in Flooded Phase and Paddy Phase.

Chironomidae (25 per cent), Naididae (16 per cent), Ceratopogonidae (4 per cent), Tipulidae (1 per cent) and Mysticidae (1 per cent). The benthic families in paddy phase consisted of Chironomidae (60 per cent), Tubificidae (11 per cent), Naididae (10 per cent), Ceratopogonidae (7 per cent), Hydrophilidae (3 per cent), Gyrinidae (3 per cent); other benthic families such as Tipulidae, Dysticidae, Aphelecherinidae, Chlorocyphidae, Calopterygidae, Corydalidae, Bithinidae, Lymnaeidae and Palaemonidae were less than 1 per cent. Chironomidae, Tubificidae and Naididae were the most common benthic family observed throughout the seasons (Table 6.1). The highest number of benthic families were observed in January, the beginning of the paddy crop season. Apart from Chironomidae, Tubificidae and Naididae which were common throughout the study period, macroinvertebrate families including, Ceratopogonidae, Tipulidae. Gyrinidae, Aphelecherinidae, Corydalidae, Lymnaeidae, Bithinidae and Palaemonidae were present. The composition of benthic families in the flooded phase and paddy phase was compared using ANOSIM. The dissimilarity between both the phases was not very strong (Global R=0.37, p>0.05).

Twelve oligochaete species were present in the study area, 9 species were observed in flooded phase including *Aulodrilus pluriseta, Pristinella jenkinae, Pristinella minuta, Aulodrilus* sp., *Branchodrilus semperi, Pristina breviseta, Haemonais waldvogeli, Nais* sp., *Aulodrilus pigueti;* Similarly in paddy phase also 9 oligochaete species were recorded including *Aulodrilus pluriseta, Pristina breviseta, Pristinella minuta, Aulodrilus* sp., *Branchodrilus semperi, Stephensoniana trivandrana, Branchodrilus hortensis, Aulodrilus pigueti* and *Haemonais sp.* The most abundant oligochaete species in flooded and paddy phases were *Aulodrilus pluriseta* (66 per cent) and *Aulodrilus* sp. (23 per cent) respectively.

Table 6.1: List of Macrobenthic Families in Flooded Phase and Paddy Phase in Maranchery Kole Paddy Fields

	Flooded Phase (July–December 2010)						Paddy Phase (January–May 2011)				
	July, 2010	Aug	Sep	Oct	Nov	Dec	Jan	Feb	Mar	Apr	May 2011
Oligochaeta											
Tubificidae	+	+	+	+–	+	+	+	+	+	+	+
Naididae	+	+	+	+	+	+	+	–	+	+	+
Insecta											
Chironomidae	+	+	+	+	+	–	+	+	+	+	+
Ceratopogonidae	–	+	+	–	+	+	+	+	–	+	–
Calopterygidae	–	–	+	–	–	–	+	–	–	–	–
Hydrophilidae	–	–	–	–	+	–	–	–	–	–	–
Gyrinidae	–	–	–	–	–	–	+	–	–	–	–
Tipulidae	–	–	–	–	–	–	+	–	–	+	–
Chlorocyphidae	–	–	–	–	–	–	–	–	–	+	+
Aphelocheiridae	–	–	–	–	–	–	+	–	–	–	–
Corydalidae	–	–	–	–	–	–	+	–	–	–	–
Dysticidae	–	–	–	–	–	–	+	–	–	–	–
Mollusca											
Bithinidae	–	–	–	–	–	–	–	–	+	–	–
Lymnaeidae	–	–	–	–	–	–	–	–	+	–	–
Crustacea											
Paleomonidae	–	–	–	–	–	–	+	–	–	–	–
Pisces											
Mysticidae	–	–	–	–	+	–	–	–	–	–	–

+: Represents present; –: Represents absent.

A difference in macrobenthic abundance between flooded and paddy phase was observed but was not statistically significant. The mean abundance of benthic macrofauna was 335.79±460.37 in flooded phase and 277±189.82 ind./m^2 in paddy phase (ANOVA $F_{5,27}$)=1.50 P> 0.05. A higher abundance of oligochaetes was observed in flooded phase (214.79±429.64 ind./m^2) than in paddy phase (102.21±160.62 ind./m^2) (ANOVA $F_{5,27}$)=1.97 P> 0.05 whereas abundance of insects was more in paddy phase (168.87±109.43 ind./m^2) compared to flooded phase (119.74±193.43 ind./m^2) (ANOVA $F_{5,27}$)=1.01 P> 0.05 (Figure 6.1).

In spite of the reduced numerical abundance in paddy phase, diversity analysis of benthic families revealed similar values in richness, diversity and dominance in flooded phases of (d=0.5± 0.22, H'=1.68±0.35, λ=0.45±0.14) and paddy phase (d=0.40±0.18, H'=1.40±0.47, λ=0.42±.06). The monthly pattern of benthic family richness and diversity showed peaks in November 2010 in flooded phase (d=0.66, H'=2.17) whereas in paddy phase it was in January 2010 (d=1.51, H'=2.28) (Figure 6.3). Temporal changes in benthic family assemblages between the phases was evaluated with multi dimensional scaling (MDS), where no marked discrete clusters of paddy and flooded phase was evident, the stress value was 0.14 (Figure 6.4).

The interactions between macrobenthic abundance and environmental parameters were analyzed using correlation analysis, but no significant correlation emerged between macrobenthic abundance and environmental parameters

Discussion

Due to the difference in ecotones, many physicochemical parameters showed a marked difference among flooded and paddy phases. Monsoon was a major parameter influencing the difference in depth which in turn influenced many other parameters. Flooding a soil results in the consumption of electrons and protons, continuous consumption of protons during consumption of electrons resulted in an increase in pH in flooded phase (Reddy and Delune 2008). Redox potential that measures electron activity showed that flooded phase was more reduced than paddy phase. Normally anoxic conditions prevail in wetlands but when water was drained as the preparation for paddy cultivation, the reduced compounds were oxidized which would have resulted in a less reduced nature of sediments in paddy phase (Jackel *et al.*, 2001; Kruger *et al.*, 2001). Available phosphorus level was less in the paddy phase; the transfer of phosphorus through the plants could be the reason.

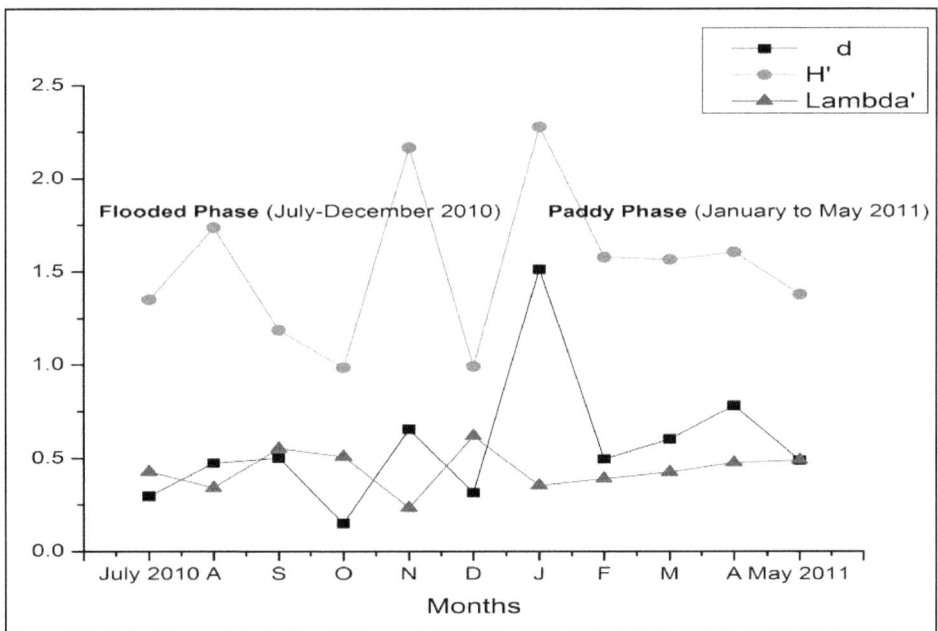

Figure 6.3: Temporal Variation in Diversity Indices of Macrobenthic Families in Maranchery Kole Paddy Fields in Flooded Phase and Paddy Phase.

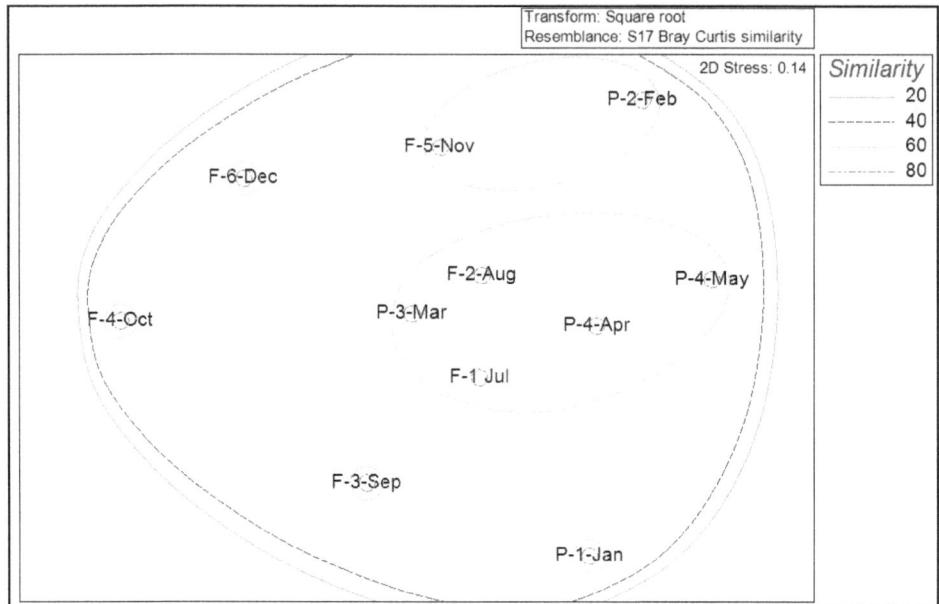

Figure 6.4: The Multi Dimensional Scaling (MDS) Ordination Plot of Bray-Curtis Similarities Showing Temporal Changes in Benthic Family Assemblages in Flooded and Paddy Phases. F1 to F6 represents consecutive months in flooded phase; P1 to P5 represents consecutive months in paddy phase.

Phosphorus assimilation and storage in plants depends on vegetative type and growth characteristics. Floating and submerged vegetation has limited potential for long-term phosphorus storage but emergent macrophytes have an extensive network of roots and rhizomes and have great potential for phosphorus storage. As paddy is an emergent plant it accumulated more phosphorus than the submerged plants in flooded phase.

In the case of aquatic segment in this wetland, there is flow through, drying off and stagnation in the aquatic habitat consecutively as the seasons progress corresponding respectively to the flooding or monsoon period, draining period as the preparation for paddy cultivation and maintaining a constant water level for paddy cultivation. As aquatic biota is characterized by adaptations to an existence in water the drying could stress or even eliminate these biota. But the organisms which were present in Maranchery Kole wetlands were capable of surviving in dry conditions by the specific mechanisms such as in Oligochaeta by diapausing eggs; resistant cysts enclosing young, adults or fragments of individuals; that in Diptera: Chironomidae (insecta) by diapausing eggs, resistant late instar larvae, sometimes in cocoons of silk or mucus; that in Ephemeroptera (insecta) by diapausing eggs; that in Odonata (insecta) by resistant nymphs, recolonising adult; that in Hemiptera (insecta) by recolonising adults; that in Trichoptera (insecta) by diapausing eggs, resistant gelatinous egg mass, terrestrial pupae in some species, recolonising adults, larvae deep in substrate, that in Coleoptera (insecta) by semi-terrestrial pupae, burrowing adults, recolonising adults; that in Bivalvia by diapausing eggs and adult stages; that in Gastropoda by adults form a protective epiphragm of dried mucus across shell opening, adults and young survive in moist air/soil under algal mats on pond/stream bed (Williams *et al.*, 1987). Apart from pisces and crustaceans whose representation was nominal, all the other benthic organisms were found to have survival mechanisms against dry periods, such survival mechanisms maintained the benthic populations in this wetland in both the phases and resulted in a similarity in composition of benthic families among the flooded and paddy phases.

During the rainy season the rice field was flooded and the aquatic medium was in communication with surrounding wetlands whereas it was compartmented or isolated during the other periods. When the continuous water body in the flooded phase provided a freedom of movement or dispersal, the discrete patches in the paddy phase prevented the same. The shift in benthic composition from oligochaete majority in flooded phase to insect majority in paddy phase was mainly due to the difference in dispersal between oligochaetes and insects. The insects adopt a flight mode of dispersal or have active dispersal but oligochaetes were benthic crawlers or have passive dispersal (Bilton *et al.*, 2001). As insects were characterized by flight dispersal mode, they were less affected by habitat fragmentation. Species that have greater adult migration abilities can disperse more easily between habitats and were less likely to be effected by habitat isolation (Smith and Brumsickle, 1989). Previous studied documented that insect population predominates in shallow water compared to oligochaetes (Che Salmah and Abu Hassan, 2002). In previous studies also the predominance of insects in paddy fields were reported such as in Srilanka (Bambardaneniya, 2000), in Bako, Ethiopia (Desta *et al.*, 2014) and Chapra, Bihar (Ojha *et al.*, 2010). The availability of more protected habitable niche by paddy plants would have favored the insect to thrive. The vegetative and reproducing growth stages of the rice plant such as tillering, booting and flowering stages attract a variety of insects (Edirisinghe and Bambaradeniya, 2006).

The variation in benthic abundance between flooded and paddy phases could be explained by the area of available habitat. The increase in habitable area could result in an increase in the number of organisms (Sommer and Horwitz, 2009). The bottom of the paddy fields were compartmented by paddy root structures providing insufficient space for the proper development of benthic fauna, which could be a probable reason for less macrobenthic abundance in paddy phase agreeing to the findings of Ojha *et al.* (2010) from paddy fields of Bihar. In paddy phase, the highest abundance was observed in January 2011, the beginning of paddy cultivation. Previous studies also showed that insect (Chironomid) abundance peaked when rice plants were young and small, as paddy grows up shading by larger plants slows weed decomposition (Bambaradeniya and Amerasinghe, 2003) or reduces the amount of detritus available (Che Salmah and Abu Hassan, 2002).

Diversity and richness in both the phases were comparable revealing that inspite of various agricultural practices that disrupt the aquatic habitat in rice fields, benthic fauna were well adapted to this temporary and highly manipulated ecosystem agreeing to the observations of Fernando (1996). Previous studies also indicate that as most rice fields are converted marshes, they have inherited the aquatic fauna of these marshes so the aquatic organisms in rice fields cover the entire spectrum of fresh water fauna (Fernando *et al.*, 1979, Fernando 1993) thus maintaining a similar composition and diversity in flooded and paddy phases. According to Heckman (1979) and Bahaar and Bhat (2011), long standing cultivation of rice over several millennia have evolved the biota through centauries to periodic disturbances and unique conditions of rice field aquatic system which sustain their diversity and richness throughout the seasons.

The results of correlation analysis between environmental parameters and the numerical abundance of the benthic fauna evolved no significant correlation among themselves. In paddy fields agricultural practices were more important driving forces for the macroinvertebrate structure unlike the usual factors in wetlands (Stenert *et al.*, 2009). Similar to the findings of this study the absence of correlation between macrobenthos and environmental variables over the cultivating cycle was documented in paddy fields in Brazil also (Stenert *et al.*, 2009). The availability of habitable area also would have been an important factor determining the benthic abundance. In spite of the physico chemical parameters analyzed in the study, the other unmeasured factors would have determined the abundance structure such as area of the habitable patch, hydrological stability, life history strategy, macrophyte structure, proximity and size of the neighboring habitat *etc*. which are known to play a key role.

The vital concern for agriculture in the future would be to manage and optimize biodiversity, stability, and productivity within agro ecosystems which is emphasized by concepts such as integrated biodiversity management (IBM) that focuses integrated pest management (IPM) and conservation approaches (Kiritani 1975). Hence the baseline information from the field is essential for management and conservation aspects as benthos play a key role in aquatic food web and as better sensors of environmental perturbations, further more macrobenthic community has the potential for enhanced nutrient replenishment in the paddy fields which could be utilized as the consequences of agrochemical are revealed gradually. Since the rice field ecosystem satisfy the interests of both agro ecologists and conservation biologists, integrated efforts can formulate biodiversity based strategies for sustainable management of the paddy field ecosystems.

References

Al-Shami, S.A., Che Salmah M.D. Rawi, Siti Azizah Mohd Nor and Abu Hassan Ahmad. 2008.

Angelini, R., Ferrero, A. and Ponti, I. 2008. II riso Bayer crop science/script ed. Bologna 680 pp.

Anonymous. 1989. Scheme for studying the possible changes in the ecosystem consequent on the conservation of Thannermukkam Bund, Thrissur, Kerala. Kerala Agricultural University.

Bahaar, S.W.N. and Bhat, G.A., 2011. Aquatic biodiversity in paddy fields of Kashmir Valley (J&K) India. *Asian Journal of Agricultural Research*, 5(5): 269-276.

Balachandran, P.V., 2007. Rice scenario of Kerala and the future strategies. *Proceedings of XIX Kerala Science Congress*, Kannur, Kerala, 22-32 pp.

Bambaradeniya, C.N.B. and Amerasinghe, F.P., 2003. Biodiversity associated with the rice field agro-ecosystem in Asian countries: a brief review. Working Paper 63. International Water Management Institute (IWMI), Colombo, Sri Lanka.

Bambaradeniya, C.N.B., 2000. Ecology and biodiversity in an irrigated rice field ecosystem in Srilanka. *Ph.D. Thesis*, University of Peradeniya, Sri Lanka, 525 pp.

Bilton, D.T., Freeland, J.R. and Okamura, B., 2001. Dispersal in Freshwater Invertebrates. *Annual Review of Ecology and Systematics*, 32: 159-181.

Brinkhurst, R.O. and Jamieson, B.G.M., 1971. *Aquatic Oligochaeta of the World*. Oliver and Boyd, Edinburgh.

Che Salmah, M.R. and Abu Hassan, A, 2002. Distribution of aquatic insects in relation to rice cultivation phases in a rain fed rice field. *Journal Biosains*, 13(1): 87-107.

Clarke, K.R. and Gorley, R.N., 2006. *PRIMER v6: User Manual/Tutorial*. PRIMER-E: Plymouth.

Covinch, A.P., Palmer, M.A. and Crowl, T.A., 1999. The role of benthic invertebrate species in freshwater ecosystems. *BioScience*, 49(2): 119-127.

Desta, L., Prabha Devi, L., Sreenivasa, V. and Amede, T., 2014. Studies on the ecology of the paddy and fish co-culture system at Dembi Gobu microwater shed at Bako, Ethiopia. *International Journal of Fisheries and Aquatic Studies*, 1(3): 49-53.

Edirisinghe, J.P. and Bambaradeniya, C.N.B., 2006. Rice fields: An ecosystem rich in biodiversity. *Journal of National Science Foundation Sri Lanka*, 34(2): 57-59.

Fernando, C.H., 1993. A bibliography of references to rice field aquatic fauna, their ecology and rice-fish culture. SUNY Geneseo - University of Waterloo, N.Y., V and 110 pp.

Fernando, C.H., 1996. Ecology of rice fields and its bearing on fisheries and fish culture. *In: Perspectives in Asian Fisheries*, (Ed.) S.S. de Silva, p. 217-237.

Fernando, C.H., Furtado J.I. and Lim, R.P., 1979. Aquatic fauna of the world's rice fields. Wallaceana Suppl. (Kuala Lampur), 2: 1-105.

Henry, R. and Stripari, N.L., 2005. The invertebrate colonization during decomposition of *Eichhornia crassipes* in the mouth zone of Guarei River into Jurumirim Reservoir (Sao Paulo, Brazil). *The Ekologia*, 3(2): 01-12.

Holme, N.A. and Mc Intyre, A.D., 1971. *Methods for Study of Marine Benthos*. IBP Handbook No. 6, Blackwell Scientific Publications.

Jäckel, U., Schnell, S. and Conrad, R., 2001. Effect of moisture, texture and aggregate size of paddy soil on production and consumption of CH_4. *Soil Biol. Biochem.*, 33: 965–971.

Jackson, M.L. 1973. *Soil Chemical Analysis*. Prentice-Hall India Ltd., New Delhi.

Johnkutty, I. and Venugopal, V.K., 1993. Kole wetlands of Kerala, Kerala Agricultural University, Thrissur, 68 p.

Kiritani, K., 1975. Pesticides and ecosystems. *J. Pesticide Sci. Commem. Issue*, pp. 65-75.

Krüger, M., Frenzel, P. and Conrad, R., 2001. Microbial processes influencing methane emission from rice fields. *Glob. Change Biol.*, 7: 49–63.

Kurup, P.G. and Varadachar, V.V.R.G., 1975. Hydrography of Purakkad mud bank region. *Indian Journal of Marine Sciences*, 4: 18-20.

Margalef, R., 1968. *Perspective in Ecological Theory*. University of Chicago Press, Chicago, 111.

McIntyre, A.D. and Antasious Eleftheriou, 2005. *Methods for the Study of Marine Benthos*, 3rd edn. Blackwell Scientific Publications.

McNeely, J.A. and Scherr, S., 2001. Common ground common future: How ecoagriculture can help feed the world and save wild biodiversity. *IUCN - The World Conservation Union*, 24 p.

Morse, C.J., Yang Lianfang and Tian Lixin (Eds.), 1994. *Aquatic Insects of China Useful for Monitoring Water Quality*. Hohai University Press, Nanjiing People's Republic of China, pp. 569.

Naidu, K.V., 2005. *The Fauna of India and the Adjacent Countries – Aquatic Oligochaeta*. Zoological Survey of India, Kolkata.

Nathuhara, Y., 2013. Ecosystem services by paddy fields as substitutes of natural wetlands in Japan. *Ecological Engineering*, 56: 97-106.

Ojha, N.K., Pandey, M.K. and Yadav, R.N., 2010. Studies on macrobenthic community of paddy field of rural Chapra. *Bioscan*: Special Issue, 2: 579-585.

Pereira, T., Cerejeira, M.J., Brito, F. and Viana, P., 2000. Rice crop superficial water effects and exposure pesticides. Final Repoet. DGA, ISA. (Exposicao e Efeitos de Pesticidas em Aguas superficiais de ecossistemas Orizicolas (1998-2000) - Relatorio Final. Direccao Geral do Ambiente, Instituto Superior de Agronomia).

Pimental, D., Stachow, U., Takacs, D., Brubaker, H.W., Dumas, A.R., Meaney, J.J., O'Neil, J.A.S., Onsi, D.E. and Corzilius, D.B., 1992. Conserving biological diversity in agricultural/forestry systems. *BioScience*, 42(5): 354-362.

Raj, N. and Azeez, P.A., 2009. The shrinking rice paddies of Kerala. *The India Economy Review*, 6: 176-183.

Reddy, K.R. and DeLaune, R.D., 2008. *Biogeochemistry of Wetlands: Science and Applications*. CRC Press, Taylor and Francis Group.

Roger, P.A., 1989. Biology and management of the floodwater ecosystem in tropical wetland ricefields. IRRI. In: *International Network on Soil Fertility and Sustainable Rice Farming (INSURF)*, Handout for the 1989 training course.

Roger, P.A., Simpson, I., Official, B., Ardales, S., Jimenez, R. and Cagauan, A.G., 1995. An experimental assesent of pesticide impacts on soil and water fauna and microflora in wetland ricefields of the Philippine. In: *Impact of Pesticides on Farmer Health and the Rice Environment*, (Eds.) P. Pingali and P.K.A.P. Roger. London, pp. 309-345.

Roger, P.A., Grant, I.F., Reddy, P.M. and Watanabe, I., 1987. The photosynthetic aquatic biomass in wetland rice fields and its effect on nitrogen dynamics. Efficiency of nitrogen fertilizers for rice. International Rice Research Institute, Los Banos, Philippines 43-68 pp.

Shannon, C.E. and Wiener, W., 1963. *The Mathematical Theory of Communication.* University of Illinois Press, Urbana, 117 pp.

Simpson, E.H., 1949. Measurement of diversity. *Nature,* 163: 688.

Smith, C.M. and Brumsickle, S.J., 1989. The effects of patch size and substrate isolation on colonization modes and rates in an intertidal sediment. *Limnology and Oceanography,* 34: 1263–1277.

Sommer, B. and Horwitz, P., 2009. Macroinvertebrate cycles of decline and recovery in Swan Coastal Plain (Western Australia) wetlands affected by drought-induced acidification. *Hydrobiologia,* 624: 191–203.

Stenert, C., Bacca, R.C., Maltchik, L. and Rocha, O., 2009. Can hydrologic management practices of rice fields contribute to macroinvertebrate conservation in southern Brazil wetlands? *Hydrobiologia,* 635: 339–350.

Strickland, J.D.H. and Parsons, T.R., 1972. *A Practical Handbook of Seawater Analysis,* 2nd edn. *Bull Fish Res Bd Can,* 167: 310.

Western D. and Pearl, M.C., 1989. *Conservation for the Twenty-first Century.* Oxford.

Williams, D.D., 1987. *The Ecology of Temporary Waters.* Timber Press, Portland OR. 205 p.

Yule, C.M. and Sen, Y.H., 2004. *Freshwater Invertebrates of the Malaysian Region* (Ed). Kuala Lumpur, Malaysia: Akademi Sains Malaysia.

Chapter 7
Demography of the Ostracod *Heterocypris incongruens* (Ramdohr, 1808) Fed Alga and Organic Wastes

Marissa F. Juárez-Franco, S.S.S. Sarma* and S. Nandini

ABSTRACT

We quantified the demographic responses of the ostracod Heterocypris incongruens using two diets (single-celled alga Scenedesmus acutus and organic matter from wastewater treatment plant) at three densities (5.8, 11.6 and 23.2 µg ml^{-1} as dry weight). Using newly hatched H. incongruens cohort life tables selective demographic variables were derived (age-specific survivorship, fecundity, average lifespan, life expectancy at birth, gross reproductive rate, net reproductive rate, generation time and the rate of population increase). The age specific survival of H. incongruens was affected by the concentration and type of food; ostracods fed algae lived longer with increasing food availability. H. incongruens showed higher reproductive peaks (up to 8 offspring female^{-1} d^{-1}) and for longer duration with an increase in the algal concentration. However, when ostracods were fed organic waste, the offspring production was much reduced (< 2 offspring female^{-1} d^{-1}) and at 23.2 µg ml^{-1}, the reproduction was inhibited. Regardless of food type and concentration, the average lifespan and life expectancy at birth of H. incongruens varied from 30-47 days. Gross and net reproductive rates varied from 10-54 and 6-32 offspring female^{-1}, respectively. Generation time (27-38 d) varied much less among the treatments. Lowest rate of population increase (0.04-0.06 d^{-1} per day) was observed in treatments containing organic waste as diet while it was nearly thrice when alga was used as food.

Keywords: Food type, Life history, Population growth rate, Survivorship, Reproduction, Crustacean.

---

* Corresponding Author.

Introduction

The ostracod *Heterocypris incongruens* (Ramdohr, 1808) (Ostracoda: Podocopida) is a cosmopolitan species found in both clean waters and organically contaminated waterbodies such as sewage waters. Both laboratory and field studies have shown that this species is omnivorous feeding mostly on algae and detritus (Schmit *et al.*, 2007). Because of its sensitivity to changes in the environment, its wide distribution and ease of culture under laboratory conditions, this species has been used as test organism in toxicological evaluations (Oleszczuk, 2007; Fernandez *et al.*, 2016). Long reproductive lifespan and post-hatching somatic growth of *H. incongruens* are well-suited to feed on particulate organic matter (Spencer and Blaustein, 2001). A long reproductive span permits ostracods to exploit resources for a greater period of time, while the presence of differently sized individuals of a population allows them feed on different size fractions of particles from the organic wastes. Earlier works on population growth of *H. incongruens* have revealed that this species is capable of growing on suspended particles from organic wastes of domestic effluents (Juárez-Franco *et al.*, 2009).

In large cities, wastewater treatment is an expensive process because of the quantity of organic load and the volume of the water generated. For example in Mexico City, every second 75 m^3 of wastewater is generated from domestic uses (Monroy *et al.*, 2000) and the number of treatment plants is not sufficient to deal with the growing quantity of domestic effluents. In treatment plants removal of particulate and suspended organic substances through microbial decomposition is time consuming and costly. Since crustaceans, including ostracods, grow in such environment, it is generally considered as a viable method to harvest the biomass from domestic effluents (Nandini *et al.*, 2004; Juárez-Franco *et al.*, 2007).

One of the difficulties in the use of suspended organic wastes from domestic effluents is the inconsistency of the water quality, which not only differ from the geographical area but also through seasons (Butler *et al.*, 1995). In spite of such large variations in the water quality, ostracods occasionally found in domestic wastewater imply some degree of adaption to such conditions (Mukhopadhyay *et al.*, 2007). Compared to many other crustaceans of comparable body size such as *Daphnia*, ostracods in general have a longer long lifespan (Delorme, 2001). Therefore various life history variables such as age-specific reproductive output can be quantified using different diets which will reveal the possible adaptations (Fernandez *et al.*, 2016). *H. incongruens* is particularly well-suited for demographic studies because of its predominantly parthenogenetic mode of reproduction (Chaplin *et al.*, 1994; Schön *et al.*, 2009). However, in literature only few life table demographic works have been undertaken using this species, especially using a single diet (Rossi and Menozzi, 1993). The aim of this present work was therefore to quantify the demographic responses of *H. incongruens* using two food types offered at three concentrations.

Materials and Methods

The ostracod *Heterocypris incongruens* was originally obtained from a temporary waterbody in the City of Guanajuato (Guanajuato, Mexico). From a single parthenogenetic individual, clonal cultures were established in 10 L aquaria using reconstituted moderately hardwater as medium (EPA medium) and the green alga *Scenedesmus acutus* (Strain no. 72, University of Texas collection center, Austin, USA) as food offered daily at a density of 1.0×10^6 cells/ml. The EPA medium was prepared by dissolving 0.9 g of $NaHCO_3$, 0.6 g of $CaSO4$, $MgSO_4$ 0.6 g and 0.002 g of KCl in one liter of distilled water (Weber, 1993). *S. acutus* was batch-cultured in 2L transparent bottles using defined nutrients (the Bold´s basal medium, Borowitzka and Borowitzka, 1988). *S. acutus* during the exponential phase was centrifuged at 3000 rpm for 5 min., rinsed and resuspended in a small quantity of distilled water. Later the density of algae was estimated using a haemocytometer. The algal cell density was converted to dry weights (0.024 ng per cell) using earlier report (García *et al.*, 2003) and using this we prepared three algal food densities (5.8, 11.6 and 23.2 µg ml^{-1} dry weight) using EPA medium. The stock cultures as well as the experimental jars of *H. incongruens* were maintained at $23\pm1°C$, pH 7.1 -7.3 and diffuse fluorescent illumination for 12h daily (L:D = 12:12 h).

In order to offer particulate organic wastes, we obtained wastewater from the treatment plant El Rosario (Mexico City). Every alternate day approximately 12 litres (in quadruplets of 3 litres each) of the wastewater containing suspended organic particles were collected from the treatment plant. This was filtered through 200-µm mesh to remove particles larger than this size. Later this was concentrated by centrifugation as in the case of alga. In order to offer diet dry weight equivalents of algae, we quantified the concentrate of organic waste, as dry weight using Cahn balance, Model C-33. For this a known quantity of organic waste was dried at 65° C for 24 h in an oven and then weighed again. In order to offer the organic waste as diet (without drying), we took into account of water content present (from the difference between wet weight and dry weight) and calculated the quantity of organic waste as equivalent to algal food. The proximate composition of organic wastes was done by the Department of Animal Nutrition and Biochemistry of the Faculty of Verternary Medicina of our university and that of *Scenedesmus acutus* was obtained from literature for the purpose of interpretation of the data.

The life table demography experiments were conducted using three concentrations of S. *acutus* (5.8, 11.6 and 23.2 µg ml^{-1} dry weight) or their equivalent of organic wastes as food. The experiments were conducted in 150 ml capacity transparent jars, each with 100 ml of test medium and with one of the chosen diet type and concentration. For each treatment we maintained four replicates. Thus, in all, we used 24 test jars (= 2 food types X 3 concentrations X 4 replicates). Into each test jar, we individually introduced 10 neonates (24 old) of *H. incongruens* using Pasteur pipette under a stereomicroscope (Nikon SMZ645, Japan). Following the initiation of the experiments, we daily counted by the number of living individuals in each jar. Later the surviving individuals of the original cohort were transferred to new jars containing appropriate food type and concentration. The dead individuals, eggs and neonates if present were separately enumerated and discarded. We continued this process until the last individual from each cohort died. For the purpose of offspring production, we combined the egg number and the neonates of a given day, although the egg hatching was never 100 per cent.

From the data of ostracod survival and the reproduction, we derived the following life history variables: age-specific survivorship, life expectancy, fecundity; average lifespan, life expectancy at birth, gross reproductive rate, net reproductive rate, generation time and the rate of population increase were obtained. The following formulae were used for deriving the life history variables (Krebs, 1985):

l_x = Proportion of surviving to start of age x

m_x = Proportion of offspring produced per female at age x

Life expectancy at the start of age x: $e_x = \dfrac{T_x}{n_x}$

where, T_x = cumulative number of individuals for further live from age x

n_x = number of living individuals at the initiation of age x (days)

Gross reproductive rate: $= \sum\limits_{0}^{\infty} m_x$

Net reproductive rate: $R_o = \sum\limits_{0}^{\infty} l_x \cdot m_x$

Generation time: $T = \dfrac{\sum l_x \cdot m_x \cdot x}{R_o}$

Rate of population increase (r), Euler-Lotka equation (solved iteratively):

$$\sum_{x=0}^{\infty} e^{-rm\,x} l_x.m_x = 1$$

In order to test if the differences among the treatments were significant we used two-way analysis of variance (ANOVA) and post-hoc Tukey tests using standard statistical software (Sigma Plot 11).

Results

Chemical analysis of the particulate waste from the treatment plant indicated very low percent of dry weight and more than 92 per cent moisture content. Crude fibre (13 per cent) and ash content (35 per cent) constituted nearly 5half of the dry weight. Carbohydrates (19 per cent), proteins (15 per cent) and lipids (18) are nearly in equal proportions.

The age specific survival (l_x) of *H. incongruens* was affected by the concentration and type of food, where the ostracods fed algae lived longer even as the food availability increased. In addition, there was little mortality during the first three weeks of *H. incongruens* when fed alga. On the other hand, ostracods fed particulate organic waste showed higher mortality especially at the lowest food level but lived longer than those fed algae of at comparable food density (Figure 7.1). Age specific reproduction showed large differences between those fed algae and those cultured on organic waste. *H. incongruens* showed higher reproductive peaks (up to 8 offspring female^{-1} day^{-1}) and for longer duration with an increase in the algal

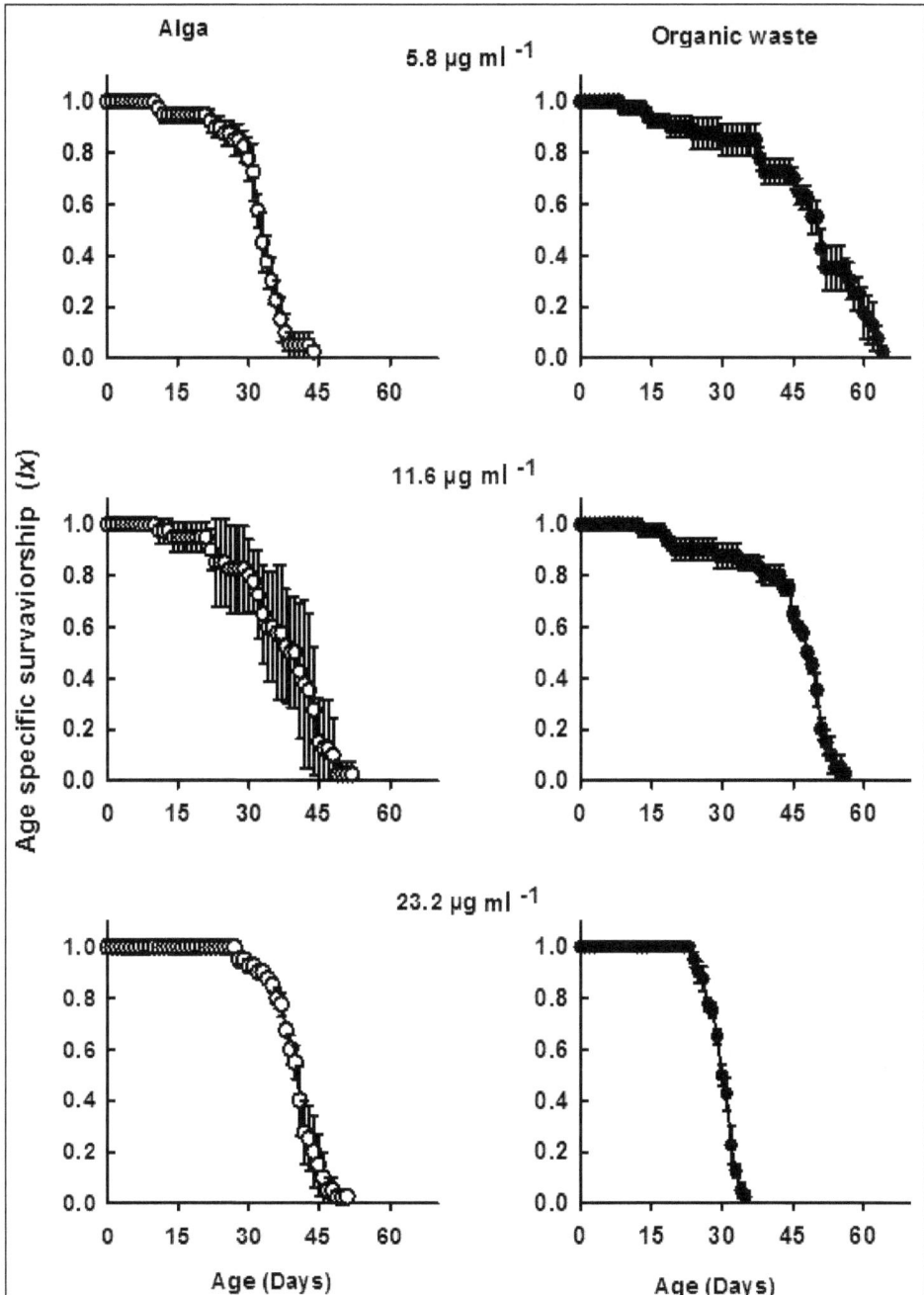

Figure 7.1: Age-specific Survivorship (l_x) Curves of *Heterocypris incongruens* Fed *Scenedesmus acutus* (Open Circles) or Organic Wastes (Closed Circles) at different Concentrations (As dry weights, µg ml^{-1}). Shown are the mean±standard error based on four replicates (Cohorts of 10 individuals each).

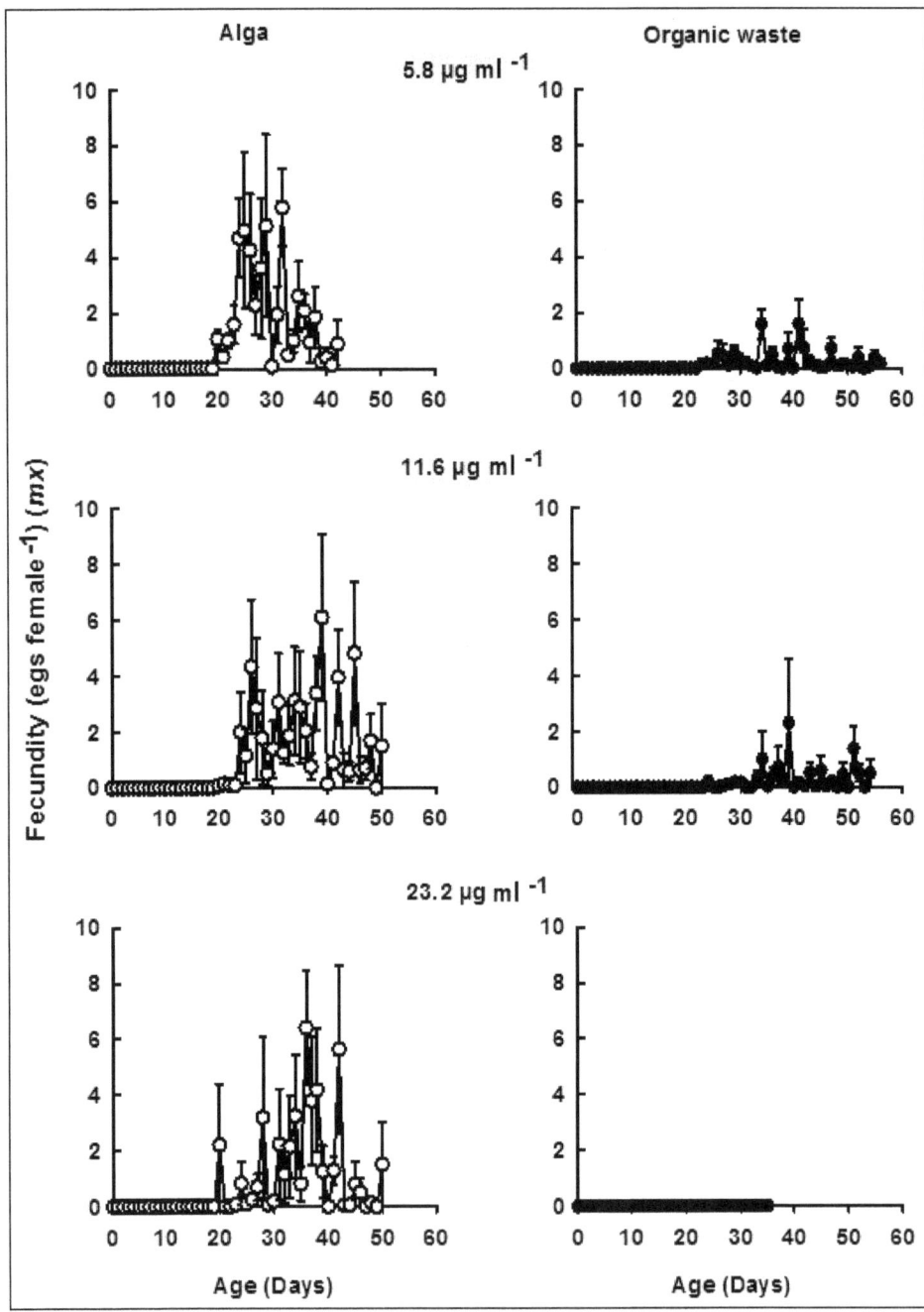

Figure 7.2: Age-Specific Fecundity (m_x) Curves of *H. incongruens* Fed *S. acutus* (Open circles) or Organic Wastes (Closed circles) at different Concentrations (as dry weights, µg ml^{-1}). Shown are the mean±standard error based on four replicates (Cohorts of 10 individuals each).

concentration. However when ostracods were fed organic waste, the offspring production was much reduced (usually < 2 offspring female^{-1} day^{-1}) and at the highest offered concentration the reproduction was inhibited (Figure 7.2).

Data on the selected demographic variables of *H. incongruens* fed alga and organic waste are presented in Table 7.1. Regardless of food type and concentration, the average lifespan varied from 30-47 days, the highest being on organic wastes offered in lowest concentration. Gross and net reproductive rates varied from 10-54 and 6-32 offspring female^{-1}, respectively. Compared to H. incongruens fed organic wastes, those cultured using algae showed significantly higher gross and net reproductive rates. Generation time (27-38 days) varied much less among the treatments. In general the generation time of H. incongruens was longer on organic wastes than on algae. Lowest rate of population increase (0.04-0.06 per day) was observed in treatments containing organic waste as diet while it was nearly 3-times when alga was used as food.

Table 7.1: Data on the Selected Life History Variables
(Average lifespan, days; Gross reproductive rate, offspring/female; Net reproductive rate, survival-weighted offspring/female; generation time, days and rate of population increase per day of *H. incongruens* in relation to different algal (*Scenedesmus acutus*) and organic wastes as food (dry weight, μg/ml). Shown are the mean±standard error based on 4 replicates. Under a given diet type, for each variable, data carrying similar alphabets are not statistically significant ($p>0.05$, Tukey test). At the highest level of organic waste, no reproduction occurred and hence data not available (-).

Food Level (μg/ml, dry weight)	Life History Variables				
	Average Lifespan	Gross Reproduction Rate	Net Reproduction Rate	Generation Time	Rate of Pop. Increase
	Alga				
5.8	32.4±1.2[a]	47.2±3.8[a]	32.0±1.1[a]	27.4±0.4[a]	0.13±0.01[a]
11.6	37.1±2.5[a,b]	54.1±10.9[a]	29.2±6.4[a]	32.4±1.1[b]	0.11±0.01[b]
23.2	40.2±0.9[b]	42.6±7.4[a]	31.2±3.3[a]	33.1±2.3[b]	0.11±0.01[b]
	Organic waste				
5.8	47.4±2.4[a]	13.3±2.0[a]	8.3±1.1[a]	38.9±2.0[a]	0.06±0.01[a]
11.6	44.8±0.7[a]	10.1±4.1[a]	6.1±2.9[a]	38.2±0.7[a]	0.04±0.02[a]
23.2	30.2±0.3[b]	–	–	–	–

For treatments containing algal diet, except for gross and net reproductive rates, rest of the derived variables was statistically significant ($p<0.01$, ANOVA). However, for treatments containing organic wastes as diet, both survivorship and reproductive variables were significantly affected by the food level ($p<0.01$, ANOVA). The interaction of food type x density was significant ($p<0.05$) for the derived variables, except for gross and net reproductive rates ($p>0.05$) (Table 7.2).

Table 7.2: Results of the Two-way ANOVA Performed on the Selected Life History Variables of *Heterocypris incongruens* Fed Organic Wastes or the Green Alga *Scenedesmus acutus* at different Concentrations

Variable/Food Type and Density	DF	SS	MS	F-ratio	P
Average lifespan					
Organic waste (A)	1	107.9	107.9	11.0	0.004
Alga (B)	2	149.5	74.7	7.6	0.004
Interaction of A x B	2	658.9	329.4	33.6	<0.001
Error	18	176.7	9.8		
Gross reproductive rate					
Organic waste (A)	1	9700.5	9700.5	70.0	<0.001
Alga (B)	2	534.0	267.0	1.9	0.174
Interaction of A x B	2	121.5	60.8	0.4	0.652
Error	18	2493.7	138.5		
Net reproductive rate					
Organic waste (A)	1	4053.4	4053.4	96.4	<0.001
Alga (B)	2	84.0	42.0	1.0	0.388
Interaction of A x B	2	80.9	40.4	1.0	0.401
Error	18	757.1	42.0		
Generation time					
Organic waste (A)	1	168.4	168.4	23.2	<0.001
Alga (B)	2	1679.8	839.9	115.6	<0.001
Interaction of A x B	2	2355.9	1177.9	162.2	<0.001
Error	18	130.7	7.2		
Rate of pop. Increase per day					
Organic waste (A)	1	0.041	0.041	184.1	<0.001
Alga (B)	2	0.006	0.003	13.9	<0.001
Interaction of A x B	2	0.002	0.001	4.8	0.022
Error	18	0.004	0.0002		

DF: Degrees of freedom, SS: Sum of squares; MS: Mean square; F: F-statistic; P: p-value.

Discussion

Heterocypris incongruens in our test jars reproduced through parthenogenesis as also recorded earlier (Fernandez *et al.*, 2016). Chaplin *et al.* (1994) also observed parthenogenetic mode of reproduction for this species. It has been reported that in some cases, the same individual of *H. incongruens* produces two different types of eggs: those that hatch within a few days and those that take up to 4 months to hatch (Angell and Hancock, 1989; Havel and Talbott, 1995). However in the present study the majority of the eggs usually hatched within a few days after being released by the female.

Ostracods have longer duration of lifespan (>10 months) (and higher age at first reproduction (> 3 weeks) (Mezquita *et al.*, 2002). In this study, *H. incongruens* had a maximal lifespan of about 60 days. In addition, this species has a long lag phase lasting from 2-3 weeks, similar to that reported for the same species ranging from 15 to 23 days (Rossi and Menozzi, 1993; Havel and Talbott, 1995).

The patterns of survivorship curves presented in most treatments appeared to be nearly ideal where there was little mortality in the cohorts during the major part of their lifespan with a rapid decline of the population due to physiological effects. In previous laboratory experiments, mortality commences after about 3 weeks following hatching (Rossi and Menozzi, 1993) as also observed in this work. Higher density of organic wastes shortened the lifespan of *H. incongruens* by about 36 per cent, while there was no such clear trend in treatments containing algal diet suggesting that this ostracod is capable of tolerating high abundances of algae rather than organic wastes (Juárez-Franco *et al.*, 2009).

At both low and intermediate levels of organic wastes in the test jars, the ostracod was able to reproduce but at a much lower rate as compared to that on algae. At the highest level of organic wastes, reproduction was completely inhibited but the population survived for about 4 weeks. Regular microscope observations revealed guts of the ostracods were indeed full with organic wastes. This suggests deficiency of nutrients in the domestic wastes, rather than non-availability of or non-consumption of diet. This is further confirmed by the chemical analysis of the organic wastes from the domestic effluents. For example, the suspended organic matter has not only high content of inorganic component but also low levels of proteins.

Scenedesmus is considered an appropriate algal diet for crustacean zooplankton based on its nutritional quality (Morris *et al.*, 1999). It has been reported that *Scenedesmus* has up to 36 per cent protein in (Dobberfuhl and Elser, 1999), which is more than twice in the organic wastes used as diet for *H. incongruens*. As for lipids, *Scenedesmus* also contains high levels of lipids (21 per cent) and these values are not much different from levels (18 per cent) found in organic wastes; however the fatty acids that are nutritionally important for ostracods are probably deficient in wastewaters (Lettinga, 1995; Ahlgren *et al.*, 1997). Wastewaters may also carry various kinds of substances including toxicants. However, a previous work conducted using the organic wastes from the same water treatment plant did not affect the population growth of *H. incongruens* suggesting that the levels of toxic substances in the organic wastes from the domestic wastes are probably not high (Juárez-Franco *et al.*, 2009). *H. incongruens* fed organic wastes in this study probably did not obtain enough proteins and appropriate fatty acids for offspring production, although survival was not drastically affected. This may suggest a trade-off between survival and reproduction (Lynch, 1980). At the highest concentration of organic wastes (23.2 µg ml^{-1}), *H. incongruens* failed to reproduce possibly due to handling of agglomerates of organic waste formed at the bottom of the test jars and energy spent in cleaning the head appendages clogged with unconsumed food and/or excreted wastes (Downing and Rigler, 1984). When *H. incongruens* was fed *Scenedesmus acutus* at the highest concentration (23.2 µg ml^{-1}), there was no formation of agglomeration and possibly this did not result in high energy costs in handling, consumption and cleaning of the appendages.

In iteroparous zooplankton species there exists a linear relation between lifespan and generation time. Based on the demographic data of rotifers, King (1982) hypothesized that for mean lifespan is twice the generation time for iteroparously reproducing species. In anostracans, this relation holds valid (Anaya-Soto et al., 2003), but varies from 0.5 to 4.3 in other crustaceans such as cladocerans (Sarma et al., 2005). In our study too we derived such a relation (Figure 7.3) where average lifespan was not twice the generation time but much less (60 per cent lower).

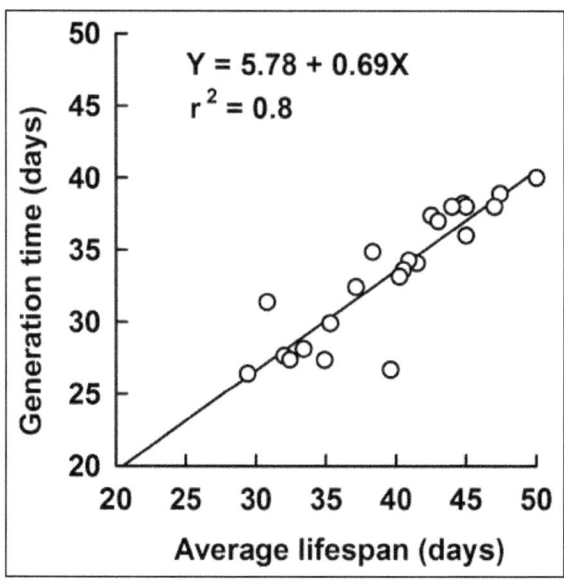

Figure 7.3: Relation between Average Lifespan (d) and Generation Time (d) of *H. incongruens* Fed *Scenedesmus acutus* or Organic Wastes at different Concentrations. Plotted are the replicated data points for each treatment.

When *H. incongruens* was fed alga at concentration (11.6 µg ml^{-1}) the difference between the two parameters was less compared with other treatments, suggesting a lower rate of mortality at this test condition. Juarez-Franco et al. (2009) have reported that with an increase in algal density from 5.8 to 11.6µg DW ml^{-1}, the population abundance of *H. incongruens* increased, although there were no significant differences when fed *Scenedesmus* from 11.6 to 23.2 µg ml^{-1}. This suggests that algal food levels close to 11.6 µg ml^{-1} are sufficient for optimal reproductive output in *H. incongruens*. The growth patterns of populations of ostracods vary greatly depending on the species (Fenwick, 1984) and the growth rates are generally low (0.03 per day) (Heip, 1977). In this work *H. incongruens* has growth rates ranged from 0.03 to 0.07 d^{-1}). The low growth rates of ostracods in general is due to longer age at maturity and the lower rate of offspring production per female, although the time taken to hatch eggs is fairly rapid (Mezquita et al., 2002).

In conclusion, it is evident that *H. incongruens* is not as efficient as some other microcrustaceans such as Moina macrocopa (Nandini *et al.*, 2004) in utilizing organic wastes as food. Not only is survival inhibited at high (23.2 µg ml^{-1}) concentrations of organic matter but also reproduction is lower on such a diet as compared to an algal diet. It remains to be seen whether ostracods, being efficient bacterivores (Monakov, 2003) are capable of reducing the load of pathogenic bacteria in domestic wastewaters.

Acknowledgements

Fernando Álvarez Noguera, Ruth Cecilia Vanegas Pérez, Pedro Ramírez García and Gloria Vilaclara Fatjo offered valuable suggestions. The first author (MFJF) received a scholarship from CONACYT (CVU-225912) for her Master's programme.

References

Anaya-Soto, A., S.S.S. Sarma and S. Nandini, 2003. Longevity of the freshwater anostracan *Streptocephalus mackini* (Crustacea: Anostraca) in relation to food (*Chlorella vulgaris*) concentration. *Freshwater Biology*, 48: 432–439.

Butler, D., E. Friedler, K. Gatt, 1995. Characterising the quantity and quality of domestic wastewater inflows. *Water Science and Technology*, 31: 13–24.

Ahlgren, G., W. Goedkoop, H. Markensten, L. Sonesten and M. Boberg, 1997. Seasonal variations in food quality for pelagic and benthic invertebrates in Lake Erken – the role of fatty acids. *Freshwater Biology*, 38: 555–570.

Angell, R.W. and J. W. Hancock, 1989. Response of eggs of *Heterocypris incongruens* (Ostracoda) to experimental stress. *Journal of Crustacean Biology*, 9: 381–386.

Borowitzka, M.A. and L. J. Borowitzka, 1988. Micro-algal biotechnology. Cambridge University Press, London. 477 pp.

Chaplin, J.A., J. E. Havel and P. D. N. Hebert, 1994. Sex and Ostracods. *Trends in Ecology and Evolution*, 9: 435–439.

Delorme, L.D. 2001. Ostracoda. In: Thorp, J.H. and A. Covich (eds). Ecology and Classification of North American Freshwater Invertebrates (2nd Edition), Academic Press, San Diego: 811–848.

Dobberfuhl, D.R. and J.J. Elser, 1999. Use of dried algae as a food source for zooplankton growth and nutrient release experiments. *Journal of Plankton Research*, 21: 957–970.

Downing, J.A. and F.H. Rigler (Editors), 1984. A manual for the methods of assessment of secondary productivity in fresh waters. 2nd edition. IBP Handbook 17. Blackwell Scientific Publications, London. 501 pp.

Fenwick, G.D. 1984. Life history and population biology of the giant ostracod *Leuroleberis zealandica* (Baird, 1850) (Myodocopida). *Journal of Experimental Marine Biology and Ecology*, 77: 255–289.

Fernandez, R., S. Nandini, S.S.S. Sarma and M. E. Castellanos-Páez, 2016. Demographic responses of *Heterocypris incongruens* (Ostracoda) related to stress factors of competition, predation and food. *Journal of Limnology*, 75(s1): 31–38.

García, C. E., S. Nandini and S.S.S. Sarma, 2003. Food type effects on the population growth patterns of littoral rotifers and cladocerans. *Acta Hydrochim. Hydrobiol.*, 31: 120–133.

Havel, J.E. and B.L. Talbott, 1995. Life history characteristics of the freshwater ostracod *Cyprinotus incongruens* and their applications to toxicity testing. *Ecotoxicology*, 4: 206– 218.

Heip, C. 1977. On the evolution of reproductive potentials in a brackish water meiobenthic community. *Mikrofauna Meeresboden*, 61: 105–112.

Juárez-Franco, M.F., S.S.S. Sarma and S. Nandini, 2009. Population dynamics of *Heterocypris incongruens* (Ramdohr, 1808) (Ostracoda, Cyprididae) in relation to diet type (algae and organic waste) and amount of food. *Crustaceana*, 82: 743–752.

King, C. E. 1982. The evolution of life span. In: Dingle, H. and J. P. Hegmann (Editors). Evolution and genetics of life histories. Springer, New York: 121–128.

Krebs, C.J. 1985. Ecology: The experimental analysis of distribution and abundance. 3rd Edition. Harper and Row, New York. 800 pp.

Lettinga, G. 1995. Anaerobic digestion and wastewater treatment systems. *Antonie van Leeuwenhoek*, 67: 3–28.

Lynch, M. 1980. The evolution of cladoceran life histories. *Quarterly Reviews of Biology*, 55: 23–42.

Mezquita, F., M.D. Boronat and M.R. Miracle, 2002. The life history of *Cyclocypris ovum* (Ostracoda) in a permanent karstic lake. *Archiv für Hydrobiologie*, 155: 687–704.

Mukhopadhyay, S.K., B. Chattopadhyay, A. R. Goswami and A. Chatterjee, 2007. Spatial variations in zooplankton diversity in waters contaminated with composite effluents. *Journal of Limnology*, 66: 97–106.

Monroy, O., G. Fama, M. Meraz, L. Montoya and H. Macarie, 2000. Anaerobic digestion for wastewater treatment in Mexico: state of the technology. *Water Research*, 34: 1803–1816.

Morris, H.J., M.M. Quintana, A. Almarales and L. Hernández, 1999. Composición bioquímica y evaluación de la calidad proteíca de la biomasa de *Chlorella vulgaris*. *Revista Cubana de Alimentación y Nutrición*, 13: 123–128.

Nandini, S. and S.S.S. Sarma, 2000. Life table demography of four cladoceran species in relation to algal food (*Chlorella vulgaris*) density. *Hydrobiologia*, 435: 117–126.

Nandini, S. and S.S.S. Sarma, 2003. Population growth of some genera of cladocerans (Cladocera) in relation to algal food (*Chlorella vulgaris*) levels. *Hydrobiologia*, 491: 211–219.

Nandini, S. and S.S.S. Sarma, 2007. Effect of algal and animal diets on life history of the freshwater copepod *Eucyclops serrulatus* (Fischer, 1851). *Aquatic Ecology*, 41: 75–84.

Nandini, S., D. Aguilera-Lara, S.S.S. Sarma and P. Ramírez-García, 2004. The ability of selected cladoceran species to utilize domestic wastewaters in Mexico City. *Journal of Environmental Management*, 71: 59-65.

Oleszczuk, P. 2007. The evaluation of sewage sludge and compost toxicity to Heterocypris incongruens in relation to inorganic and organic contaminants content. *Environmental Toxicology*, 22: 587–596.

Rossi, V. and P. Menozzi, 1993. The clonal ecology of *Heterocypris incongruens* (Ostracoda): life history traits and photoperiod. *Functional Ecology*, 7: 177–182.

Sarma, S.S.S., S. Nandini and R.D. Gulati, 2005. Life history strategies of cladocerans: comparisons of tropical and temperate taxa. *Hydrobiologia*, 542: 315–333.

Schmit, O., G. Rossetti, J. Vandekerkhove and F. Mezquita, 2007. Food selection in *Eucypris virens* (Crustacea: Ostracoda) under experimental conditions. *Hydrobiologia*, 585: 135–140.

Schön, I., K. Martens and P. van Dijk, 2009. Lost sex: The evolutionary biology of parthenogenesis. Springer, Dordrecht. 615 pp.

Spencer, M. and L. Blaustein, 2001. Risk of predation and hatching of resting eggs in the ostracod *Heterocypris incongruens*. *Journal of Crustacean Biology*, 21: 575–581.

Weber, C. I. 1993. Methods for measuring the acute toxicity of effluents and receiving waters to freshwater and marine organisms. 4th Edition. United States Environmental Protection Agency, Cincinnati, Ohio. 293 pp.

Chapter 8

Benthic Faunal Diversity in Indian Mangroves

Philomina Joseph, S. Sreelekshmi, Rani Varghese, C.M. Preethy and S. Bijoy Nandan*

ABSTRACT

Indian mangroves are noted for their inherent biotic diversity along with providing the nourishment and energy to the coastal system for maintaining the trophic processes. The deltaic and nutrient rich soils of the east coast support rich mangrove vegetation compared to west coast of India. Benthic fauna, the ecosystem engineers, helps in remineralisation, transformation and release of nutrients and also balancing the secondary production, thus forming the inevitable part of mangrove ecosystem. This study attempts a detailed review on various works in Indian mangroves along with collating information on the benthic fauna along the east coast and west coast of India. Out of various mangrove sites, Sundarbans ranks first in benthic abundance and diversity followed by Pichavaram mangroves in Tamil Nadu in east coast whereas lower diversity was observed in the west coast. This study also highlights the biodiversity status of Indian mangroves with emphasis on its benthic faunal segments in maintaining the trophic system.

Keywords: Benthic fauna, Mangroves, Biodiversity, Pichavaram.

Introduction

India ranks one among the 12 mega biodiversity countries of the world and enjoys warm tropical climatic conditions suitable for flourishing of mangrove vegetation and related faunal segments that adds to the radiance of the country.

---

* Corresponding Author.

India with a coastline of 7517 km is divided into east coast, west coast and bay islands. Mangroves in India account for about 3 per cent of the world's mangrove vegetation, 8 per cent of the Asian mangrove area and 0.14 per cent of the country's land area, and are spread over an area of about 4740 sq km (Forest Survey of India, 2015) along the coastal states of the country with 38 true mangrove species (Bijoy Nandan *et al.*, 2015a).There has been a net increase in mangrove cover of 112 sq km as compared to 2013 assessment (Forest Survey of India, 2015). About 58.09 per cent of the total mangrove area is recorded on the east coast of India (Bay of Bengal region) and 28.87 per cent on the west coast (Arabian Sea region) and rest of 13.03 per cent in Bay islands (Andaman and Nicobar). Benthic fauna forms the basement of any sustainable ecosystem. Mangroves form a major supporter for flourishing of benthic fauna within their complex structural characteristics such as root structures, pneumatophores and loose muddy substratum. Besides, a well-designed concomitance can be seen between benthic fauna and mangroves. Mangroves, the producer of the aquatic as well terrestrial system, fuel the underlying soil fauna. Benthos in turn modify the mangrove's physical and vegetation structure through their burrowing activities and grazing on detritus (Berry, 1972; Smith, 1987; Smith *et al.*, 1991) and helps in remineralization of organic matter (Gallep *et al.*, 1978). They run the detritus food chain and play focal role in the storage, transformation and release of nutrients to the overlying water column (Coull, 1999; Cummins *et al.*, 2004). They function as ecosystem engineers - directly or indirectly modulating the availability of resources to other species especially crabs (Jones *et al.*, 1994) and also as sediment destabilisers due to the bioturbation activities and thus helps in mixing of the soil (Day *et al.*, 1987; Bird *et al.*, 1999). The benthic faunal community has been considered as main criteria for judging the success of mangrove restoration programs (Lewis, 2005; Ye *et al.*, 2006; Field, 1998) and are useful bio-indicators providing more accurate understanding of changing aquatic conditions (Ravera, 1998, 2000; Ikomi *et al.*, 2005). Benthic realm in Sundarbans and Tamil Nadu mangroves of east coast are well studied. The distribution, diversity and community structure of benthic fauna in mangrove ecosystem is flimsy especially in west coast. To unravel the benthic diversity of mangrove ecosystems of India and the less studied west coast especially Kerala, a fast developing state with a fast degrading mangroves is inevitable. This study shoots out the comparative status of benthic fauna in east coast and west coast with emphasis on Kerala mangrove habitats particularly on Cochin mangroves, Kerala.

Benthic Diversity in East Coast

The deltaic mangroves and nutrient rich alluvial soils of east coast always prospered with higher benthic diversity. Various studies undertaken in east coast reported 616 species of benthic fauna from mangrove ecosystem (Kathiresan and Qasim, 2005). East coast is endowed with 32 true mangrove species (Bijoy nandan *et al.*, 2015a). The major mangrove areas in east coast include Sundarbans in Gangetic delta of West Bengal, Bhitarkanika in Mahanadhi delta of Odisha, Coringa in Godavari delta of Andhra Pradesh, Pichavaram in Cauvery delta of Tamil Nadu.

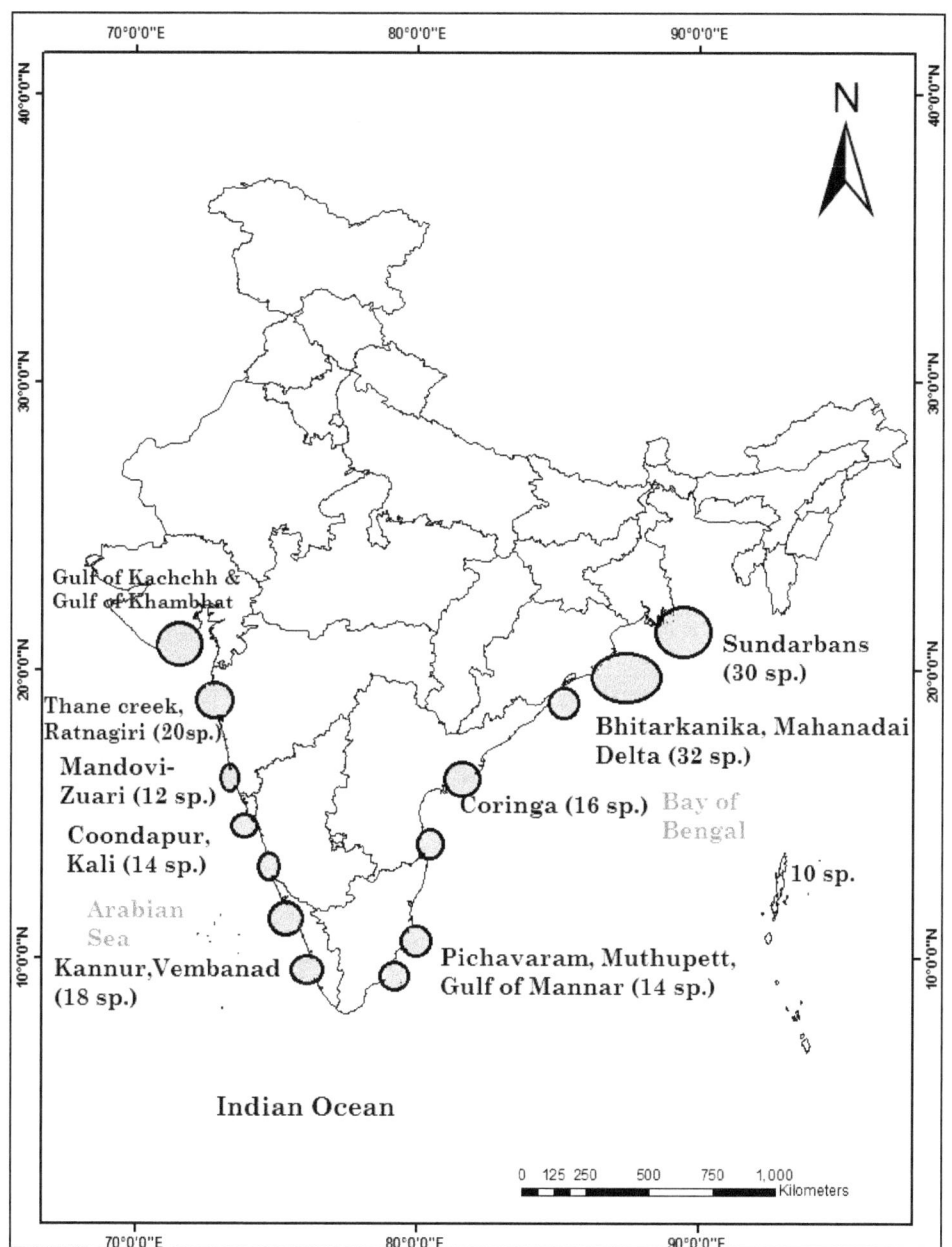

Figure 8.1: Map Showing Major Mangrove Ecosystems and Distribution of Mangrove Species in India.
Source: Manual on Mangroves (Bijoy Nandan et al., 2015a).

Sundarbans

Sundarbans forms the largest mangrove region in the world and a UNESCO world heritage site, a National Park, Tiger Reserve and a Biosphere Reserve of India with a mangrove cover of 2106 sq km (Forest Survey of India, 2015). The deltaic Sundarbans mangroves of West Bengal is the earth's most extensive mangrove ecosystem named after the governing mangrove species, *Heritiera fomes*, locally known as 'Sundari' and is the single mangrove ecoregion that harbours the Royal Bengal tiger (*Panthera tigris*). About 30 true mangrove species were found in Indian Sundarbans (Figure 8.1). The dominant mangrove species include *Heritiera fomes, Avicennia sps, Xylocarpus granatum, Sonneratia apetala, Bruguiera gymnorrhiza, Ceriops decandra, Aegiceras corniculatum, Rhizophora mucronata* and *Nypa fruticans*.

Table 8.1: Studies of Benthic Arthropods and Molluscs Reported from Sundarbans Mangroves

Sl.No.	Type of Organism	Type of Study	Authors
1.	Brachyuran crabs	Diversity and distribution	Chakraborty *et al.*, 1986; Chakroborthy, 1994
2.	Fiddler crabs (*Uca* sps.) and brachyuran crabs (*Metaplex intermedia*)	Zonation pattern, distribution and population ecology	Chakraborthy and Choudhury, 1985; 1992a,b,c
3.	Red ghost crab (*Ocypode macrocera*)	Biology, sexual dimorphism	Dubey *et al.*, 2014
4.	Mud crab (*Scylla serrata*)	Morphometrics and habitat ecology	Poovachiranom, 1992; Mahapatra *et al.*, 1996; Nandi *et al.*, 1996
5.	*Penaeus monodon*	Production	Chakraborti *et al.*, 2002
6.	Molluscs (Dominant sps - *Cellana radiata, Telescopium telescopium, Meretrix meretrix, Enigmonia aenigmatica* and *Dicyathifer manni*)	Diversity and distribution	Subba Rao *et al.*, 1983; Jahan *et al.*, 1990; Sing and Choudhury, 1995a; Ghosh *et al.*, 1995; Dey *et al.*, 2005; Dey, 2006
7.	Tellinid bivalve (*Macoma birmanica*)	Biocalcification and physico-chemical characteristics	Saha and Jana, 1999; Saha *et al.*, 2000
8.	Oyster (*Saccostrea cucullata*)	Biochemical composition	Mitra *et al.*, 2008

Detailed study of benthic realm was carried out by Bhunia and Choudhury, 1981; Choudhury *et al.*, 1984; Mandal and Nandi, 1989; Patra *et al.*, 1988, 1990; Chakraborti *et al.*, 1990,1992; Ghosh *et al.*, 1990; Dehadrai, 1994; Khan, 2003 on the distribution, composition and abundance of benthic fauna. Actiniarians, polychaetes, nemertines, bivalves, echiurids, decapods, isopods, amphipods and gobiids are the major macrobenthic residents with the dominance of sipunculids and gastropods. Chaudhuri and Choudhury, 1994 reported about 476 species of arthropods, molluscs (143 species), annelids (78 species) and nematodes (68 species). Colonization and community ecology of macrobenthic intertidal polychaetes was studied by Misra and Choudhury, 1985; Sarkar *et al.*, 2005; Chandra and Chakraborty, 2008 reported

30 species of polychaetes with distinctive assemblage of *Mastobranchus indicus* – *Dendronereides heteropoda* and *Lumbrinereis notocirrata* – *Ganganereis sootai*– *Glycera tesselata* in mangroves. Diversity and distribution of arthropods (Hazra *et al.*, 2005) and molluscs (Table 8.1) were extensively studied.

Sing and Choudhury (1984, 1985, 1992, 1995b) studied distribution, breeding and feeding behaviour of hemichordate worm, *Saccoglossus* sps. and reported a new record on occurrence of the holothuroid, *Protankyra similis*. Das (2016) studied the macrobenthic burrowing animals and their bioturbation activities, whereas biomonitoring studies using bivalves were carried out by Sarkar *et al.*, 2008. Mangroves also harbour a rich abundance of benthic insects. A total of 14 dipteran species belonging to 5 families and 11 genera (Ray and Choudhury, 1985) and about 14 coleopteran species (Poddar *et al.*, 1990) have been encountered.

Rao and Misra (1983) studied meiofaunal abundance in Sagar Island, Sundarbans with predominance of nematodes followed by copepoda, polychaeta, ostracoda. Sinha *et al.*, 1987 identified a new species of nematode, *Anoplastoma macrospiculum* and a new record of 17 stylet bearing nematode from Sundarbans mangroves (Sinha and Choudhary, 1988). Ghosh *et al.*, 2014; Majumder *et al.*, 1996 identified 44 species of foraminifera dominated by the family Rotaliidae. Habitat preference showed maximum density in the sandy flats with moderate mangroves than in mudflats with least mangrove.

Bhitarkanika and Coringa Mangroves

Bhitarkanika has the second single largest block of mangrove formations in India next to Sundarbans situated on Brahmani-Baitrani delta of Odisha with a mangrove cover of 231 sq km (Forest Survey of India, 2015). There are about 62-67 species of mangroves in this region, of these 32 are true mangroves (Figure 8.1) and ranks first among the Indian state with highest number of true mangrove species. In Bhitarkanika, *Heritiera fomes* exhibited highest density followed by *Excoecaria agallocha* and *C. ramiflora* and these three species together accounted for 77 per cent of the total mangrove plants (Misra *et al.*, 2005). Coringa mangrove wetland, the largest mangrove in Andhra Pradesh (367 sq km), occupies the northern portion of the Godavari delta. It is home to as many as 35 species of mangroves, of which 16 are true mangroves (Figure 8.1), the rest being associated species. The three communities of mangroves making up the Coringa mangrove forest are *Excoecaria-Avicennia*, *Avicennia-Sonneratia* and *Avicennia* community. A rare mangrove species, *Scyphiphora hydrophyllacea* (Rubiaceae) was reported from Andhra Pradesh (Venkanna, 1991).

Bhitarkanika provides all conducive environmental situations to nourish Indian horse shoe crabs. Among the 4 species of horse shoe crabs of the world, two are present in Bhitarkanika -*Tachypleus gigas* and *Carcinoscorpious rotundicauda*. About 8 species of crabs were found in Bhitarkanika that includes fiddler crabs (*Uca* sps.) and hermit crabs. Commercially important species of prawns such as *Penaeus indicus, Penaeus monodon, Metapenaeus affinis* and crabs mainly *Scylla serrata* are residents of mangrove ecosystem of Bhitarkanika (http:www.bhitarkanika.org). Other mud dwellers include amphibious fish *Periopthalmus* sps. and *Bolephthalmus* sps. and a variety of molluscs. Twenty species of molluscs have been reported from the

Mahanadi estuary (Subba Rao and Mukherjee,1969). Meiofaunal study showed 11 major faunal taxa, of which nematodes are dominant in Bhitarkanika mangrove sediments (Sarma and Wilsanand, 1994).

Benthic studies in mangroves of Andhra Pradesh dates back to earlier studies on marine borers (Molluscs) in mangrove ecosystem by Ganapati and Rao,1959 in Godavari mangroves (11 species) and Krishna mangroves (9 species) by Radhakrishnan and Janakiram,1975. Murthy and Rao, 1977; Ramanamurthy and Kondala Rao, 1993 studied the ecology and diversity of mangrove molluscs and reported 10 to 23 species of molluscs. Recent studies displayed a reduction in diversity of molluscan fauna to 9 species (Chakravarty and Uday Ranjan, 2014). Radhakrishna and Ganapati (1969) recorded two species of polychaetes, namely, *Eurythoe parvecarunculata* and *Micronereis* sps from the mangrove zones of the Kakinada Bay which has been replaced by principal species *Diopatra neopolitana* and gastropod *Cerithidia cingulata* during the last four decades, besides the disappearance of echinoderms, crustaceans and molluscs were also observed (Dipti Raut *et al.,* 2005). The commercial fishery is supported by more than 20 species of penaeids and non-penaeids of these *Penaeus monodon* contributes around 4-6 per cent to the annual fishery followed by bivalves (*Placuna placenta, Anadara granosa, Macoma sp, Meretrix sp.*) and gastropods (*Cerithidea cingulata, Telescopium telescopium*) (Rajyalakshmi, 1991). Rambambu *et al.,* 1987 reported the preference of mangrove substratum and mangrove plants by mud snail, *Terebralia palustris*. Stable carbon isotope ratios recognised mangroves as primary carbon source for benthic fauna (Mohan *et al.,* 1997; Bouillon *et al.,* 2002). A new record of mangrove clam *Geloina erosa* was reported from Coringa mangroves (Srinivasulu, 2001). Critical Habitat Information System for Coringa Mangroves (Andhra Pradesh), 2001 reported 114 species of macrobenthos (41 sps of crustaceans, 26 sps of polychaetes, 21 sps each of gastropods and bivalves and 5 sps of other taxa). About 28 meiofaunal taxa comprising nematoda, copepoda, foraminifera, ostracoda, polychaeta were recorded with the predominance of nematodes followed by foraminiferans and harpacticoid copepods (Kondala Rao and Ramanamurty, 1988).

Pichavaram and Muthupet Mangroves

Tamil Nadu have a coastline of 1,076 km which is the second longest coastline in the country with a mangrove cover of 47 sq km (Forest Survey of India, 2015). The major mangrove zones includes Pichavaram, Muthupet and Gulf of Mannar. Pichavaram mangrove, part of Vellar-Coleroon estuarine complex on the northern side of Cauvery delta, with a mangrove cover of 21 sq km, is one of the exquisite mangrove forests in India. It represents 14 true mangrove species (Figure 8.1) of which dominant mangroves includes *Rhizophora* sps., *Avicennia marina, Excoecaria agallocha, Bruguiera cylindrica, Lumnitzera racemosa, Ceriops decandra* and *Aegiceras corniculatum*. Muthupet mangrove forest is located in the southern end of the Cauvery delta, covering an area of around 6803.01 ha that enjoys the status of a Reserve Forest since 1880 (Meher Homji, 1991). Out of 8 true mangrove species present in the Muthupet mangroves, *Avicennia marina* is the conqueror of the forest. Gulf of Mannar, Marine National Park is a protected area of India has small patches of mangroves with 9 true mangrove species. *Pemphis acidula*, a true mangrove is endemic to these islands. Pondicherry mangroves, a minor mangrove zone, over 168 ha have 7 true mangrove species.

Major benthic study in Pichavaram was on mangrove crabs and nematodes. Pioneering works on brachyuran crab diversity was carried out by Sethuramalingam and Khan, 1991. Later Chandrasekaran and Natarajan, 1994; Ravichandran and Kannupandi, 2007; Ravichandran and Wilson, 2012, 2013) reported 38-46 species of brachyuran crabs with the dominance of *Sesarma* and *Uca* species from Pichavaram mangroves suggests that sediment suitability, effects of tidal flushing and mangrove vegetation were the possible factors that could influence zonation and abundance of the crabs. *Avicennia marina* zone support more crabs than *Rhizophora* zone (Raffi *et al.*, 2002; Ravichandran *et al.*, 2001, 2011). About 8 species of crabs from Vellar mangroves (Khan *et al.*, 2005) and 15-22 species from Pondicherry with dominance of portunids that prefer *Avicennia marina* as suitable habitat (Satheesh Kumar, 2012; Satheesh Kumar and Khan, 2013; Kamalakkannan, 2015), 18-32 species from different islands of Gulf of Mannar (Jeyabaskaran and Ajmal Khan, 2007) was reported. The biochemical changes in decomposing leaves and its influence on crab was reported by Ravichandran and Kannupandi, 2004. Sesarmid crabs prefer decaying mangrove leaves of *Avicennia marina* and *Rhizophora mucronata* as their most favourite food. Most recent study reported 33 to 36 species of crabs (Soundarapandian *et al.*, 2008; Wilson and Ravichandran, 2013) from entire mangrove ecosystems of Tamil Nadu. Crabs belonging to the family Grapsidae and Ocypodidae were the dominant forms. Roy and Nandi, 2007 reported 355 species of crabs from entire coastal ecosystems of Tamil Nadu, of these 40 species of crabs are endemic to mangroves.

Several crab species, collected from the mangrove area, have been successfully reared in the laboratory (Pasupathi and Kannupandi, 1986, 1987, 1988a, 1988b; Vijayakumar and Kannupandi, 1987; Krishnan and Kannupandi, 1987,1989; Balagurunathan and Kannupandi, 1993, 1995; Kannupandi and Pasupathi, 1994). The fishery potential of *Portunus sanguinolentus* (John, *et al.*, 2004, John and Soundarapandian, 2009) and nutritional value of *Podophthalmus vigil* (Soundarapandian *et al.*, 2014) was studied.

Density, distribution and composition of penaeid prawn were assessed in a *Rhizophora* dominated mangrove area at Pichavaram. There was an occurrence of eight species of penaeid prawns, of these *Penaeus indicus, Metapenaeus dobsoni* and *M. monoceros* form major portion (99.44 per cent) of prawn fishery (Rajendran, 1997; Rajendran and Kathiresan, 1999; Chandrasekaran, 2000) that depend on the mangrove vegetation and their structural aspects like basal area, canopy height and tree diameter (Kathiresan *et al.*, 1994). In Muthupet mangroves *P. merguiensis, M. dobsoni* and *P. indicus* show preference to the detritus-rich substratum, whereas *P. monodon* was equally abundant over different substrate types (Mohan *et al.*, 1995).

Malacofaunal studies were initiated by Kasinathan and Shanmugam (1985), 10-11 species were recorded, gastropods were the governing fauna and they prefer prop roots and pneumatophores as their substratum (Suresh *et al.*, 2012). Shanmugam and Vairamani, 2009; Kesavan *et al.* (2009 a,b) gave checklists of molluscan fauna of Vellar estuarine mangroves and Pondicherry mangroves. The larval development, age and growth pattern and allometric relationship of snails

Melampus ceylonicus (Shanmugam and Kasinathan,1987; Shanmugam,1994,1995b) and *Pythia plicata* (Shanmugam,1991a,b; 1995a,c; 1996), the salinity and temperature tolerance and reproductive cycle of gastropods *Cassidula nucleus* and *Melampus ceylonicus* (Dious and Kasinathan, 1994; Shanmugam,1998), biology and population structure of periwinkle *Littorina scabra* (Maruthamuthu *et al.*, 1985; Maruthamuthu and Kasinathan,1986) were studied in Pichavaram mangroves. Mangroves provide suitable habitat for the production of edible oyster *Crassostrea madrasensis* compared to coastal and estuarine areas (Rajapandian *et al.*, 1990).

Kasinathan and Shanmugam, 1988; Sethuramalingam and Ajmal Khan, 1991; Vijayakumar *et al.*, 2009; Pravinkumar *et al.*, 2013 observed 22 - 44 species of macrofauna from Pichavaram, while 112 species of insects, 14 species of crustaceans and 18 species of molluscs were observed in Muthupet mangroves (Oswin,1998). Khan *et al.*, 2008; Saravanan *et al.*, 2008; Satheesh Kumar and Anisa Khan, 2013 studied the macrofaunal diversity and community structure of Pondicherry mangroves, reported 76 species that comprised of molluscs, crustaceans, amphipods, polychaetes, barnacles and oligochaetes (Figure 8.2). Samidurai *et al.*, 2012; Thilagavathi *et al.*, 2013; Sekar *et al.*, 2013 compared the macrobenthic communities of developing Vellar mangroves (31-156 sps), riverine Pichavaram mangroves(35-252 sps.) and island mangroves of Gulf of Mannar(31-163 sps.) and found that more pristine zone was Pichavaram.

Devi *et al.* (1986) studied the polychaete *Ceratonereis costae* and the amphipod *Paracalliope fluviatilis* while Fernando and Rajasekaran, 2002 reported the presence of *Ceratonereis burmensis* in the sediments of Pichavaram mangroves. Amphipods of Pichavaram was studied by Lyla *et al.*, 1998 and Mondal, 2010 and found higher species composition and diversity in mangrove habitats and out of 29 amphipods,

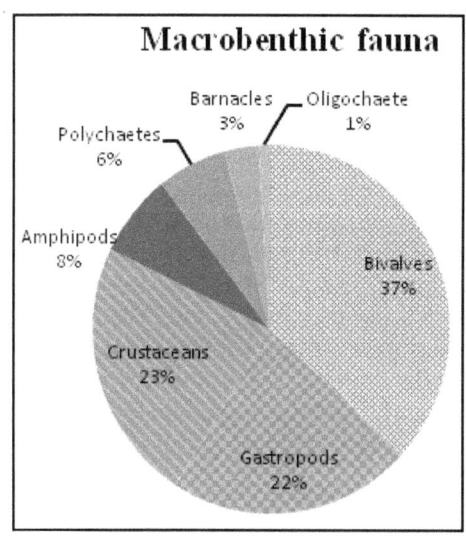 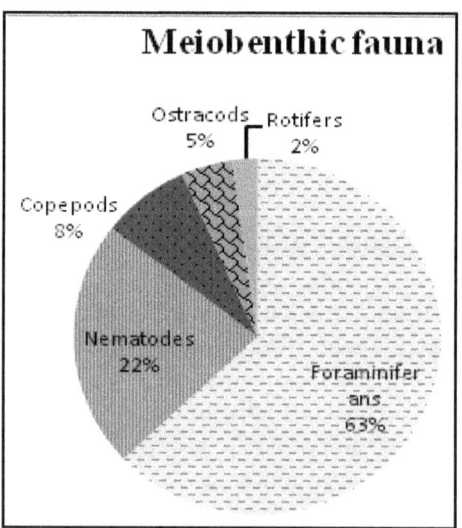

Figures 8.2 and 8.3: Macrobenthic and Meiobenthic Diversity in Mangrove Ecosystem of Tamil Nadu Mangroves. (Redrawn from Ajmal Khan *et al.*, 2013 and Thilagavathi *et al.*, 2011).

10 were exclusively from mangroves, 7 species of amphipods were reported from Pondicherry mangroves (Kumar and Khan, 2013). Sundaravarman *et al.*, 2012 compared the macro-meiofaunal composition in mangrove site lined with *Excoecaria agallocha, Avicennia marina,* land ward and coastal area without mangroves and found maximum macrofaunal counts in the mangrove-lined lagoon and the minimum in the landward site.

Meiobenthic studies were initiated by Ali *et al.* (1983) using nematodes for studying the energy flow in Pichavaram mangroves. Later Ali *et al.*, 1998; Chinnadurai and Fernando, 2003, 2006 a, 2007a studied the meiobenthic composition, nematode diversity and observed 44 species of nematodes, of these 37 species from Pichavaram and 14 from artificial Vellar mangroves (Chinnadurai and Fernando, 2006b). Meiofauna population density and the assemblage of free-living marine nematodes were higher in areas with *Avicennia marina* compared to *Rhizophora apiculata* cover (Chinnadurai and Fernando, 2007b; Ansari *et al.*, 2014). Suresh *et al.*, 2014; Thilagavathi *et al.*, 2011 identified 106 species of meiofauna from Muthupet and Sethukuda mangrove, foraminiferans, nematodes, harpacticoid copepods, ostracodes and rotifers were identified with the predominance of foraminiferans (Figure 8.3).

Andaman and Nicobar Islands

The Andaman and Nicobar Islands, located in the northeast Indian Ocean, floats on Bay of Bengal endowed with about 96 Wildlife Sanctuaries, nine National Parks and one Biosphere Reserve. Andaman and Nicobar Islands harbour 617sq km (Forest Survey of India, 2015) of dense and diverse mangrove cover (Selvam,2003) with 10 true mangrove species (Figure 8.1) with dominance of *Rhizophora mucronata, Bruguiera gymnorrhiza, Avicennia sps, Ceriops* tagal etc (Kannan, 1990).

A detailed study of benthic fauna of Andaman and Nicobar islands was done by Das and Roy (1989). He recorded higher diversity of molluscs, crustaceans, echinoderms, sipunculids and polychaetes, of these *Marphysa mossambica* is well established in highly deoxygenated soil of Andaman mangroves. Rajashekaran and Fernando, 2012 recorded 30 polychaetes belonging to eight families and 23 genera. Diversity, distribution and relative abundance of mangrove crabs (Thomas *et al.*, 2002), description of a new species *Myopilumnus andamanicus* (Deb, 1989) and biological studies of *Scylla serrata* (Poovachiranom,1992) was done in Andamans mangroves. Marine wood borers of mangroves of the Bay islands have been dealt by Dev Roy and Das, 1985; Das and Dev Roy (1980, 1981,1984a, b) and Tiwari *et al.* (1980). Among the marine borers, teredinid borer *Bactronophorus thoracites* was the most copious and detrimental species which causes considerable damage to the mangroves of these islands (Dev Roy and Das, 1985). Rao (1986) studied the meiofauna of mangrove sediments in South Andaman and recorded nematodes, copepods, gastrotrichs, kinorhynchs, archiannelids, polychaetes and ostracods. Among these, nematode contributed 80 per cent of the total fauna followed by copepods which comprise only 12 per cent. *Avicennia marina* exhibited highest meiofaunal diversity and density and the lowest by *Acrostichum aureum*. Higher carbonate and moderate organic carbon was essential for density and distribution of meiofauna (Mohan *et al.*, 2012).

The pictorial representation of the distribution of benthic fauna in various mangrove habitats along the Indian coast is depicted in Figure 8.4.

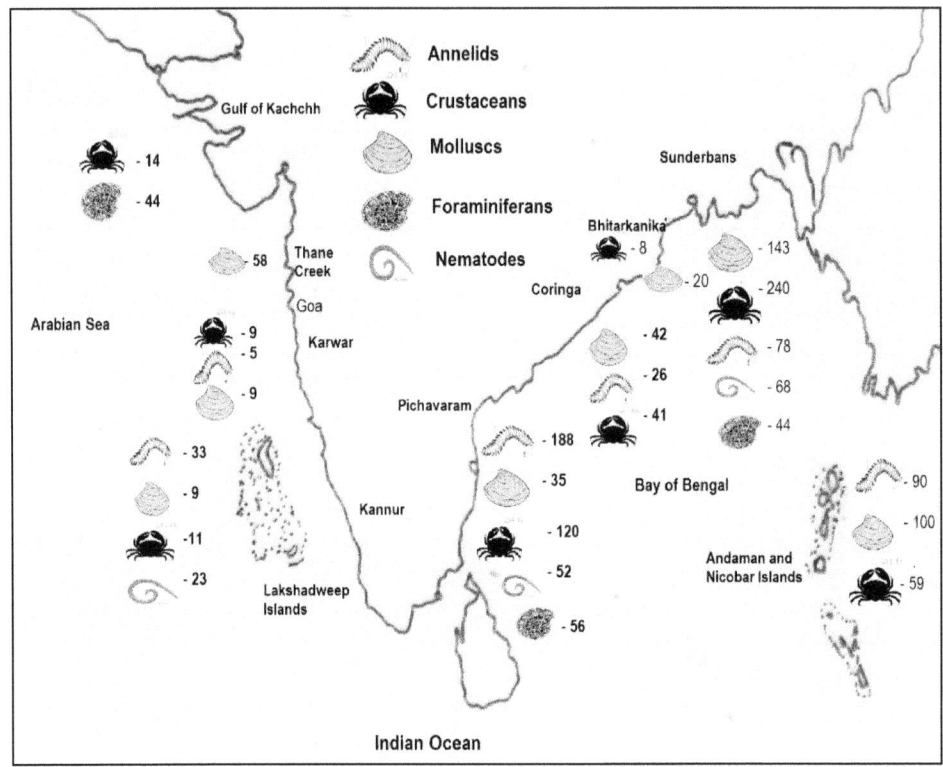

Figure 8.4: Map Showing the Distribution of Benthic Faunal Species in Indian Mangroves.

Benthic Studies in West Coast of India

The west coast covers only 28.87 per cent of mangroves of India receiving tidal flushing from Arabian sea. The west coast is characterised by backwater estuarine type of mangroves experiencing intense upwelling associated with south-west monsoon. Mangroves of west coast is distributed in five states, Gujarat with Gulf of Kachchh and Gulf of Khambhat mangroves, Maharashtra with Thane creek mangroves, Goa with Mandovi and Zuari estuarine mangroves, Karnataka with Karwar mangroves, Kerala with Kannur and Cochin mangroves along Vembanad estuary. The benthic abundance was comparatively less in west coast and about 290 species were reported (Kathiresan and Qasim, 2005).

Gulf of Kachchh and Gulf of Khambhat

Gujarat is having the largest mangrove patch on the west coast and second largest in India after Sundarbans with an area of 1107 sq km under mangrove cover (Forest Survey of India, 2015). Nearly 77 per cent of the Gujarat mangroves are confined to the Gulf of Kachchh and the remaining are found in the Gulf of Khambat region. Gujarat has 14 species of mangroves(Sahu *et al.*, 2015).The mangrove species *Avicennia officinalis* and *Rhizophora mucronata* dominate on the Gulf of Kachchh and *Avicennia marina* as a single dense species in Gulf of Khambhat, *Sonneratia apetala* has dispersed and sparse distribution.

Studies on benthic faunal assemblage (Sesh Serebiah, 2003) and brachyuran crabs was extensively studied by Saravanakumar, 2007, Shukla *et al.*, 2013, Trivedi *et al.*, 2012 reported 10-14 species from Kachchh mangroves. Major families include Grapsidae, Portunidae, Ocypodidae, Gecarcinidae and Goneplacidae. *Uca lactea annulipes, Parasesarma plictum, Scylla serrata* and *Cardisoma cranifex* were the most dominant brachyuran crabs in mangrove mud flats.

Mangroves of Goa, Karnataka and Mumbai

In Goa, mangroves exist over an area of 26 sq km (Forest Survey of India, 2015) especially in dense patches along the Mandovi and Zuari estuary. Goa has low diversity of mangroves represented by 12 true species (Figure 8.1) dominated by *Acanthus ilicifolius and Kandelia kandel* followed by *Avicennia officinalis and Sonneratia caseolaris*. Mangroves are only sparsely distributed along the Karnataka coast that comes around 3 sq km (Forest Survey of India, 2015). The mangrove forest occurs along the northern coast of Karnataka, particularly in Karwar and Ullal. About 14 species of mangroves belonging to 9 genera are extensively distributed in the state (Figure 8.1). The dominant mangrove flora includes *Acanthus ilicifolius, Rhizophora mucronata* and *Excoecaria agallocha*. Mangroves of Maharashtra existed largely in the Thane creek, Mahim, Versova, Gorai and Ghodbunder spreading around 222 sq km (Forest Survey of India, 2015) with 20 true mangrove species (Figure 8.1).

Untawale and Parulekar (1976) conducted extensive studies on ecological aspects of estuarine mangrove area of Goa. The penaeid prawn stock of the Mandovi and Zuari estuaries comprises 13 species with *Metapenaeus dobsoni* and *M. monoceros*, together accounting for 80 per cent of the total harvest (Parulekar and Achuthankutty, 1993). Habitat preference of mangrove ecosystem has been strikingly observed in *Penaeus merguiensis, Metapenaeus dobsoni* and *M. monoceros*. A mangrove clam, *Gelonia erosa*, a new record to west coast and its population characteristics (Ingole *et al.*, 1994, 2002) and marine wood-borers (Santhakumaran, 1983) was reported. Ansari *et al.* (1993) studied the meiobenthic fauna and reported that the nematodes, turbellarians and harpacticoids were reduced due to vertical gradients such as redox potential, organic matter in the environment and was positively correlated with interstitial water of the sediment and also to the microbial density in mangrove mudflats. Life cycle studies of the free-living marine nematode *Innocuonema tentabunda* was done by Singh *et al.*, 2009.

Marakala *et al.,* 2005 studied the ecology and biodiversity of macrofauna of Karnataka mangroves and observed higher diversity index in dense mangrove area while evenness was higher in riverine stretch. About 46 species were identified with the dominance of crustaceans and molluscs. Vasanth Kumar *et al.*, 2013 identified

14 taxa including foraminifera, coelenterata, polychaeta, gastropoda, bivalvia, harpacticoida, cumacea, tanaidacea, isopoda, amphipoda, mysidacea, shrimps, decapoda and pisces, observed higher species diversity at the upper surface and decreased with sediment depth (Gowda et al., 2008), found the dominance of polychaetes followed by molluscs and crustaceans. Boominathan et al., 2012 studied the molluscan fauna of mangroves of India with special reference to Karnataka mangroves, about 215 species of molluscs were reported from mangrove areas of east and west coasts. It includes 133 gastropods, 77 bivalves, four cephalopods, and one polyplacophores. Pradnya et al., 2011, studied the biodiversity of crabs in Karwar mangrove environment.

Macrobenthos from the mudflats of Thane Creek, Mumbai includes polychaetes, gastropods, bivalves and sea anemones that inhabit the mangrove systems (Athalye and Gokhale, 1998). Padmakumar (1984) investigated the benthos of mangroves in Mumbai with reference to sewage pollution. Diversity of bivalve and gastropod molluscs from mangrove habitat, rocky substrata, sandy beach, and muddy habitat was compared by Khade and Mane, 2012 a,b,c. About 12-19 species of bivalves and 13-39 species of gastropods were encountered from Ratnagiri and Raigad coast of Maharashtra. A preliminary study on dominant fauna of mangroves at Versova (Pereira et al., 2002), tidal ponds with kandalvan (Deepalakshmi and Sunanda, 2013) was carried out. Biology and biochemistry of mangrove bivalve *Gelonia proxima* and their fishery status (Kale and Pawar, 2002), biology of mud crab *Scylla serrata* (Funde et al., 2009), diversity of decapod fauna (Pawar, 2012) and wood borers of mangroves were studied (Yerangi and Yerangi, 2002) from Maharashtra mangrove area. Influence of copper on polychaete *Lycastis merukensis* (Mukherji and Gokhale, 2002) and a biofouler sea anemone, *Acontiactis gokhaleae* in mangrove mudflats along Thane creek (Mishra et al., 1994) was explored. The most commonly seen macrofauna in mangroves of Mumbai are the fiddler crabs and the gastropod *Telescopium telescopium*. Studies on the meiobenthos of intertidal zone of mangrove mudflats of Maharashtra revealed dominance of nematodes (Goldin et al., 1996).

Mangroves of Kerala

Kerala once had a lush mangrove cover now reduced to just about 9 sq km (Forest Survey of India, 2015) in isolated bits at Kannur, Cochin, Alappuzha and along the Vembanad backwaters. Kerala eventhough have less stretch of mangroves has the highest diversity with 18 species of true mangroves and 38 species of associate forms (Bijoy Nandan et al., 2015 a, b). *Acanthus ilicifolius* ranks first in its density followed by *Avicennia officinalis*. Other major mangroves include *Rhizophora mucronata, Bruigera cylindrica* and *B.gymnorrhiza*.

Benthic studies were scanty in mangrove ecosystem of Kerala. Kurian, 1984 studied the benthic fauna in Cochin Mangroves. Later Sunil Kumar worked extensively on mangrove fauna. Community structure and distribution of macrobenthic fauna in mangrove sediments of Cochin mangroves has been studied (Sunil Kumar,1993;1995a,b,c) and compared the mangrove macrobenthic fauna with estuarine fauna of Vembanad estuary (Sunil kumar, 2002). Studies on environmental and sediment influence in diversity and distribution of polychaete fauna have been extensively studied from Cochin Mangroves (Sunil Kumar and Antony, 1993, 1994a,b).Thirty-three species belonging to 20 genera under 10 families have been reported, of these five polychaetes were new record from mangroves of Cochin (Sunil Kumar,1999). Among the polychaetes, *Dendronereis aestuarina, Paraheteromastus tenuis, Nereis glandicincta, Marphysa gravelyi, Dendronereides heteropoda* are found to be the most prevailing species. Sunil Kumar and Antony (1994c) reported the existence of pollution indicator polychaete worm, *Paraheteromastus tenuis* from Cochin mangroves. Depth wise distributional variations in macrofauna especially polychaetes in mangrove sediments showed their persistence in top 0-5 cm substratum and decreased towards deeper layers (Sunil Kumar, 1997). Macrobenthos in a traditional prawn field 'Chemmen Kettu' enclosed by mangrove vegetation in Cochin backwaters have been assessed (Sunil Kumar, 1998). Ecological distribution and population structure of mud dwelling actinarian *Edwardsia* sps. and isopoda *Sphaeroma terebrans* in a mangrove habitat of Cochin area was reported (Sunil Kumar, 2001a,b).A check list of polychaetous annelids of Indian mangrove environment confirmed 62 species of Indian record and it showed higher diversity among the Asian countries (Sunil Kumar,2001c). Population dynamics studies of *Eriopisa chilkensis* was reported from mangrove swamps of Cochin (Aravind *et al.*, 2007). Rajagopalan *et al.* (1986) studied the molluscs and crabs of Cochin mangroves and observed distinctive zonation pattern: *Uca* sps. in the upper littoral zone, hermit crabs and *Nautica* sps. in the mid-littoral zone, *Cerethidium* sps. and *Terebralia* sps. on the mud-flats; *Littorina* sps. on the trunks and leaves of mangroves and juveniles of prawns and fishes in mangrove waters. Dalia Susan Vargis, 2005 compared the macrobenthos at the intertidal zones of seagrass and mangrove ecosystems of Minicoy Island of Lakshadweep. Benthic diversity was lower in mangroves (17species) compared to sea grass (62-137sps). The mangroves showed an unstable ecosystem, having a diversity index less than 3, resulting in less species diversity. The species diversity was governed by prey-predator relationships and food resource availability. Gastropods were the dominant fauna, then crustaceans and crab. *Littorina undulata* and *Terebralia palustris* were the abundant gastropods in mangroves.

Author studied the benthic realm of mangrove ecosystems in various districts of Kerala and observed higher macrofaunal density in Kannur district (av.256 ind/m^2) followed by Ernakulam (av.251ind/m^2), Alappuzha (av.218 ind/m^2) while least abundance was observed in Thiruvananthapuram (av.10 ind/m^2) (Figure 8.5). Higher abundance of benthic fauna in Kannur might be due to luxuriant mangrove vegetation which covers almost 80 per cent of the total mangrove forests of the state. While in Thiruvananthapuram, lower benthic density might be due to the low diversity and declining mangrove vegetation. Species diversity was found to be high in Ernakulam district, evenness index in Alappuzha and dominance index in Kottayam. In Kasaragod, Kozhikode and Kottayam dominant fauna were polychaetes while in Kannur, Thrissur, Ernakulam, Alappuzha, Kollam and Thiruvanathapuram amphipods were the pioneering fauna. Bivalves were the dominant fauna in Malappuram. A detailed study was done on Cochin mangroves by the author, revealed the abundance of Amphipods followed by polychaetes, tanaids, bivalves (Figure 8.6). Diversity index was comparatively low and disappearance of many of species was noted. A decline in species composition of polychaetes from 33 species as reported by Sunil kumar,1993 to 11 species in the present study dominated by *Dendronereis aestuarina, Prionospio cirrifera, Paraheteromastus tenuis* and *Capitella capitata*.

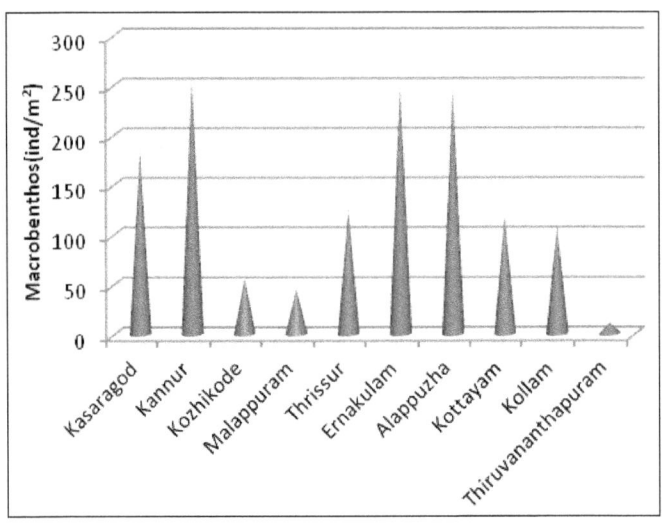

Figure 8.5: Mean Density of Macrobenthic Fauna in different Districts of Kerala.

Chinnadurai and Fernando, 2006c studied the inter-relationship between the meiofauna and mangrove vegetation of Cochin, out of 7 major taxa of meiofauna (nematodes, copepods, foraminifera, polychaetes, oligochaetes, ostracods and turbellarians) as in other areas nematodes (Comesomatidae) was abundant in *Avicennia marina* (48.2 per cent) and *Sonneratia caseolaris* (30.3 per cent) with 23 species belonging to 16 genus. *Daptonema oxycerca* was the most common species that existed in all the stations due to high mud concentration in the sediments. Habitat preference studies indicate more assemblage in *A.marina* zone than *Rhizophora* zone. Meiofaunal studies by the author exhibited similar results as higher dominance of nematodes as well as foraminiferans (Figure 8.7) in Puthuvype mangrove area, only zone where *Avicennia marina* present in Ernakulam district.

Conclusion

The present paper points out the paucity of benthic studies in mangrove ecosystem in many of Indian states both in east and west coast. But state of Tamil Nadu, stands different from this, as most of mangrove benthic realm was fully exploited by the continuous research in mangrove areas such as Pichavaram, Muthupet. Hence maximum number of species were identified from these ecosystems. Sunderbans stands second in benthic diversity studies and most of the studies were concentrated on crabs. Least number of benthic species was reported from mangrove ecosystems of west coast might be due to difficulty in sampling or

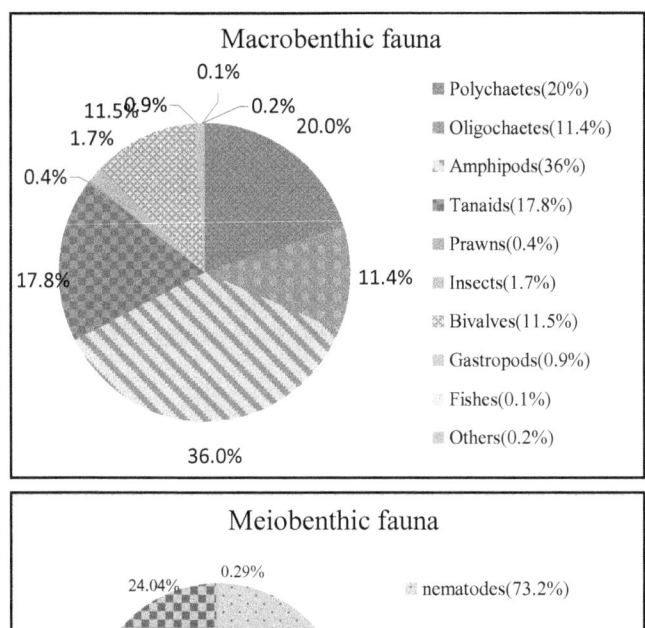

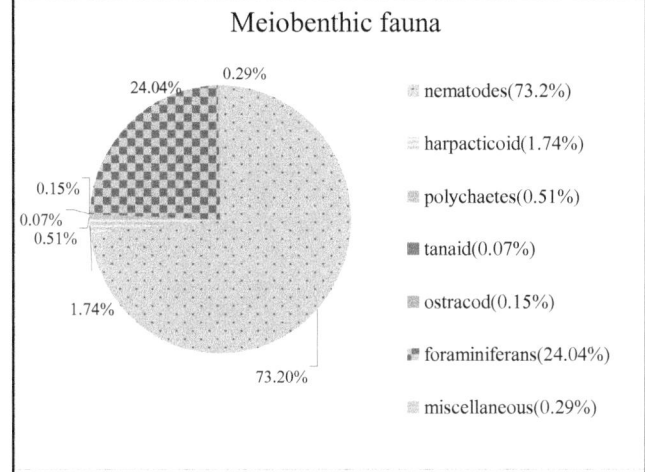

Figures 8.6 and 8.7: Percentage Abundance of Macrobenthic and Meiobenthic Fauna in Mangroves of Cochin.

lack of knowledge on importance of mangrove ecosystems and its supporting fauna in ecosystem balance resulting in gradual disappearance of many fauna. Besides most of the areas are cleared for developmental activities especially in Cochin, recent developmental projects, urbanisation has resulted in decline of many benthic fauna that ruled the substratum. But Gujarat state stands unique compared to other west coast states as their mangrove cover has increased from 936 sq km (Forest Survey of India, 2005) to 1107 sq km (Forest Survey of India, 2015). By considering the importance of mangrove ecosystem and related fauna, necessary actions should be taken to increase the mangrove vegetation all over India and thereby reintroduce the benthic faunal components for the functioning and balancing of the trophic system.

Acknowledgement

Authors are thankful to UGC Maulana Azad National Fellowship and Directorate of Environment and Climate Change for the financial Support and also to the Head, Department of Marine Biology, Microbiology and Biochemistry for the facilities provided. We are extremely thankful to Dr. K.J. Joseph for his immense guidance and support.

References

Ajmal Khan, S., Raffi, S.M. and Lyla, P. S., 2005. Brachyuran crab diversity in natural (Pichavaram) and artificially developed mangroves (Vellar estuary). *Current Science*, 88: 1316–1324.

Ali, M.A.S., Ajmal Khan, S. and Balasubramanian, T., 1998. Nematodes of Pichavaram mangroves. GIS based information system for Pichavaram Mangroves sponsored by DODICMAM, Govt. of India, CAS in Marine Biology, Parangipettai, 48 pp.

Ali, M.A.S., Krishnamurthy, K. and Jeyaseelan, M.J.P., 1983. Energy flow through the benthic ecosystem of the mangrove with special reference to nematodes. Mahasagar – *Bulletin of the National Institute of Oceanography*, 16: 317-325.

Ansari, K.G.M.T., Manokaran, S., Raja, S., Lyla, P.S. and Ajmal Khan, S., 2014. Interaction of free-living marine nematodes in the artificial mangrove environment (southeast coast of India). *Environmental Monitoring and Assessment*, 186: 293–305.

Ansari, Z.A., Rodriguez, C.L., Chatterji, A. and Parulekar, A.H., 1993. Distribution, abundance and ecology of the meiofauna in a tropical estuary along the west coast of India. *Hydrobiologia*, 262: 115-126.

Aravind, N.P., Sheeba, P., Nair, K.K.C. and Achuthankutty, C.T., 2007. Life history and population dynamics of an estuarine amphipod, *Eriopisa chilkensis* Chilton (Gammaridae). *Estuar. Coast. Shelf Sci.*, 74(1): 87-95.

Athalye, A.P. and Gokhale, K.S., 1998. Macrobenthos from the mudflats of Thane Creek, Maharastra, India. *J. Bombay Nat. Hist. Soc.*, 95: 258-266.

Balagurunathan, R. and Kannupandi, T., 1993. Effect of salinity on larval survival and development of the mangrove crab *Metaplax elegans*. *J. Mar. Biol. Ass. India*, 35(1 and 2): 193 - 197.

Balagurunathan, R. and Kannupandi, T., 1995. Biochemical changes during larval development of mangrove crab *Metaplax elegans* (De Man). *J. mar. biol. Ass. India*, 37(1 and 2): 35 - 38.

Berry, A.J., 1972. The natural history of west Malaysian mangrove faunas. *Malay. Nat. J.*, 25: 135–162.

Bhunia, A.B. and Choudhury, A., 1981. Observations on the hydrology and the quantitative studies on benthic macrofauna in a tidal creek of Sagar Island, Sunderbans, West Bengal. *Proc. Indian Nat. Sci. Acad.*, 47: 398-407.

Bijoy Nandan, S., Sreelekshmi, S., Preethy, C.M., Rani Varghese and Philomina Joseph (eds.)., 2015a. *Manual on Mangroves*. Directorate of Public Relations and Publications, CUSAT, Kochi, India, 133 pp.

Bijoy Nandan, S., Sreelekshmi, S., Rani Varghese, Preethy, C.M. and Philomina Joseph, 2015b. Ecology of the mangroves of south west coast of India. In: *Proceedings of the National Seminar on Conservation and Management of Mangrove Ecosystem with Special Reference to its Faunal Diversity*, 3-5 pp.

Bird, F.L., Ford, P.W. and Hancock, G.J., 1999. Effect of burrowing macrobenthos on the flux of dissolved substances across the water-sediment interface. *Marine and Freshwater Research*, 50: 523-532.

Boominathan, M., Ravikumar, G., Subash Chandran, M.D. and Ramachandra T.V., 2012. Mangrove associated molluscs of India. LAKE 2012: *National Conference on Conservation and Management of Wetland Ecosystems*.

Bouillon, S., Raman, A.V., Dauby, P. and Dehairs, F., 2002. Carbon and nitrogen stable isotope ratios of subtidal benthic invertebrates in an estuarine mangrove ecosystem (Andhra Pradesh, India). *Estuarine Coastal and Shelf Science*, 54: 901–913.

Chakraborti, R.K., Naskar., Chattopadhyay, G.N., Nath, D. and Bhowmik, M.L., 1990. Ecology and faunal association of intertidal mangrove habitats in the Hooghly - Matlah estuarine system. *J. Inland Fish. Soc. India*, 22(1 and 2): 31-37.

Chakraborti, R.K., Sundaray, J.K., and Ghoshal, T.K., 2002. Production of *Penaeus monodon* in the tide fed ponds of Sundarbans. *Indian J. Fish.*, 49(4): 419-426.

Chakraborty, K., 1994. Biodiversity of mangrove ecosystem of Sunderbans. In: S. Maity, and N. Mukherjee (eds.), Bidhan Chandra Krishi Viswavidyalaya Publ, 57-64 pp.

Chakraborty, S.K. and Choudhury, A., 1992c. Population ecology of *Metaplex intermedia* (Brachyura: Grapsidae) of Sagar Island, Sunderbans, India. *Proceedings of Zoological Society, Calcutta*, 47(1): 41-45.

Chakraborty, S.K., Chaudhury, A. and Deb, M., 1986. Decapod brachyura from Sundarbans mangrove estuarine complex, India. *Journal of Bengal Natural History Society*, 5: 55–68.

Chakraborty, S.K. and Choudhury, A., 1985. Distribution of fiddler crabs in Sunderbans mangrove estuarine complex, India. In: *Proc. Nat. Symp. Biol. Util. Cons. Mangroves* (Ed. L.J. Bhosale), 467- 472 pp.

Chakraborty, S.K. and Choudhury, A., 1992a. Ecological studies on the zonation of brachyuran crabs in a virgin mangrove Island of Sunderbans, *India. J. Mar. Biol. Ass. India*, 34(1 and 2): 189-194.

Chakraborty, S.K. and Choudhury, A., 1992b. Population ecology of fiddler crabs (*Uca* sps.) of the mangrove estuarine complex of Sunderbans, India. *Trop. Ecol.*, 33(1): 78-88.

Chakraborty, S.K., Poddar, T.K. and Choudhury, A., 1992. Species diversity of macrozoobenthos of Sagar island, Sunderbans, India. *Proc. Zool. Soc., Calcutta,* 45(A): 435-444.

Chakravarty, M.S. and Joseph Uday Ranjan, T., 2014. A check list of malacofauna from the Nuvvalarevu backwaters of Srikakulam district, Andhra Pradesh, India. *International Journal of Research in Marine Sciences,* 3(1): 11-15.

Chandra, A. and Chakraborty, S.K., 2008. Distribution, density and community ecology of macrobenthic intertidal polychaetes in the coastal tract of Midnapore, West Bengal, India. *J. Mar. Biol. Ass. India,* 50 (1): 7–16.

Chandrasekaran, V.S. and Natarajan, R., 1994. Seasonal abundance and distribution of seeds of mud crab *Scylla serrata* in Pichavaram mangrove, southeast cost of India. *J. Aquacult. Trop.,* 9(4): 343-350.

Chandrasekaran, V.S., 2000. Relationship between plankton and finfish and shellfish juveniles in Pichavaram mangrove waterways, southeast coast of India. *Seaweed Res. Utiln.,* 22 (1 and 2): 199 - 207.

Chinnadurai, G. and Fernando O.J., 2003. Meiofauna of Pichavaram mangroves along southeast coast of India. *J. Mar. Biol. Ass. India,* 45(2): 158-165.

Chinnadurai, G. and Fernando O.J., 2006 b. New records of five free-living marine nematodes from an artificial mangrove of India. *J. Mar. Biol. Ass. India,* 48(1): 105-107.

Chinnadurai, G. and Fernando, O.J., 2006 a. New records of free-living marine nematodes from India. *Records of Zoological Survey of India,* 106: 45-54.

Chinnadurai, G. and Fernando, O.J., 2006c. Meiobenthos of Cochin mangroves (southwest coast of India) with emphasis on free-living marine nematode assemblages. *Russian Journal of Nematology,* 14(2): 127-137.

Chinnadurai, G. and Fernando, O.J., 2007a. Meiofauna of mangroves of the southeast coast of India with special reference to the free-living marine nematode assemblage. *Estuar. Coast. Shelf. Sci.,* 72(1-2): 329-336.

Chinnadurai, G. and Fernando, O.J., 2007b. Impact of mangrove leaves on meiofaunal density: An experimental approach. *Journal of Life Sciences,* 1: 62-70.

Choudhury, A., Das, A., Bhattacharya, S. and Bhunia, A.B., 1984a. A quantitative assessment of benthic macrofauna in the intertidal mudflats of Sagar Island, Sunderbans, India. In: *Proc. Symp. Mangr. Enviorn: Res. Manage.,* (Eds.) E. Soepadmo, A.N. Rao and D.J. Macintosh, 298-310 pp.

Choudhury, A., Bhunia, A. and Nandi. S., 1984b. Preliminary survey on macrobenthos of Prentice Island, Sunderbans, West Bengal. *Rec. Zool. Sur. India,* 81 (2 and 4): 81-92.

Coull, B.C., 1999. Role of meiofauna in estuarine soft-bottom habitats. *Aust. J. Ecol.,* 24: 327–343.

Cummins, S.P., Roberts, D.E. and Zimmerman, K.D., 2004. Effects of the green macroalgae *Enteromorpha intestinalis* on macrobenthic and seagrass assemblages in a shallow coastal estuary. *Marine Ecology Progress Series*, 266: 77-87.

Dalia Susan Vargis, 2005. Macrobenthos of Minicoy island, Lakshadweep. *Ph.D Thesis*, Cochin university of science and Technology.

Das, A.K. and Dev Roy, M.K., 1980. On the wood-boring molluscs of South Andaman, India. *Rec. Zool Surv. India.* 77: 179-187.

Das, A.K. and Dev Roy, M.K., 1984a. Marine borers of mangroves of Little Andaman, India. *Bull. Zool. Surv. India*, 6: 95-98.

Das, A.K. and Dev Roy, M.K., 1984b. Report on the marine wood borers from the mangroves of Neil, Havelock and Peel Island, Ritchie's Archipelago, Andaman, India. *Bull. Zool. Surv. India*, 6: 327-329.

Das, A.K. and Dev Roy, M.K., 1981. On the teredinid borers of mangroves of Camorta Island, Nicobar, India. *Bull. Zool. Surv. India*, 4(3): 391-393.

Das, A.K. and Dev Roy, M.K., 1989. A general account of the mangrove fauna of Andaman and Nicobar Islands. Conservation area series 4, *Zoological Survey of India*.

Day, J. W., Hall, C.A.S., Kemp, W. M. and Yáñez-Arancibia, A.,1987. *Estuarine Ecology*. John Wiley and Sons, Brisbane, 558 pp.

Deb, M., 1989. *Myopilumnus andamanicus*, a Xanthid crab from Andamans. *J. Andaman Sci. Assoc.* 5(2): 113-116.

Deeplaxmi Satam and Sunanda Deshmukh., 2013. Macrobenthos of Tidal Ponds at Kandalvan along Eastern Suburb of Mumbai. *National Conference on Biodiversity: Status and Challenges in Conservation - 'FAVEO' 2013*.

Dehadrai, P.V., 1994. Mangrove fauna of Sunderbans: Ecological features and utilisation. In: Deshmukh, S.V. and Balaji, V. (Eds.), Conservation of Mangrove Forest Genetic Resources: A Training Manual. ITTO- CRSARD Project, M.S. Swaminathan Research Foundation, Madras, India, 287- 293 pp.

Dev Roy, M.K. and Das, A.K., 1985. Marine wood-borers from the mangrove ecosystem of Great Nicobar Island, India. *Bull. Zool. Surv., India*, 7(2-3): 251-254.

Devi, L.P., Padmakumar, K. and Ayyakkannu, K., 1986. Qualitative and quantitative study of gut microflora of *Ceratoneries costae* (polychaete) and *Paracalliope fluviatilis* (amphipod) associated with the sediments of Pichavaram mangroves. *Natl. Sem. Microbial Ecology*, 53 p.

Dey, A., 2006. *Handbook on Mangrove Associate Molluscs of Sunderbans*. (Eds. Director, ZSI, Kolkata), 96 pp.

Dey, M., Yusuf Ali, J. and Abhijit, M., 2005. Distribution of intertidal malacofauna at Sagar Island. *Rec. Zool. Surv. India*, 105(Part 1-2): 25-35.

Dious, S.R.J. and Kasinathan, R., 1994. Tolerance limits of two pulmonate snails *Cassidula nucleus* and *Melampus ceylonicus* from Pichavaram mangroves. *Envir. Ecol.*, 12: 845–849.

Dipti Raut., Ganesh, T., Murty, N.V.S.S. and Raman A.V., 2005. Macrobenthos of Kakinada Bay in the Godavari delta, East coast of India: comparing decadal changes. *Estuarine, Coastal and Shelf Science*, 62: 609–620.

Fernando, O.J. and Rajasekaran, R., 2002. On the occurrence of the polychaete *Ceratonereis burmensis* Monro, from Pichavaram mangrove. *J. mar. boil. Ass. India*, 44(1 and 2): 237-238.

Field, C.D., 1998. Rehabilitation of mangrove ecosystems: an overview. *Mar. Poll. Bull.* 37: 383–392.

Funde, A.B., Naik, S.D. and Mohite, S.A., 2009. Contribution to the biology of the mudcrab, Scylla serrata (Forslal) of Ratnagiri, Maharashtra. *Aquacult*, 10(1): 73-79.

Gallep, G.W., Kitchell, J.F. and Bartell, S.M. 1978. Phosphorous release from Lake Sediments as affected by *Chironomid. Ver. Inter. Vere Limnologic*, 20: 458-465.

Ganapati, P.N. and Rao, M.V.L., 1959. Incidence of marine borers in the mangrove of Godavari estuary. *Current Science*, 28(8): 332.

Gautam Kumar Das, 2016. Occurrence of Bioturbation Structures at Estuarine Environment of the Sunderbans, Eastern India. *Earth Science India*, 9 (I): 1 – 20.

Ghosh, A., Chakraborti, P.K., Naskar, K.R., Chattopadhyay, G.N., Nath D. and Bhowmik, M.L., 1990. Ecology and faunal association of intertidal mangrove habitats in the Hoogly-Matlah estuarine system. *J. Inland Fish. Soc. India*, 22(1-2): 31 - 37.

Ghosh, A., Sing, B.N. and Choudhury. A., 1995. Studies on the distribution of Gastropoda (Mollusca) in a mangrove forest (Prentice Island) of Sunderbans, India. *J. Mar. Biol. Asso. India*, 37(1 and 2): 283-286.

Ghosh, D., Majumdar, S. and Chakraborty, S.K., 2014. Taxonomy and distribution of meiobenthic intertidal foraminifera in the coastal tract of Midnapore (East), West Bengal, India. *Int. J. Curr. Res. Aca. Rev.*, 2(3): 98-104.

Goldin, Q., Mishra, V., Ullal, V., Athalye, R.P. and Gokhale, K.S., 1996. Meiobenthos of mangrove mudflats from shallow region of Thane Creek, Central West coast of India. *Indian Journal Marine Sciences*, 25: 137–141.

Gowda, G., Rajesh, K.M. and Mridula, R. M., 2008. Abundance and vertical distribution of macrobenthos in a mangrove - fringed brackish water pond in Mangalore, India. *The Ecoscan*, 2 (2): 181 – 186.

Hazra, A.K., Dey, M.K. and Mandal, G.P., 2005. Diversity and distribution of Arthropod fauna in relation to mangrove vegetation on a newly emerged island on the river Hooghly, West Bengal. *Rec. Zool. Surv. India*, 104(Part 3-4): 99 – 102.

Ikomi, R.B., Arimoro, F.O. and Odihirin, O.K., 2005. Composition, distribution and abundance of macroinvertebrates of the upper reaches of river Ethiope, Delta State, Nigeria. *The Zoologist*, 3: 68-81.

India State of Forest Report, 2005. Forest Survey of India (FSI), Dehradun, India.

India State of Forest Report, 2013. Forest Survey of India (FSI), Dehradun, India.

India State of Forest Report, 2015. Forest Survey of India (FSI), Dehradun, India.

Ingole, B., Naik, S., Furtado, R., Ansari, Z. and Chatterji, A., 2002. Population characteristics of the mangrove clam *Polymesoda* (*Geloina*) *erosa* (Solander, 1786) in the Chorao mangrove, Goa. In: *National Conference on Coastal Agriculture*, 6-7 April 2002. Indian Council of Agricultural Research, Old.

Ingole, B.S., Kumari, L.K., Ansari, Z.A. and Parulekar, A.H., 1994. New record of mangrove clam *Geloina erosa* (Scholander, 1786) from the west coast of India. *J. Bombay Nat. Hist. Soc.*, 91(2): 338-339.

Jahan, M.S., Mannan, M.A. and Mandal, K.N., 1990. Intertidal molluscs of Sunderbans, Bangladesh. *Enviorn. Ecol.*, 8(2): 603-607.

Jeyabaskaran, R. and Ajmal Khan, S., 2007. Diversity of brachyuran crabs in Gulf of Mannar (Southest coast of India). In: *Biodiversity Conservation of Gulf of Mannar Biosphere Reserve*, (Eds.) S. Kannaiyan and K. Venkataraman. National Authority, Chennai, India. 68-82pp.

John Samuel, N. and Soundarapandian, P., 2009. Fishery potential of commercially important crab *Portunus sanguinolentus* (Herbst) along Parangipettai coast, south east coast of India. *International Journal of Animal and Veterinary Advances*, 1: 99-104.

John Samuel, N., Thirunavukkarasu, N. and Soundarapandian, P., 2004. Fishery potential of commercially important portunid crabs along Parangipettai coast. In: *Proceeding of Ocean Life Food and Medicine Expo*, p. 165-173.

Jones, C.G., Lawton, J.H. and Shachak, M., 1994. Organisms as ecosystem engineers. *Oikos*, 69, 373–386.

Kale, R.B. and Pawar, B.K., 2002. Survey of mangrove clam (*Gelonia proxima*) fishery in Ratnagiri district. In: *Proceedings of the National Seminar on Creeks, Estuaries and Mangroves, Pollution and Conservation*, (Ed.) Goldin Quadros. Zoology Department, Thane, 120-124 pp.

Kamalakkannan, P., 2015. Studies on habitat distribution and diversity of brachyuran crabs in Pondicherry mangrove environments, Southeast coast of India. *International Journal of Fisheries and Aquatic Studies (IJFAS)*, 2(4): 370-373.

Kannan, L., 1990. Mangroves-their importance and need for conservation. *Biol. Education*, 7(2): 93 - 102.

Kannupandi, T. and Pasupathi, K., 1994. Laboratory reared larval stages of a mangrove crab Sesarma edwardsi De Man 1887 (Decopods: Grapsidae). *Mahasagar*, 27: 105–115.

Kasinathan, R. and Shanmugam, A., 1988. Benthic macrofauna of the Pichavaram mangroves, South India. *J. Annamalai Univ.*, Part-B, 34: 109 - 119.

Kasinathan, R. and Shanmugam, A., 1985. Molluscan fauna of Pichavaram mangroves, Tamil Nadu. In: *Proc. Nat. Symp. Biol. Util. Cons. Mangroves*, (Ed.) L.J. Bhosale. 438-443 pp.

Kathiresan, K. and Qasim, S.Z., 2005. *Biodiversity of Mangrove Ecosystems*. Hindustan Publication Corporation, New Delhi, 251 pp.

Kathiresan, K., Rajendran, N., Palaniselvam, V. and Ramanathan, T., 2000. Macrofauna population in the Mangrove nursery of *Rhizophora apiculata* Blume. *Environ. and Ecol.*, 18(1): 230-232.

Kathiresan, K., Ramesh, M.X. and Venkatesan, V., 1994. Forest structure and prawn seeds in Pichavaram mangroves. *Environment and Ecology* 12, 465-468.

Kathiresan, K., 2000. A review of studies on Pichavaram mangroves, Southeast India. *Hydrobiol.*, 430: 185-205.

Kesavan, K., Babu, A., Ravi, V. and Rajagopal, S., 2009a. A checklist of malacofauna from Pondicherry mangroves. *AES Bioflux*, 1(1): 31-36.

Kesavan, K., Palpandi, C. and Shanmugam, A., 2009b. A checklist of malacofauna of the Vellar estuarine mangroves, India. *Journal of Threatened Taxa*, 1(7): 382-384.

Khade, S.N. and Mane, U.H., 2012a. Diversity of Bivalve and Gastropod Molluscs in Mangrove ecosystem from selected sites of Raigad district, Maharashtra, West coast of India. *Recent Research in Science and Technology*, 4(10): 16-20.

Khade, S.N. and Mane, U.H., 2012b. Diversity of Bivalve and Gastropod Molluscs from selected localities of Raigad district, Maharashtra, West coast of India. *World Journal of Science and Technology*, 2 (6): 35-41.

Khade S.N., and Mane U.H., 2012c. Diversity of edible Bivalve and Gastropod Molluscs from Ratnagiri, Maharashtra. *IJSPER*, 8: 1-4.

Khan, A.B., Saravanan, K.R. and Ilangovan, K., 2008. Floristic and macro faunal diversity of Pondicherry mangroves, South India. *Tropical Ecology* 49(1): 91-94.

Khan, S.A., Raffi, S.M. and Lyla, PS., 2005. Brachyuran crab diversity in natural Pichavaram and artificially developed mangroves Vellar estuary. *Curr. Sci.*, 88: 1316–1324.

Kondalarao, B. and Ramanamurty, K.V., 1988. Ecology of intertidal meiofauna of the Kakinada Bay (Gautami- Godvari estuarine system), East coast of India. *Indian Journal of Marine Sciences*, 17: 40-47.

Krishnan, T. and Kannupandi, T., 1987. Larval development of the mangrove crab *Sesarma bidens* (De Hann, l853) in the laboratory (Brachyura: Grapsidae: Sesarminae). *Mahasagar*, 20: 171–181.

Krishnan, T. and Kannupandi, T., 1989. Laboratory cultured zoeae and megalopa of the mangrove crab *Metaplax distincta* H. Milne Edwards, 1852 (Brachyura: Sesarminae). *J. Plankton Res.* 11: 633–648.

Kurian, C.V., 1984. Fauna of the mangrove swamps in Cochin estuary. In: Soepadmo, E., Rao, A.N., and Macintosh, D.J. (Eds.), *The Asian Symposium on Mangrove Environment Research and Management*, 25-29 Aug 1980. University of Malaya, Kuala Lumpur, 226- 230 pp.

Lewis, R.R., 2005. Ecological engineering for successful management and restoration of mangrove forests. *Ecol. Eng.*, 24: 403–418.

Lyla, P.S., Velvizhi and Khan, S.A., 1998. *Brackishwater Amphipods of Parangipettai Coast*. India: Annamalai University, 1–80pp.

Mahapatra, B.K., Pal, B.C., Chattopadhyay, P., Saha, D. and Datta, N.C., 1996. Some aspects of biology and fishery of mud crab, *Scylla serrata* (Forskal) with a note on its cultural aspects in the Sundarbans. *J. Mar. Biol. Ass. India*, 38(1-2): 8 - 14.

Majumder, S., Choudhary, A., Naidu, T.K. and Bandyopadhyay, S., 1996. A Reconnaissance Survey of Recent Benthic Foraminifera from the Mangrove-Estuarine Sector of Indian Sunderbans. *V.U.Jr. of Biological Sciences*, 2: 40-46.

Mandal, A.K. and Nandi, N.C., 1989. Fauna of Sundarban mangrove ecosystem, West Bengal, India. Zoological Survey of India.

Marakala, C., Rajesh, K.M., Ganapathi Naik, M. and Mridula, R.M., 2005. Ecology and biodiversity of macrofauna in a mangrove fringed lagoon, South-West coast of India. *Indian J. Fish.*, 52(3): 293-299.

Meher Homji, V.M., 1991. Mangroves of the Kaveri delta. In: *Coastal Zone Management*, (Eds.) R. Natarajan, S.N. Dwivedi and S. Ramachandran, 236–248 pp.

Mishra, P.K., Sahu, J.R. and Upadhyay, V.P., 2005. Species diversity in Bhitarkanika Mangove ecosystem in Odisha, India. *Lyonia, Journal of Ecology and Application*, 8(1).

Mishra, V., Quadros, G., Ullal, V., Gokhala, K.S. and Athalye, R.P., 1994. Sea anemone, *Acontiactis gokhaleae* as biofouler in the mangrove mudflats along Thane creek. *Mahasagar*, 27(1): 73-78.

Misra, A. and Choudhury, A., 1985. Polychaetous annelids from the mangrove swamps of Sunderbans, India. In: *Proc. Nat. Symp. Biol. Utili. Cons. Mangroves*, (Ed.) L.J. Bhosale, 448-452 pp.

Mitra, A., Basu, S. and Banerjee K., 2008. Seasonal variation in biochemical composition of edible oyster (*Saccostrea cucullata*) from Indian Sundarbans. *Fish. Technol. Soc. Fish. Technol. (India)*, 45(2): 209-216.

Mohan, P.C., Rao, R.G. and Dehairs, F., 1997. Role of Godavari mangroves (India) in the production and survival of prawn larvae. *Hydrobiologia*, 358: 317-320.

Mohan, P.M., Dhivya, P., Sachithanandam, V. and Ragavan, P., 2012. Distribution of mangrove meiofaunal composition in relation to organic carbon and carbonate in Port Blair, South Andaman, India. Tropical ecosystems structure, function and services, Published by Institute of Forest Genetics and Tree Breeding (IFGTB), Coimbatore, 47-54 pp.

Mohan, R., Selvam, V. and Azariah, J., 1995. Temporal distribution and abundance of shrimp post larvae and juveniles in the mangroves of Muthupet, Tamil Nadu, India. In: *Proc. Asia Pacific Symp. Mangrove Ecosystems*, (Eds.) W.Y. Shan and N.F.Y. Tam, 295(1-3): 183 - 191.

Mondal, N., Rajkumar, M., Sun, J., Kundu, S., Lyla, P.S. and Ajmal Khan, S., 2010. Biodiversity of brackishwater amphipods (Crustacean) in two estuaries, southeast coast of India. *Environmental Monitoring and Assessment*, 17: 471–486.

Mukherji, M.N. and Gokhale, K.S., 2002. Study of copper in polychaeta *Lycastis merukensis* from Thane creek. In: *Proceedings of the National Seminar on Creeks, Estuaries and Mangroves, Pollution and Conservation*, (Ed.) Goldin Quadros. Zoology Department, Thane, 201-202 pp.

Murty, A.S. and Balaparameswara Rao, M., 1977. Studies on the ecology of mollusks in a South Indian Mangrove swamp. *J. Moll. Stud.*, 43: 223–229.

Nandi, N.C., Dev Roy and M.K., Pal, S., 1996. Biometrical studies on the Mud crab, *Scylla serrata* Forskal from Sundarban, west Bengal. *Seafood Export J.*, 27(6): 17-22.

Nandi, S. and Choudhury, A., 1983. Qualitative studies on the benthic macrofauna of Sagar Island, intertidal zones, Sunderbans, India.

Nigam, R. and Chaturvedi, S.K., 2000. Foraminiferal study from Kharo creek, Kachchh (Gujarat), northwest coast of India. *Indian J. Mar. Sci.*, 29: 133-138.

Oswin, S.D., 1998. Biodiversity of the Muthupet mangroves, southeast coast of India. *Seshaiyana*. 6(1): 9-11.

Padmakumar, K.G., 1984. Ecology of a mangrove swamp near Juhu Beach, Bombay with reference to sewage pollution. *Ph.D. Thesis*, University of Bombay,Panchanadikar.

Parulekar, A.H. and Achuthankutty, C.T., 1993. Resource potential of juvenile marine prawns in the estuaries of Goa, 85 p.

Pasupathi, K. and Kannupandi, T., 1986. Laboratory reared larval stages of the mangrove grapsid crab, *Metopograpsus maculatus* H. Milne Edwards. *Mahasagar*, 19: 233–244.

Pasupathi, K. and Kannupandi, T., 1987. Laboratory culture of a mangrove crab *Sesarma pictum* De Haan, 1853 (Brachyura: Grap-sidae). *In*: *Proc. Fifth Indian Symposium of Invertebrate Reproduction*, (Ed.) S. Palanichamy, Palani, 294–307 pp.

Pasupathi, K. and Kannupandi, T., 1988a. The zoeae, megalopa and first crab of the mangrove crab *Metaplax elegans* De Man, cultured in the laboratory. *Mahasagar*, 145–160 pp.

Pasupathi, K. and Kannupandi, T., 1988b. The complete larval development of the mangrove ocypodid crab *Macrophthalmus depressus* Ruppell, 1830 (Brachyura: Macrophthalminae) reared in the laboratory. *J. Nat. Hist.* 22: 1533–1544.

Patra, K.C., Bhunia, A.B. and Mitra, A., 1988. Ecology of macrobenthos from a coastal zone of West Bengal, CMFRI Spec, Publi. 40: 45 pp.

Patra, K.C., Bhunia, A.B. and Mitra, A., 1990. Ecology of macrobenthos in a tidal creek and adjoining mangroves in West Bengal, India. *Envir. Ecol.*, 8(2): 539-547.

Pawar, P.R., 2012. Diversity of Decapod fauna from mangrove ecosystem of Uran (Raigad) Navi Mumbai, Maharashtra, West coast of India. *Indian J. Sci. Res.*, 3(1): 87-90.

Pereira, C., Rao, C.V. and Krishnan, S., 2002. Study of the dominant fauna of mangroves at Seven Bungalows beach, Versova, Mumbai - A preliminary study. *In*: *Proceedings of the National Seminar on Creeks, Estuaries and Mangroves*

- *Pollution and Conservation*, (Ed.) G. Quadros, 28-30 Nov 2002. VidyaPrasarak Mandal's B.N. Bandodkar College of Science, Thane, India, 196-200 pp.

Poddar, T.K., Chakrabory, S.K. and Choudhury, A., 1990. Littoral buttes of the sand and mudflats of Hoogly Estuary. *Annals of Entomology*, 8(1): 31-35.

Poovachiranom, S., 1992. Biological studies of the mud crab *Scylla serrata* (Forskal) of the mangrove ecosystem in the Andaman Sea. *Report of the Seminar on the Mud Crab Culture and Trade*, Surat Thani, Thailand.

Pradnya, D.B., Kusuma, Neelkantan and Kakati, V.S.,2011. Biodiversity of crabs in karwar mangrove environment west coast of India. *Recent Research in Science and Technology*. 3(4): 01-05.

Radhakrishna, Y. and Ganapati, P.N., 1969. Fauna of Kakinada Bay. Proc. Symp. Indian Ocean. *Bull. Nat. Inst. Sci. India*, 38 (2): 689-699.

Radhakrishna, Y. and Janakiram, K., 1975. The mangrove molluscs of Godavari and Krishna estuaries. In: *Recent Researches in Estuarine Biology*, (Ed.) R. Natarajan, 20-24 Jan 1972. Porto Novo. Hindustan Publishing Corporation, Delhi, India, 177-184 pp.

Raffi, S.M., Thomas, J.K., Lyla, P.S. and Ajmal Khan S., 2002. Species composition, distribution and abundance of mangrove crabs in an artificially created mangrove ecosystem. In: *Proceedings of the National Seminar on Marine and Coastal Ecosystems : Coral and Mangrove – Problems and Management Strategies*, (Eds.) J.K. Patterson Edward, A. Murugan and Jamila Patterson. SDMRI Res. Pub., 2: 29-36.

Rajagopalan, M.S., Pillai, C.S.G., Gopinathan, C.P., Selvaraj, G.S.D., Pillai, P.P., Aboobaker, P.M. and Kanagam, A., 1986. An appraisal of the biotic and abiotic factors of the mangrove ecosystem in the Cochin backwater, Kerala. In: *Proc. Symp. on Coastal Aquaculture, MBAI*, 4: 1068-1073.

Rajasekaran, R. and Fernando, O.J., 2012. Polychaetes of Andaman and Nicobar Islands. In: *Ecology of Faunal Communities on the Andaman and Nicobar Islands*, (Eds.) K. Venkataraman *et al*. 16: 340.

Rajendran, N., 1997. Studies of mangrove associated prawn seed resources of the Pichavaram, Southeast coast of India. *Ph.D Thesis*, Annamalai University, India, 135pp.

Rajendran, N. and Kathiresan, K., 1999. Seasonal occurrence of juvenile prawn and environmental factors in a Rhizophora mangal, southeast coast of India. *Hydrobiologia*, 394: 193-200.

Rajyalakshmi, T., 1991. The prawn fisheries of the Godavari estuarine system Kakinada Bay complex. *J. Inland Fish. Soc. India*, 23(2): 50 - 59.

Ramanamurty, K.V. and Kondala Rao, B., 1993. Studies on mangrove ecosystems of Godavari and Krishna estuaries Andhra Pradesh, India. UNESCO Curriculum workshop on Management of Mangrove Ecosystem and Coastal Protection. Andhra University, Visakhapatnam, 21pp.

Rambabu, A.V.S. and Prasad, B.V., Rao, M.B., 1987. Response of the mangrove mudsnail *Terebralia palustris* (Prosobranchia: Potamididae) to different substrata. *J. Mar. Biol. Ass. India*, 20(1 and 2): 140-143.

Rao, C.G., 1980. On the zoogeography of the interstitial meiofauna of the Andaman and Nicobar Islands, Indian Ocean. *Records of Zoological Survey of India*,77: 153-178.

Rao, G. C., 1986. Meiofauna of the mangrove sediments in South Andaman. *Journal of Andaman Science Association*, 2: 23–32.

Rao, G.C. and Misra, A., 1983. Meiofauna of Sagar Island. *Proceedings of Indian Academy of Sciences (Animal Science)* 92: 73-86.

Ravera, O., 2000. Ecological monitoring for water body management. Proceedings of Monitoring Tailor- Made III. *International Workshop on Information for Sustainable Water Management*, 157-167 pp.

Ravichandran, S. and Kannupandi, T., 2004. Biochemical changes in decomposing leaves and crabs of Pichavaram mangroves. *Biochem. Cell. Arch.*, 4(2): 79- 86.

Ravichandran, S. and Kannupandi, T., 2007. Biodiversity of crabs in Pichavaram mangrove environment. Zoological Survey of India. *National Symposium on Conservation and Valuation of Marine Biodiversity*, 331-340 pp.

Ravichandran, S. and Soundarapandian, P., Kannupandi, T., 2001. Zonation and distribution of crabs in Pichavaram mangrove swamp, southeast coast of India, *Indian Journal of Fish*, 48(2): 221-226.

Ravichandran, S. and Sylvester Fredrick, W., Ajmal Khan, S., Balasubramanian, T., 2011. Diversity of Mangrove Crabs in South and South East Asia. *Journal of Oceanography and Marine Environmental System*, 1 (1): 01-07.

Ravichandran, S. and Wilson, F.S., 2012. Variations in the crab diversity of the mangrove environment from Tamil Nadu, Southeast coast of India. *Proceedings of the International Conference 'Meeting on Mangrove ecology, functioning and Management - MMM3'*, Galle, Sri Lanka. VLIZ Special Publication, 57: 152 pp.

Ray, S. and Choudhury, A., 1985. Ecology of tabanid larvae and pupae (Diptera : Tabanidae) in Sunderbans mangrove ecosystem, Sagar Island. In: *Proc. Nat. Symp. Biol. Util, Cons. Mangroves* (Ed. L.G. Bhosale), 516-512.

Saha, A. and Jana, T.K., 1999. Biocalcification of aragonite by tellinid bivalve *Macoma birmanica* (Philippi) on the tidal mudflat in the Sundarban mangrove forest, north-east coast of India. *Indian J. Mar. Sci.*, 28: 404-407.

Saha, A., Mukhopadhyay, S.K. and Jana, T.K., 2000. Physico-chemical characterization of extrapallial fluid of a common tellinid bivalve *Macoma birmanica* (Philippi) in mudflats of Sundarbans mangrove, Bay of Bengal. *Indian J. Mar. Sci.*, 29: 158 - 164.

Sahu, S.C., Suresh, H.S., Murthy, I.K. and Ravindranath, N.H., 2015. Mangrove Area Assessment in India: Implications of Loss of Mangroves. *J Earth Sci.Clim. Change*, 6: 280.

Samidurai, K., Saravanakumar, A. and Kathiresan, K., 2012. Spatial and temporal distribution of macrobenthos in different mangrove ecosystems of Tamil Nadu coast, India. *Environmental Monitoring and Assessment*, 184: 4079–4096.

Santhakumaran, L.N., 1983. Incidence of marine wood-borers in mangroves in the vicinity of Panaji Coast, Goa. *Mahasagar*. 16(3): 299-307.

Saravanakumar, A., Sesh Serebiah, J., Thivakaran, GA. and Rajkumar, M., 2007. Benthic macrofaunal assemblage in the arid zone mangroves of Gulf of Kachchh-Gujarat. *J. Ocean Univ. of China*, 6: 303–309.

Saravanan, K.R., Ilangovan, K. and Khan, A.B., 2008. Floristic and macrofaunal diversity of Pondicherry mangroves, South India. *Tro Eco*, 49(1): 91–94.

Sarkar, S.K., Cabral, H., Chatterjee, M., Cardoso, I., Bhattacharya, A. K., Satpathy, K.K. and Alam, M. A., 2008. Biomonitoring of heavy metals using the bivalve molluscs in Sunderban mangrove wetland, northeast coast of Bay of Bengal (India): Possible risks to human health. *Clean-Soil, Air, Water*, 36(2): 187-194.

Sarkar, S.K., Bhattacharya, A., Giri, S., Bhattacharya, B., Sarkar, D., Nayak, D.C. and Chattopadhaya, A.K., 2005. Spatiotemporal variation in benthic polychaetes (Annelida) and relationships with environmental variables in a tropical estuary. *Wetlands Ecol. Manage.*, 13: 55–67

Sarma, A. L. N. and Wilsanand, V., 1994. Littoral meiofauna of Bhitarkanika mangroves of river Mahanadi systems. East coast of India. *Indian Journal of Marine Sciences*, 23: 221–224.

Satheesh Kumar, P. and Anisa Basheer Khan., 2013. The distribution and diversity of benthic macroinvertebrate fauna in Pondicherry mangroves, India. *Aquatic Biosystems*, 9: 15 pp

Satheeshkumar, P., 2012. Mangrove vegetation and community structure of brachyuran crabs as ecological indicators of Pondicherry coast, South east coast of India. *Iran J Fish Sci.*, 11(1): 184–203.

Sekar, V., Prithiviraj, N., Savarimuthu, A. and Rajasekaran, R., 2013. Macrofaunal assemblage on two mangrove ecosystems, southeast coast of India. *International Journal of Recent Scientific Research*. 4(5): 530- 535

Selvam, V., 2003. Environmental classification of mangrove wetlands of India. *Current Science*, 84(6): 757-765.

Sesh Serebiah, J., 2003. Studies on Benthic faunal assemblage on Mangrove environment of Jakhau, Gulf of Kachchh-Gujarat. *Ph.D Thesis*, Annamalai University, India, 147 pp.

Sethuramalingam, S. and Ajmal Khan, S., 1991. Brachyuran crabs of Parangipettai coast. CAS in Marine Biology publication, Annamalai University, Tamil Nadu, India, 92 pp.

Sethuramalingam, S. and Khan, S.A., 1991. Brachyuran crabs of Parangipettai coast. CAS in Marine Biology publication. Annamalai University, Tamil Nadu, India.

Shanmugam, A. and Kasinathan. R., 1987. Larval development of salt marsh snail *Melampus ceylonicus* (Ellobiidae: Pulmonata) from Pichavaram mangroves, Tamil Nadu. *J. Mar. Biol. Ass. India*, 29(1 and 2): 69-73.

Shanmugam, A. and Vairamani, S., 2009. Molluscs in Mangroves: A Case Study. *In*: *Training Course on Mangroves and Biodiversity*, 371-382 pp.

Shukla, M.L., Patel, B.K., Trivedi, J.N. and Vachhrajani, K.D., 2013. Brachyuran Crabs Diversity of Mahi and Dhadhar Estuaries, Gujarat, India. *Res. J. Marine Sci.*, 1(2): 8-11.

Sing, B.N. and Choudhury, A., 1984. Occurrence of an Enteropneust Hemichordate worm in the mangrove swamps of Sunderbans, India. *Bull. Zool. Surv. India*, 1(1-3): 1-4.

Sing, B.N. and Choudhury, A., 1992. A new record of *Protankyra similis* (Semper) (Holothurioidea: Apoidida) from Indian brackish water environment. *Obelia*, (18): 109-119.

Sing, B.N. and Choudhury, A., 1995 a. Studies on the distribution of Gastropoda (Mollusca) in a mangrove forest (Prentice Island) of Sunderbans, India. *J. Mar. Biol. Asso. India*, 37(1 and 2): 283-286.

Sing, B.N. and Choudhury, A., 1995 b. Seasonal distribution of *Saccoglossus* sp. in relation to abiotic parameters in the mangrove swamps of Sunderbans, West Bengal, India. *J. Mar. Biol. Asso. India*, 37(1 and 2): 143-146.

Sing, B.N. and Choudhury, A., 1985. Morphological excellence, feeding and breeding behaviour of *Saccoglossus* sp. (Hemichordata: Enteropneusta) from mangrove mudflats of Sunderbans, India. In: *Proc. Nat. Symp. Biol. Util. Cons. Mangroves* (Ed. L.J. Bhosale), 505-510 pp.

Singh, R., Ingole, B.S. and Nanajkar, M.R., 2009. The life cycle of the free-living marine nematode *Innocuonema tentabunda* De Man. *Nematol Medit.*, 37: 235-238.

Sinha, B. and Choudhury, A., 1988. On the occurrence of stylet-bearing nematodes associated with mangroves of Gangetic estuary, West Bengal, India. *Current Science*, 57(23): 1301-1302.

Sinha, B., Choudhury, A. and Barqri, Q.H., 1987. Studies on the nematodes from mangrove swamps of deltaic Sundarbans, West Bengal, India, III. *Anoplostoma macrospiculum*. sp. (Anoplostomatidae: Nematoda). *Current Science*, 56: 539-540.

Smith, T.J., 1987. Seed predation in relation to tree dominance and distribution in mangrove forests. *Ecology*, 68: 266–273.

Smith, T.J., Boto, K.G., Frusher, S.D. and Giddins, R.L., 1991. Keystone species and mangrove forest dynamics: the influence of burrowing by crabs on soil nutrient status and forest productivity.

Soundarapandian, P., John Samuel, N., Ravichandran, S. and Kannupandi, T., 2008. Biodiversity of crabs in Pichavaram Mangrove environment, South east coast of India. *International Journal of Zoological Research*, 4(2): 113-118.

Soundarapandian, P., Varadharajan, D. and Ravichandran, S., 2014.Mineral composition of edible crab *Podophthalmus vigil* Fabricius (Crustacea: Decapoda). *Arthropods*, 3(1): 20-26.

Srinivasulu, C., 2001. Mangrove clam *Geloina erosa* (Solander, 1786) from Coringa (Godavari) estuary: A new record for Andhra Pradesh. *J. Bombay Natural History Society*, 98(1): 144.

Subba Rao, N. V., Dey, A. and Baruna, S., 1983. Studies on the malacofauna of Muriganga estuary, Sunderbans, West Bengal, *Bull Zool Surv India*, 5(1): 47–56.

Subba rao, N.V. and Mukherjee, H.P., 1969. On a collection of molluscs from the Mahanadi estuary, Odisha. *Recent Researches in Estuarine Biology*. Ed. B. Natarajan, Marine Biological Association of India, Pt. I: 85-93.

Sundaravarman, K., Kathiresan, K., Saravanakumar, A. and Balasubramanian, T., 2012. Studies on a mangrove lagoon at Muthupet, southeast coast of India. *International Journal of Current Research*, 4(9): 15-22.

Sunil Kumar, R., 1998. A critique on the occurrence and distribution of macrozoobenthos in a traditional prawn field and adjacent mangroves in Cochin backwaters. *J. Mar. Biol. Ass. India*, 40(1 and 2) : 11-15.

Sunil Kumar, R., 1993a. Studies on the benthic fauna of the mangrove swamps of Cochin area. *Ph.D. Thesis*, Cochin University of Science and technology,Cochin.

Sunil Kumar, R., 1995a. Comparative study on the community structure and distributional ecology of benthos in two mangrove swamps of Cochin estuary. In: *Proc. Seventh. Kerala Sci. Congr.* (Ed. P.K. Iyengar), 121-122 pp.

Sunil Kumar, R., 1995b. Macrobenthos in the mangrove ecosystem of Cochin backwaters, Kerala (Southwest coast of India). *Indian J. Mar. Sci.*, 24: 56-61.

Sunil Kumar, R., 1995c. Animal-Sediment interaction with respect to the distribution pattern of polychaetous annelids in the mangrove ecosystem of Cochin backwater. *J. Zool. Soc. Kerala*, 5: 43-48.

Sunil Kumar, R., 1997. Vertical distribution and abundance of soil dwelling macro invertebrates in an estuarine mangrove biotope. *Indian J. Mar. Sci.*, 26: 26-30.

Sunil Kumar, R., 2001a. Ecological distribution and population structure of mud dwelling *Edwardsia* (Cnidaria: Actinaria) in a mangrove habitat of Cochin area, Kerala. *J. Bombay Natural History Society*, 98(2): 308 – 311.

Sunil Kumar, R., 2001b. Habitat selection, distribution and population density of *Sphaeroma terebrans* (crustacea: isopoda) in the littoral subsoil of a tropical estuarine mangrove ecosystem. *Zoos' Print Journal*, 16(6): 509-513.

Sunil Kumar, R., 2001c. A check list of polychaetous annelids from some Indian Mangroves. *Zoos' Print Journal*, 16: 439-441.

Sunil Kumar, R., 2002. Comparison of the macrofaunal benthic assemblages of an estuarine mangrove habitat and adjacent area in Cochin backwaters, Kerala. In: J.K. Patterson Edward, A. Murugan and Jamila Patterson (eds.), *Proceedings of the National Seminar on Marine and Coastal Ecosystems: Coral and Mangrove – Problems and Management Strategies.* SDMRI Res. Pub., 2: 37-41.

Sunil Kumar, R. and Antony, A., 1993. Influence of substratum on the polychaetous annelids in the mangrove swamps of Cochin area. In: *Environmental issues of Water Resources Projects* (Ed. P.N. Unni), p. 43 pp. (Abstract).

Sunil Kumar, R. and Antony, A., 1994b. Impact of environmental parameters on polychaetous annelids in the mangrove swamps of Cochin, South West Coast of India. *Indian J. Mar. Sci.,* 23: 137-142.

Sunil Kumar, R. and Antony, A., 1994c. *Paraheteromastus tenuis* Monro (Annelida: Polychaeta), an Indicator species of pollution in Cochin Backwater. In: *Proc. Third. Nat. Symp. Envir. with Special Emphasis on High Background Radiation Areas* (Eds.) N.B. Nair, C.D. Eapen, V.N. Bapat, S. Sadasivan and P. Gangadharan, 107-109 pp.

Sunil Kumar, R. and Antony. A., 1994a. Preliminary studies on the polychaete fauna of the mangrove areas of Cochin. In: *Proc. Sixth Kerala Sci. Congr.* (Ed.) R. Ravi Kumar, 74-77 pp.

Sunilkumar, R., 1999. New record of five amnelids (Class: Polychaeta) from the mangrove habitat of the southwest coast of India. *J. Mar. Biol. Ass. India*. 41(1 and 2): 116-118.

Suresh Gandhi, M., Jisha, K. and Rajeshwara Rao, N., 2014. Recent Benthic Foraminifera and its ecological condition along the surface samples of Pichavaram and Muthupet Mangroves, Tamil Nadu, East Coast of India. *Int. J. Curr. Res. Aca.Rev.*, 2(9): 252-259.

Suresh, M., Arularasan, S. and Ponnusamy, K., 2012. Distribution of molluscan fauna in the artificial mangroves of Pazhayar back water canal, Southeast Coast of India. *Advances in Applied Science Research*, 3(3): 1795-1798.

Thilagavathi, B., Das, B., Saravanakumar, A. and Raja, K., 2011. Benthic meiofaunal composition and community structure in the Sethukuda mangrove area and adjacent open sea, east coast of India. *Ocean Science Journal* 46: 63-72.

Thilagavathi, B., Varadharajan, D., Babu, A., Manoharan, J., Vijayalakshmi, S. and Balasubramanian, T., 2013, "Distribution and diversity of macrobenthos in different mangrove ecosystems of Tamil Naducoast, India," *Journal of Aquaculture Research and Development*, 4(6).

Thomas, J.K., Raffi, S.M., Ajmal Khan, S. and Kannan, L., 2002. Diversity, distribution and relative abundance of mangrove crabs in nullahs of Campbell Bay, Great Nicobar island. In: J.K. Patterson Edward, A. Murugan and Jamila Patterson (eds.), *Proceedings of the National Seminar on Marine and Coastal Ecosystems : Coral and Mangrove – Problems and Management Strategies.* SDMRI Res. Pub. 2: 42-47.

Tiwari, K.K., Das, A.K., Dev Roy, M.K. and Khan, T.N., 1980. On the wood-borers of mangroves of Andaman and Nicobar Islands, with a note on the gallery pattern of some insect borers. *Rec. Zool. Surv. India*, 77: 357-362.

Trivedi, J.N., Gadhavi, M.K. and Vachhrajani, K.D., 2012. Diversity and habitat preference of brachyuran crabs in Gulf of Kachchh, Gujarat, India. *Arthropods* 1(1): 13-23.

Untawale, A.G. and Parulekar, A.H., 1976. Some observations on the ecology of an estuarine mangrove of Goa. *Mahasagar*, 9 (1 and 2): 57-62.

Vasanth Kumar, B., Roopa, S.V. and Gangadhar, B.K., 2013. Distribution and Abundance Of Macrobenthos In Mangroves Ecosystem Of Kali Estuary, Karwar Karnataka. *Int. J. of Life Sciences*, 1(4): 313-316.

Venkanna, P., 1991. Present status of the estuarine flora of the rivers Godavari and Krishna. *J. Bombay. Nat. Hist. Soc.*, 88 (1): 47-54.

Vijayakumar, G. and Kannupandi, T., 1987. Laboratory-reared zoeae and megalopa of the mangrove crab *Sesarma brockii* de Man. *Ind. J. Fish*, 34: 133–144.

Vijayakumar, N., Sakthivel, D. and Anandan, V., 2009. Studies on mangroves, crustaceans and molluscs in the Ngaithittu estuary, Puducherry, South India. *J. Aquat. Biol.*, 24: 13-16.

Wilson, S.F. and Ravichandran. S., 2013. Diversity of Brachyuran Crabs in the Mangrove Environment of Tamil Nadu. *World Journal of Fish and Marine Sciences*, 5(4): 441-444.

Ye, Y., Weng, J., Lu, C.Y. and Chen, G.C., 2006. Mangrove biodiversity restoration. *Acta Ecol. Sin*. 26: 1243–1250.

Yeragi, S.G. and Yeragi, S.S., 2002. Studies on wood borers from the mangroves of Mithbav creek. In: Goldin Quadros (ed.), *Proceedings of the National Seminar on Creeks, Estuaries and Mangroves, Pollution and Conservation*. Zoology Department,Thane. 293-295pp.

Chapter 9
Water Bugs as Forage Base in World Fishery
P. Venkatesan

ABSTRACT

Water bugs are of great ecological importance in fresh water bodies, since they are not only polyphagous but also voracious predators on mosquito larvae, fish fingerlings and frog tadpoles. Among them, Belostomatid bugs play a prime role in the relationships among various strata of fresh water ecosystem. Diplonychus is the genus which is well represented globally and are known to feed on prey that are surface dwellers, column dwellers and also bottom dwellers. They contribute substantially to world fishery production by providing the foraging base for many fresh water fish populations.

Keywords: Water bugs, Forage base, Freshwater.

Introduction

The relationship between freshwater fishes and insects is an intimate one. Water bugs may be said, therefore, to contribute substantially to world fishery production by providing the forage base for many freshwater fish populations. Understanding the interactions between the fish and insect communities of freshwater habitats is fundamental both to the dynamics of natural ecosystems and to the management of aquatic resources for food production.

Three aspects of the interaction between fish and water bugs are functional adaptations of fish as predators, methods for calculating the number of insects eaten by fish in natural populations and the impact of fish predation on the abundance and composition of the insect community. Among various water bugs, the candidate group is belostomatid due to their uniqueness in various aspects that go in coincidence in fish production.

Characteristics of Belostomatid Bugs

Bugs belonging to the family Belostomatidae are called giant water bugs, electric light bugs or toe biters. Members of this family include dorso-ventrally flattened bugs with strong, thick forelegs which are used for grasping, mid and hind legs which are broad, flat and fringed with swimming hairs, and tarsi which are 2-3 segmented. Ocelli are absent. Antennae are usually four segmented and are concealed in pockets beneath the head. The most distinctive feature in the nymphs and adult is a pair of retractable, strap-like appendages at the abdominal apex which is used to obtain air. These air straps are homologous with the respiratory siphon of nepids, being derived from the eighth abdominal tergum. Each air strap bears a spiracle at the base.

Although belostomatids are strong swimmers, most are sedentary hunters, preferring to perch on submerged vegetation or any other support and wait for prey to swim by. The raptorial forelegs are readily held out to seize any moving object that passes nearby. The diet of belostomatids is varied and these bugs will suck dry anything that they can subdue.

Seven genera of belostomatids in the three sub-families- Lethocerinae, Horvathininae and Belostomatinae have been recognized by Lauck and Menke (1961). Approximately 150 species are known in the family, their size ranging from 1- to 110 mm.

Sub-family Belstomatinae

This subfamily comprises relatively small to large elongate oval bugs with short air straps. Due to sex role reversal, females glue their eggs on the dorsum of males who carry and care them until they hatch. Such encumbered males float horizontally on the water surface and engage in a variety of brooding behaviour that include keeping the eggs wet, frequently exposing them to atmospheric air and maintaining an intermittent flow of water over them by stroking them with the hind legs, when they move below the water surface. Eggs fail to develop, if kept submerged in water or are left in open air. Sub family Belostomatinae is represented by five genera which display great diversity of form and development.

Genus *Limnogeton* are restricted to the northern part of Africa. These are the most primitive members of the family, with legs poorly developed for swimming and the front femur poorly developed for grasping. They possess unspecialized, long and slender air straps.

Genus *Hydrocerius* (Water King) are restricted to Africa and Madagascar. These are large elongate bugs, whose body size ranges from 42-70 mm. They are also primitive bugs with a bifurcate phallobase and short arms.

Genus *Belostoma* is distinct in having dart mouth, that refers to the painful bite inflicted by these bugs). These are large bugs ranging from 9 – 50 mm in length. This is the largest genus in the sub-family and it includes nearly 60 species which are very diverse in form. A few species are found in N. America but most are distributed in South America. They are found along the vegetated margins of ponds and lakes and occasionally along banks of streams.

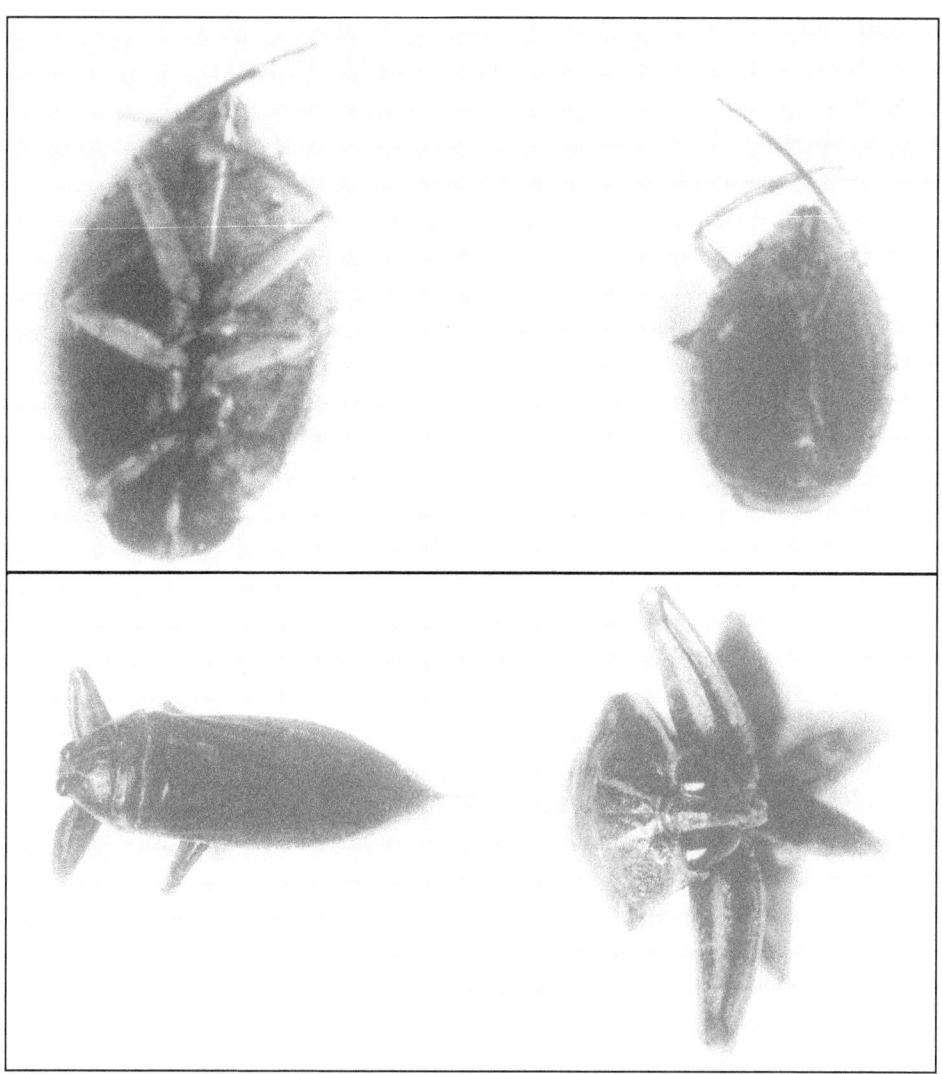

Figure 9.1: Water Bugs.

Genus *Abedus* (toe eater) refers to the voracious appetite exhibited by these bugs. The genus is restricted to southern part of United States, Mexico and Central America. Species of *Abedus* are almost invariably found in streams/running water, clinging on to rocks and aquatic vegetation. Although they occur in the same ecological niche as *Belostoma*, both the genera are rarely found to occur together in the same body of water because of their different habitat preference. These are medium sized bugs (13-40 mm) in length. *A. herberti* secretes a mixture containing four pregnanes which are deterrent to fish.

Genus *Diplonychus* is characterized by the presence of 2 claws in the front tarsus. These bugs are very common and are collected in large numbers in quiet bodies of fresh water with dense vegetation of water hyacinth and twigs of submerged *Hydrilla* plants. They prefer habitat with vegetation and hide themselves among their adventitious roots. There are nearly 20 species under this genus. They measure 11 to 30 mm. The Indian species - D. *indicus* Venk. and Rao is ocharaceous brown and range in size from 13.5 – 18.5mm. Legs are covered with minute spinules. Claws are small and equal in size.

They exhibit distinct migratory flight periodicity correlated with the lunar cycle as in *D. nepoides*

Fabr. *D. indicus* has been much studied in southern India.

Distribution

Members of the genus Diplonychus are cosmopolitan in distribution. They occupy a variety of fresh water habitats- both lentic and lotic.

Habitat Structure

The substratum determines to a large extent the microenvironmental conditions, under which the insects live and it affects their growth and survival. The vegetation structure and its association with the species distribution of water striders determine the ecological separation of gerrid species, whereas the stone structure and nature of pebbles influence the occurrence and abundance of naucorids.

Abiotic Factors

Abiotic factors such as temperature, rainfall, alkalinity and dissolved oxygen of the habitat exert an influence over the population of aquatic bugs. Corixids have been found in habitats of several types, occasionally even in saline water. The microenvironmental condition includes water temperature which affects the time of emergence of aquatic insects. Such an impact affects the population size, distribution and dispersal. In gerrids, the developmental response to temperature formed a basic aspect of adaptive strategies which varied within a species depending upon local climatic conditions.

Significance of the Water Bugs

Predation is generally regarded as an important factor in structuring aquatic communities. A negative correlation between the presence of the predator and the potential prey species has been used as an evidence of predation, although such a correlation in the absence of experimental manipulation is due to other causes. The impact of predators on the stability and the structure of their prey population varies from one species to another. Though most insect orders with aquatic life stages contain species that are predaceous, the predatory activities of some aquatic insects have a strong influence in shaping the evolution of aquatic insect communities. The study of predator-prey interactions are effectively accomplished by use of behavioral experimentation.

Based on the mode of feeding, aquatic insect predators can be classified into four general categories:

1. Active searchers of prey (*e.g.*) some odonates, plecopterans, trichopterans, dipterans, hemipterans and coleopterans
2. Ambush predators or sit and wait predators, who wait for prey to encounter them (*e.g.*) some odonates, trichopterans, dipterans, hemipterans and coleopterans.
3. Engulfers which include predators that swallow whole prey or pieces of prey *e.g.* plecopterans, some odonates, trichopterans and dipterans.
4. Piercers, which include the predatory insects that penetrate their prey, inject toxins and proteolytic or paralytic enzymes and suck out part or all the contents of the prey body. *e.g.* Some coleopterans, dipterans and hemipterans.

The notion that predators can have profound effect on prey communities is one of the paradigms of ecology. The ecological significance of these effects has only recently begun to be quantitatively studied. Aquatic insects also affect the dynamics of their prey. Activity time, habitat use and diet are the main parameters that determine predator-prey interactions including the effect of food quality on physiology and behavior.

Aquatic and semi-aquatic bugs are the most important predators on other aquatic animals as they are polyphagous and carnivorous predators. The large size of some aquatic heteropterans often make them top predators in fresh water systems lacking vertebrates, and their contribution to production and energy flow can be substantial in these systems. The role of aquatic heteropterans in the trophic structure of these aquatic ecosystems has been largely unexplored.

The development of predator-prey interactions has been viewed as a coevolutionary "arms race". Predator species evolve more efficient modes of prey capture and prey species encounter with defensive adaptations. The role of *Diplonychus rusticus* as an efficient predator on the larvae of vector mosquitoes in the evolution of morphology, distribution and behavior of aquatic insects has been investigated for the past two decades and more by me with my school of students who have applied various parameters in order to evaluate its predatory success that are as follows:

Taxonomy

Diplonychus indicus n.sp.

Specimens collected from Chetpet pond, Madras, India were identified to be a new species-

Diplonychus indicus based on the following features

☆ Head length more than the width between the eyes;

☆ The posterolateral margin of the respiratory strap of male with a cluster of setal tufts or spikes.

☆ Air straps not meeting at the tip of aedeagus.

Diplonychus indicus n.sp. is closely related to *D. rusticus* (Fabr.) in having hemelytra shorter than the total body length, anterior claws short and the presence of tuft of setae on the lateral sides of the basal plate in the female genitalia. It differs from *D. rusticus* in head length being more than the width between the eyes, cluster of setae forming the spike being present on the posterolateral margins of the respiratory straps, air straps not meeting at the tip of aedeagus, the pubescence of ventrolateral tergites from III to VII reaching the external margin on the segment III only and the membrane of the hemelytra with a patch of spinules at the bottom. This species is inseparable from *Diplonychus rusticus* (Fabricius) and therefore must fall as a junior synonym (New synonymy).

Population Dynamics and Abiotic Factors

The brooding behaviour of *Diplonychus indicus* is compared with that of other genera of the subfamily Belostomatinae. Surfacing is a common brooding behaviour in males of all genera. In *Diplonychus*, subsurface wandering, sub-surface resting and brood paddling are also observed. Significance of surfacing in belostomatids is attributed to the physiological need to acquire oxygen from the atmosphere and to trap air in a sub-alar reservoir. Infrequent surfacing by encumbered males may be to avoid their predators and other egg feeders, or to avoid loosening the egg pad and to minimize their oxygen requirement. This feature is compensated by sufficient storage reserve. Thus brooding behaviour in general may reflect the necessity to keep the eggs away from predators and parasites.

In female *D. indicus*, two pairs of accessory sex glands are present. Such glands are noted to be elaborate consisting of hundreds of follicles. Increased number of such follicles and their secretion may be of great importance in gluing the eggs on the back of the male bug. The belostomatid bugs exhibited male parental care in order to keep the eggs viable so as to cause an increased rate of hatching success and to multiply the population size in fresh water habitat. This brood risk would have necessitated the belostomatid bug to deviate as a specialized offshoot from the main line of family Belostomatidae.

Male dominates even in summer when population was low and environmental factors-temperature and salinity were high, suggesting a higher tolerance and resistance of males compared to females for adverse environmental conditions. The maximum survival temperature differs between starved and well fed insects, suggesting that the maximum survival temperature is dependent on the nutritional status of the insect. Utilisation of protein in the insect during starvation and thermal stress strongly argues for the existence of a protective mechanism for the detoxification and elimination of nitrogenous wastes. The tolerance and resistance met with in males and females in relation to environmental stresses by the utilisation of protein as the energy reserve, reveals their adaptability. Such a protective mechanism during adverse environmental conditions may be of high value to promote the predatory efficiency on mosquito larvae.

During development, the 3 day old eggs are not susceptible to the effect of changes in temperature or humidity when compared to the 32 day old eggs. This may be due to the utilisation of storage reserve to promote their resistance capacity to varying temperature and relative humidity. Thus, this intake of water in older eggs may coincide with the hydrolysis of old protein, the products of which may be utilised for energy requirements and for the synthesis of tissues of the embryo.

Availability of food highly influence the reproductive strategy of the insects. Under well fed conditions, male bugs showed 63 and 15.2 per cent increase in the amount of protein present in the muscle and testis respectively. Female showed 32 and 22 per cent increase of protein in their muscle and ovary. Under starvation, the amount of protein in the muscle and gonads of both the sexes decreased drastically. Protein decrease was 7.4 and 14 per cent in the muscle and testis of the male respectively. In the female, the respective decrease was 7.6 and 10.03 per cent protein in the muscle and ovary. Since the maturation of gonads need large amount of protein, thoracic muscles of the insect of the pre-reproductive phase contain only small amount of protein. During the reproductive phase, the male indulges in enforced copulation so that thoracic muscles contain higher amount of protein. After copulation, the females store the sperms in their spermatheca temporarily. Since the spermatozoan itself is a protein, the gonads of the males contain more protein than the muscle. In the post-reproductive phase, the amount of protein present in the muscle and gonads of both the sexes decrease slowly. In general, females are more active than the males resulting in more protein in the muscle and gonads of females than the males. As the male requires more amount of protein during encumberance, the thoracic muscles of encumbered male contained more protein than the gonads.

Among the post-embryonic stages, the fourth and fifth nymphal stages occurred throughout the year. But the first and second nymphal stages were poorly represented. Sex ratio revealed male dominance in the population. The population of this water bug varied considerably even when the pH value showed a marginal variation from 7.03 to 7.57. The chlorinity of water samples collected from the study site, Cooum river, showed a wide range of fluctuations from 452 ppm in July 1991 to 4068 ppm in June 1991 and during the remaining period from 565 to 2488 ppm. The bug probably adapted to such variations in salinity.

Stress Tolerance

Respiratory efficiency of belostomatid bug *Diplonycus indicus* was determined at different temperature and salinities of water media. The temperature influenced the rate of oxygen consumption of *D. indicus* which was higher at 34 C in well water than the oxygen consumption at different temperatures of water media. Similarly, salinity also influenced the rate of oxygen consumption significantly. Among the different salinities, maximum oxygen consumption of the water bug was received in well water.

Members of *Diplonychus* have extra-tracheal air store, carrying a bubble of air down into the water when it dives. Its respiratory behavior is under the stress of environmental changes. In 2 per cent concentration of the biolarvicide-Bacticide, the bugs showed good tolerance, in spite of their disturbed respiratory behavior, whereas in 4 per cent concentration of the biolarvicide- Bacticide, the bug was able to show tolerance only to certain extent.

The concentration of the pesticides- Abate and Solfac responsible for the 50 per cent mortality of the bugs was more or less equal to the concentration applied against mosquito population in the field. The female bugs were more tolerant to the pesticides than the males.

Diplonychus rusticus tolerated upto 300ug/l of the pyrethroid pesticide decamethrin (K-Othrine) than the fish- *Gambusia* which tolerated only upto 200ug/l.

Studies on the impact of pesticides such as Baytex, K-othrine, Dimilin and their effect on AchE, SDH, LDH enzyme activity of the bug showed that AchE activity in male and female exposed to Baytex was similar up to 24 hour. After that, a sharp decline in the activity was observed up to 96 hour. Baytex and K-othrine inhibited SDH activity up to 96 hour than Dimilin. The bug exposed to Dimilin maintained uniform rate of LDH activity compared to control. But enzyme activity of the bug decreased in baytex and K-othrine treated water.

Among different tissues of male bug, the reproductive tissue was less sensitive to all pesticides except dimilin. But in female, the ovary is highly affected and dimilin decreased its protein. During pesticidal exposure, much of the energy must have been used to compensate the stress. Hence the depletion of the protein content is observed. Among the three pesticides, dimilin is highly toxic to the mosquito larvae. However, the present study reveals that dimilin is the safest pesticide for non target organisms like water bugs.

The decline in protein level of the gonad of the female bug may be attributed to the influence of the pesticide on reproductive protein. These would result in higher mortality of female than males. K-Othrine treated bugs had a uniform pattern of total free sugar in various tissues as that of control suggesting thereby the lesser influence of pesticide on glycogenolysis and glycolysis. Depletion of total free sugar in relation to period of exposure apparently shows the need to meet increased energy requirements in a stress condition. Such a pattern is not shown in Dimilin or Baytex treated bugs. Besides protein and carbohydrates, the storage reservoir of energy in the form of lipid may also throw light on the pesticidal stress in water bugs. The result on lipid value reveals that the lipid level decreases in relation to the study period in the gonad and the muscle of the female. Probably, such insects utilize the energy resource stored in large quantities as lipid.

Results on physico-chemical characteristics of the water sample treated with pesticide in which bugs were exposed strongly suggested distinct variations in the water quality. Pesticides used were known to exhibit their persistence, solubility and absorptivity properties. The most strongly absorbed insecticide among the three used was noted to be synthetic pyrethroid.

Insecticidal resistance management is of central importance to an increasing number of vector control programme. Upper confidence limit of male *D. indicus* exhibited higher rate of tolerance.

Male bugs are highly resistant to the pesticide than the females towards the pyrethroid-K-othrine. The application of male *D. indicus* in insecticide treated water bodies as the bioagent may be possible due to their tolerance capacity to decamethrin.

Prey-Predator Relationship

Belostomatid bugs display great variety in form and development. Of all the known five genera, many species of *Diplonychus* contain brachypterous forms that enhance flight activity during night hours in monsoon periods. The presence of a two –segmented fore tarsus and well developed phallus make them successful predators and male brooders.

The bug is polyphagous in feeding. Though it feeds on mosquito larvae, naiads. dragon fly., frog tadpole, worms, snails and other waster insects, it shows preferential selection of mosquito larvae due to their size, shape, texture of the body and body content.

Predator's Performance

With reference to anopheline larvae, the number of prey killed by *D. indicus* increased with increasing prey density. The prey death rate is positively accelerated in the fourth nymphal stage preying on the second and fourth prey ages suggesting that larger predators often search faster and make a higher proportion of successful attack than the smaller predators when exposed to the same size of the prey. Further, the attack rate seems to be correspondingly high with the first and fourth nymphal stages than those with the intermediate age groups of the prey species. Such an observation is not in conformity with the report on the predation of *D. indicus* preying on *Culex* and *Aedes* larvae. Normally they remain clung to twigs within the water surface. Periodically, they come to the surface for trapping air bubbles. Such occasional visits may also permit the function of predation wherein the smaller predator instars kill and eat the prey but the larger nymphal stages exhibit only marked successful attack. Such a phenomenon of prey switching in these biological agents is of importance in the control of two types of mosquitoes – larvae of *Culex* and *Anopheles* living in one and the same habitat.

Role of Predator's Sex

Among the sexes, female killed more number of mosquito larvae than that of male in one day as well as in twenty day treatments. In 20 day treatment, male predator maintained uniform predation on all days except 9th and 15th day. Whereas, predation by female fluctuated widely. From the predatory rate, energy content was determined and the energy uptake of male bug from mosquito larvae was lower than the energy uptake by female bug.

Ovipositional Frequency

Study on ovipositional frequency of *D. indicus* for a fifteen day period reveals that *Culex* fed ones exhibit phenomenal rate of oviposition than those fed with *Chironomus* larvae, suggesting thereby that the nutritional value of the prey equates the ovipositional frequency through a flow of nutrients in the haemolymph to the ovary for oocyte development.The wild adults might have had optimal predation in water bodies and accumulated sufficient nutrients for the development of oocytes. It may be interpreted that *D. indicus* is a continuous feeder and the dearth of ovipositional site is very rare, since it has the adaptability of selecting the male's dorsum for oviposition with which it has last mated. The intimate relationship of increased gonad index with high count of encumbered males further confirms the phenomena of intraspecific adjustment.

D. indicus is capable of ovipositioning the eggs throughout the year. The eggs are kept viable by their unique parental behaviour. Eggs are numerous in an egg pad and they take about 9 days to emerge out as the first nymphs which take 50 days to become adults. Hence en masse rearing, the phenomenon that challenges other bioagents is possible here.

Mass eclosion of these eggs can be effective in attaining a high population density of the predator. Larvae are voracious predators when compared to adults and are not likely to leave the water body they are placed in. High attack rate and lower handling time of the prey were observed as the density of the prey increased.

Predator's Niche

The habitat of the bug population suitably overlaps that of mosquito larvae. The water bug stills shows aerial respiration by rising to the water surface to trap an air bubble. During this behaviour, they attack the surface dwelling *Anopheles* larvae; when they move in water for other activities, they depend on the *Culex* and *Aedes* larvae that exhibit wriggling movements, thus effecting a consolidation of predation on all types of mosquito larvae.

Prey species were taken in combination of twos as 75:25, 50:50, 25:75, 20:30, 30:20, and 25: 25 with larvae of *Culex* and *Chironomus*, fish fingerlings and frog tadpoles. Prey death rate of frog tadpole was the least as 1.0 in combination with fish fingerlings (50:50) by fifth predator stage. The highest predatory efficiency was with *Culex* larvae in fifth predator stage killing 41.4 in combination with *Culex* larvae in the ratio of 50:50. Relatively fifth predator stage showed the highest predatory efficiency than the others with all prey species combinations.All predator stages preferred *Culex* larvae than the other prey species.

The impact of predator on the stability and structure of their prey population varies from one species to another. Two hypotheses could explain in water bugs the size effect on the number of items killed and on the amount of dry matter ingested.

1. Predation rates decrease with increasing prey size because of increasing difficulty in capturing, subduing and handling a prey item.
2. Larger prey provide more energy per item so that a predator requires less items to become satiated.

Predator's Combinations

Among different combinations of predators- 1 male + 1 female killed more prey at the density of 150 and 200. Similar trend was observed in 2 male + 2 female and 3 male + 3 female combinations in one hour and one day exposures. In one day exposure, 3 male + 3 female killed more prey than the others, when the prey density was increased from 50 to 200. An increasing rate of predation by various combination of sexes of *D. indicus* reveals that they effectively predator *Culex* larvae at higher prey densities. Various combinations of sexes could be practiced in fresh waters where the mosquito breed enormously.

Functional Response

Five predator instars of the water bug *Diplonychus indicus* when exposed to four size classes of two different prey species of larval mosquitoes *Aedes aegypti* and *Culex fatigans* at varying densities showed the type II functional response with the increasing attack rate corresponding to decreasing handling time. Largest predator instars (V) killed maximum number of smallest prey (1st) and vice versa of both prey species. Larger predator instars showed more successful attack and shorter handling time than smaller predator instars. However, changes in functional response were observed in the instars II and III of *D. indicus* preying on 2nd and 3rd sizes classes of *Aedes aegypti* and *Culex fatigans*.

With reference to functional response of the adult predator, *Diplonychus indicus* to various combinations of different size classes of larvae of *Culex quinquefasciatus* as prey revealed that:

1. Normal as well as encumbered males showed a preferential selection for the larger rather than the small sized prey,
2. Time interval in attacking the prey was not constant with all prey size classes and
3. The predator killed more prey than those from which it sucked the prey contents.

Prey

Prey Preference

Male showed an increased rate of predation of 4th instar of *Culex* larvae as well as 2nd and 3rd instars of *Anopheles* larvae. Females predated at higher rates on 2nbd instar of *Culex* larvae and 1st instar of *Anopheles* larvae. Egg carrying males killed more of 1st instar of *Culex* larvae, 4th instar of *Anopheles* larvae and *Chironomus* larvae. Effect of prey density: when the predators were exposed to varying prey densities of all instars of *Culex* larvae, the rate of predation was higher for males and egg carrying males at the density of 64 and 128 respectively and for females at the density of 32. The predators showed a preference for 4th instar larvae of both the prey species. Irrespective of the period of starvation, males killed more of 4th instars, females the 2nd and egg carrying males the 1st instar of *Culex* larvae. In general, males killed relatively more number of *Culex* larvae than other prey. Irrespective of the sex and state of predator, the attack rate is inversely proportional to handling time *i.e.* attack rate decreases when handling time increases. Minimum attack rate of male was observed in presence of size 2 prey and minimum attack rate of female occurred in presence of size 1 prey. Maximum attack rate of both male and female was observed in the presence of size 4 prey. In egg carrying males, the handling time was maximum when the attack rate was also relatively higher.

Prey Density

The density of mosquito larvae in the breeding sites is also benefitial *i.e.*, when the density is predominantly high, the predatory performance is often disturbed by the moving prey population. This interruption results in the incomplete handling of the prey by the bug after paralysing it. The bug thus becomes restless, till it succeeds in bringing down the density of mosquito larvae by extensive attack and then attains satiation in feeding. Where mosquito menace is acute, the introduction of the bug becomes purposeful.

Members of predator's population exhibit a high rate of competition at high prey densities leading to higher attack rate and low handling time. The development of *Diplonychus rusticus* in the field proves the survival tolerance of the nymphal stages. Wherein nymphs predate vigorously on mosquito larvae to enhance moulting from nymph to adult. The efficiency of this bioagent depends on the time of its introduction into the habitat and its efficiency and adaptability are directly related to the habitat structure.

Prey Quality

Quantitative analyses of food utilization in *Sphaerodema*(= *Diplonychus*) *annulatum* by providing larvae of *Culex* mosquito as food are made. The third nymphal instar shows potential conversion efficiency than the first instar and that the conversion efficiency is maintained in subsequent instars. The rate of conversion is the highest in the second instar. The significant increase in the morphometry of the third nymphal instar from that of the second instar is reflected in the conversion efficiency and the rate of conversion of food in that stage. The duration of each nymphal instar of the bug fed with mosquito larvae in the laboratory conditions is higher than that of the nymphal instars fed with other insects under identical conditions.

In *Diplonychus indicus*, the prey quality affects the rate of predation as well as food utilization. Allometric growth of various body parts and the longevity of each nymphal instar of the bug showed distinct variation, when exposed to individual prey item- *Culex, Anopheles* and *Aedes* larvae and fish fingerlings. The high conversion rate of second nymphal instars that fed on culicine larvae is reflected in the increased body size of the penultimate instars of considerable interest. Also, relationship between conversion rate and allometric growth including total body length was noted.

Plant Extracts

Petroleum ether crude extracts of *Abutilon indicum* (leaves), *Citrullus coloycynthis* (whole plant) and acetone extract of *Ficus racemosa* (bark) exhibited high toxic effect on mosquito larvae. The active compounds were identified by TLC plate and structural characterization of pure compounds have been elucidated by various spectral analysis as B sitosterrel glauanol-3-acetate and glauanol-3-one. They were reported as effective larvicidal agents of mosquitoes.

Thus one may utilize the concept of integrated control of mosquitoes with this water bug as bioagent given in alternation with pesticides. The optimization of this bug as a bioperiodic element may well be enhanced by mapping and exploiting an anticipated broad ultradian, circadian and infradian time structure, of which the available results document at least a circasemidian component.

Suitability

Some reasons for the suitability of the aquatic bug- *D. indicus* as a bioagent follow:

1. It is a highly efficient predator both on anopheline and culicine larvae.
2. It predates anopheline larvae while surfacing and culicine larvae while wandering in the water column.
3. En masse rearing is possible due to size, using an adequate supply of food
4. Mass eclosion of the eggs from the egg pad on the dorsum of encumbered male is effective in attaining a high population density of the predator.
5. All immature stages and adults kill mosquito larvae
6. Sexual dimorphism is distinct morphologically which facilitate sorting
7. Male dominance in the population, paternal care of eggs and the males carry the eggs on their backs are benefitial factors.
8. Fluctuation in the life span of nymphal development relates to kind of mosquito larvae devoured with an apparent preference for disease spreading vectors of mosquitoes.
9. Analysis on aggressivity and blood sugar supports the time effect and the occurrence of intraspecific adjustment.
10. The timing of aggression may depend upon factors beyond the sugar that may be mobilized from fat reserves.
11. Their adaptability to the insecticide treated water is noteworthy.

Conclusion

The analysis of predatory behavior and estimation of predation rate suggest certain important consequences for the insect prey community provided the fish predators are sufficiently abundant. Without even having looked in the stomach of a fish taken from a natural population, one could reasonably predict that predation pressure would be greatest on the larger components of the aquatic insect community, on those most conspicuous by virtue of their activity pattern or habitat as noted above, and on those least protected by cases or tubes. These components of the community would decline in abundance relative to other components under the influence of fish predation. The following two types of studies are warranted to confirm the effectiveness of fish predation, namely:

1. Studies in which natural variation in abundance of fish is compared with the occurrence of particular components of the insect community

2. Studies in which the abundance of fish predators or their access to the insect community is artificially manipulated.

References

Anjali, N.U. and Venkatesan, P., 2009. Role of co-existence of water bugs on their bio-control potential with *Culex* larvae as prey. *Hexapoda*, 16(2): 103–109.

Arivoli, S., Chandramohan, G. and Venkatesan, P., 2005. Influence of abiotic factors on seasonal fluctuation of population of waterstrider *Tenagogonus fluviorum* (Fabricius) in a permanent pond. *J. Natcon.* 17: 363–370.

Hynes, H.B.N., 1984. The relationship between the taxonomy and ecology of aquatic insects. pp. 9 – 23. *In*: *Ecology of Aquatic Insects*, (Eds.) Resh, V.H. and D.M. Rosenberg. Praeger, New York, p. 625.

Lauck, D.R. and Menke, A.S. 1961. The higher classification of Belostomatidae (Hemiptera) *Ann. Ent. Soc.*, 56: 644–657.

Polhemus, J.T., 1978. Aquatic and semiaquatic Hemiptera, pp 119 – 133. *In*: *An introduction to aquatic insects of North America*, (Eds.) R.W. Meritt and K.W. Cummins. Kendall/Hunt Publishing Co. Iowa, p 441.

Schuh, R.T. and Slater, J.A., 1995. True bugs of the world (Hemiptera: Heteroptera): classification and natural history. Cornell University Press, Ithaca.

Venkatesan, P. and Rao, T.K.R., 1981. Description of a new species and a key to Indian species of Belostomatidae. *J. Bombay Nat. History*, 77: 299–303.

Venkatesan, P., Guillareme, G.C. and Halberg, F., 1986. Modelling prey predator cycles using hemipteran predators of mosquito larvae for reading world wide mosquito borne disease incidence. *Chronobiologia*, 13(4): 351–354.

Venkatesan, P. and Cloarec, A. , 1988. Density dependent prey selection in *Ilyocoris* (Naucoridae). *Aquatic Insects*, 10: 105-116.

Venkatesan, P., 2000. Giant water bugs (Belostomatidae) pp. 577–582. *In*: *Heteroptera of Economic Importance*, (Eds.) C. W. Schaefer and A.R. Panizzi, CRC Press, New York, p 828.

Venkatesan, P. and Betsy, S. 2005. Biodiversity of water bugs in a fresh water ecosystem. *J. Ecotoxicol. Environ. Monit.*, 15(1): 27–32.

Chapter 10

Induced Spawning and Seed Production of *Pangasianodon hypophthalmus* in Three different Types of Hatcheries under Agro-climatic Conditions of Raipur (Chhattisgarh), India

C.S. Chaturvedi, Rashmi S. Ambulkar,
R.K. Singh and A.K. Pandey

ABSTRACT

Sutchi catfish (Pangasianodon hypophthalmus), one of the important riverine catfish, has great potential for freshwater aquaculture. This species, a native of Mekong river of Thailand, is highly fecund, seasonal spawner and breeds once in a year during monsoon season in flooded rivers. Males attain maturity after second years while females at the end of third year. In this species, the females were found to be larger than males. In the present experiment, 15 females and 15 males (15 sets; 1:1 sex ratio) were selected and induced bred by varying doses of pituitary gland extract (PGE) depending upon the physiological status of fishes. Male brooders were also given PGE at the time of second injection to females. Fertilization of eggs varied from 30-80 per cent. After spawning, hatching of eggs (5 sets) was studied in circular hatchery, 5 sets in vertical jar hatchery and 5 sets in Thailand model hatchery. After incubation of fertilized eggs from vertical jar hatchery 17,30,400 hatchlings, in circular hatchery 14,08,000 hatchlings while in Thailand model hatchery, it gave 3,17,500 hatchlings only. The hatching percentage were observed 60 per cent in vertical jar hatchery, 50 per cent in circular hatchery and 30 per cent in Thailand model hatchery. After 2 days, yolk absorption was observed and from the three types of hatcheries - 9,88,740 fry were realised from vertical jar hatchery, 7,40,000 from circular hatchery and 65,550 from Thailand model hatchery. After rearing the fry in nursery ponds for 25-30 days, 5,93,244 fingerlings from Krundh-Liey Fish Farm, 4,22,400 fingerlings from State Fisheries Farm, Raipur and 65,550 from Deepak Mandal Fish Farm (Thailand model hatchery) (total 10,81,144) were obtained paving the way for mass seed production of the commercially important catfish under agro-climatic conditions of Raipur (Chhattisgarh).

Keywords: Induced spawning, Seed production, Hatching systems, Raipur, India.

Introduction

Pangasianodon hypophthalmus, commonly known as striped (sutchi, iridescent shark) catfish, fetches high price in markets. Culture of this species is growing day-by-day in Bangladesh (Rahman *et al.*, 2006; Ahmed and Hasan, 2007; Ahmed *et al.*, 2013), Indonesia (Griffith *et al.*, 2010), India (Lakra and Singh, 2010; Singh and Lakra, 2012; Kumar *et al.*, 2013) and Vietnam (Phan *et al.*, 2009; Bui *et al.*, 2010). Vietnam is the top producer and exporter of *P. hypophthalmus* (Phan *et al.*, 2009; Bui *et al.*, 2010). Contribution of the Indian major carps in Indian aquaculture is more as compared to those of catfishes and in Andhra Pradesh itself, major carps contribute about 85 per cent of the total freshwater fish production while catfishes and murrels show the next to them (Laxmappa, 2004). *P. hypophthalmus* is native of river Mekong Basin and Chao Phraya river in Thailand, Cambodia and Vietnam. It has been introduced in Singapore, Philippines, Taiwan, Malaysia, China, Myanmar, Bangladesh, Nepal and India. In India, it was brought in West Bengal through Bangladesh during 1997 (Mukai, 2011). Initially, its culture was carried out in Andhra Pradesh and West Bengal in private sector but the Government of India permitted aquaculture of *P. hypophthalmus* in 2010-11. Young ones of the species are bottom feeder and carnivore while the fingerlings feed on snail, worm, insects, gastropods *etc*. This species attain maturity at the end of third year while male mature in two years (Phuong and Oanh, 2009; Griffith *et al.*, 2010; Vidthayanon and Hogan, 2013; Anon, 2014).

P. hypophthalmus is a promising candidate species for freshwater catfish culture (young ones also possess ornamental values) and has captured all the markets of India in shorter period (Lakra and Singh, 2010; Singh and Lakra, 2012). There exist reports that this species is being sold in more than 100 countries, mainly in European Union (EU), Russia, South-east Asia and USA in the form of white fillets (Nguyen, 2007; Phuong and Onah, 2009; Phan *et al.*, 2009). For culture of this species in West Bengal, the seed were initially procured through Bangladesh. Though the species has been induced bred in West Bengal, Andhra Pradesh and Chhattisgarh and some hatcheries established in these states, the survival of offspring has been very poor. Sutchi catfish is highly fecund fish, seasonal spawner and breeds once in a year in flooded rivers. Recently, *P. hypophthalmus* has been bred successfully in Mekong Delta region of Vietnam by using high doses of human chorionic gonadotropin (HCG) (Bui *et al.*, 2010). Success has also been achieved in induced breeding of *P. hypophthalmus* employing GnRH-based drug and dopamine antagonist (ovaprim) at Raipur (Chaturvedi *et al.*, 2014). An attempt has been made to induce breeding in the sutchi catfish by exogenous pituitary gland extract (PGE) administration and larval rearing in different types of hatcheries for mass seed production under agro-climatic conditions of Raipur (Chhattisgarh). Since physico-chemical conditions of water like pH, dissolved oxygen, temperature, alkalinity as well as metabolites play important role in fish breeding (Dwivedi and Ravindranathan, 1982), these parameters were monitored regularly and kept optimal while undertaking induced breeding experiments.

Materials and Methods

Breeding and hatching experiments were carried out at State Fisheries Department and private Fish Farms at Raipur (Chhattisgarh). Male and female brooders of *P. hypophthalmus* (Sauvage, 1878) (Family Pangasiidae) were reared at M/S Hemant Chaudrakar Fish Farm at Dhamtari. At this farm, vertical glass jar hatchery with 15 cemented vertical jars was developed in the year 2010-11. Physico-chemical parameters of the water during the breeding experiments were analyzed as per APHA (1998). For induced breeding experiments, mature and gravid brooders of both the sexes of age group 3 (+) years were collected and induced bred by varying doses of pituitary gland extract (PGE) depending upon the physiological status of fishes. Male brooders were also given PGE at the time of second injection to females. Injected brooders were kept in cemented breeding tanks of size (3 x 2 x 1 m) with flowing water. The stripping was done in the early morning (6 am) after 10-12 hours of the final injection as female were ready for spawning. After fertilization, separated eggs were transferred for incubation to the three type of hatcheries - (i) Circular Hatchery (Figures 10.1 and 10.2), (ii) Vertical Jar Hatchery (Figures 10.3 and 10.4) and (iii) Thailand Model Hatchery (Figures 10.5 and 10.6).

Results and Discussion

Physico-chemical parameters of the water during the breeding experiments were found to be within the optimum range (Table 10.1). Details of the breeding trials conducted on *P. hypothalamus* at Raipur (Chhattisgarh) have been summarized in Table 10.2. In the present experiment, 15 females and 15 males (15 sets, 1:1 sex ratio) were selected and induced bred by varying doses of pituitary gland extract (PGE) depending upon the physiological status of the brooders (after first injection, the second injection was administered after 6 hours). In this study, male brooder were also given pituitary gland extract at the time of second injection to females (Table 10.2). The eggs of *P. hypophthalmus* were very small (diameter 1.4-1.8 mm), adhesive in nature while fertilized eggs were light creamy or brown in colour. For fertilization of one million eggs of *P. hypothalamus*, one ml milt was used. After fertilization, three type of solutions such as cow milk, multani soil (mitti) and black soil were used for removal of stickiness of eggs. Separated eggs were transferred to the three type of hatcheries - (i) Circular Hatchery (Figures 10.1 and 10.2), (ii) Vertical Jar Hatchery (Figures 10.3 and 10.4) and (iii) Thailand Model Hatchery (Figures 10.5 and 10.6) for incubation. Fertilization of eggs varied from 30-80 per cent and survival of the hatchlings varied from 30-60 per cent in all the three hatching systems. After incubation of fertilized eggs from vertical jar hatchery 17,30,400 hatchlings, in circular hatchery 14,08,000 hatchlings while in Thailand model hatchery, only

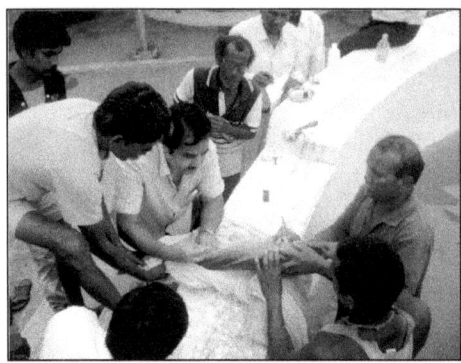

Figure 10.1: Circular Hatchery of *P. hypothalamus* at State Fisheries Department, Raipur.

Figure 10.2: Circular Hatchery of *P. hypothalamus* with Hatching Pool at State Fisheries Department, Raipur.

Figure 10.3: Vertical Jar Hatchery of *P. hypothalamus*.

Figure 10.4: Vertical Jar Hatchery of *P. hypothalamus* Owned by Private Fish Farmer at Raipur.

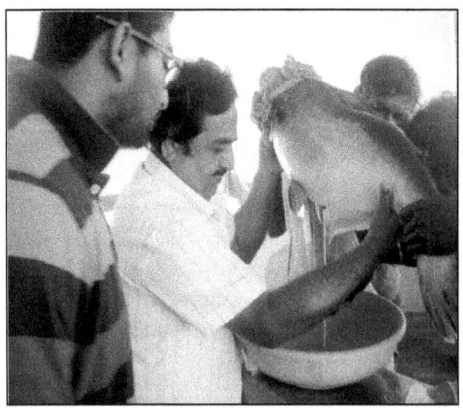

Figure 10.5: Thailand Model Hatchery of *P. hypothalamus* at Raipur.

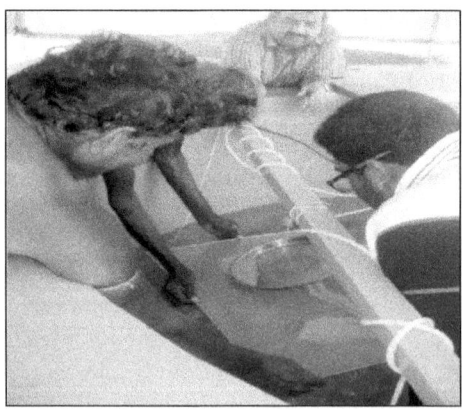

Figure 10.6: Detailed View of Thailand Model Hatchery of *P. hypothalamus* at Raipur.

3,17,500 hatchlings were obtained. The hatching percentage were observed 60 per cent in vertical jar hatchery, 50 per cent in circular hatchery and 30 per cent in Thailand model hatchery. After 2 days, yolk absorption was observed and from the three types of hatcheries - 9,88,740 fry were realised from vertical jar hatchery, 7,40,000 from circular hatchery and 65,550 from Thailand model hatchery. After rearing the fry in nursery ponds for 25-30 days, 5,93,244 fingerlings from Krundh-Liey Fish Farm, 4,22,400 fingerlings from State Fisheries Farm, Raipur and 65,550 from Deepak Mandal Fish Farm (Thailand model hatchery) (total 10,81,144) were obtained.

Table 10.1: Physico-chemical Parameters during the Breeding Experiments at Raipur

Sl.No.	Parameters	I 24.8.2014	II 25.8.2014	III 26.8.2014	IV 27.8.2014	V 28.8.2040	Remarks
1.	Dissolved oxygen (mg/l)	4.8	4.8	5.2	4.8	5.0	Hatchery water
2.	Free carbon dioxide (mg/l)	--	--	--	--	--	--
3.	Iron (mg/l)	0.2	0.1	0.2	0.1	0.2	
4.	Water temperature (Hatchery)	27.4 °C	27.6 °C	28.5 °C	28.4 °C	28.6 °C	Hatchery water
5.	pH	7.4	7.5	7.8	7.5	7.4	
6.	Salinity (ppt)	--	--	--	--	--	--
7.	Total alkalinity (mg/l)	111	114	121	120	112	
8.	Weather temperature (°C)	34.6	34.2	34.2	34.0	34.6	
9.	Weather	Sunny	Sunny	Sunny	Sunny	Sunny	
10.	Fertilization per cent (after removal of stickiness)	70 per cent	60 per cent	70 per cent	80 per cent	80 per cent	After stripping and washing of eggs

Induced breeding of the Indian major carps has been achieved successfully by administration of pituitary gland extract (PGE) and different preparation of synthetic GnRH-based drugs and dopamine antagonists (Chaudhuri and Alikunhi, 1957; Chaudhuri, 1960; Chaudhuri et al., 1966; Varghese et al., 1975; Dwivedi and Ravindranathhan 1982; Chaudhuri and Singh, 1984; Peter et al., 1988, 1993; Nandeesha et al., 1989, 1990; Lakra et al., 1996; Alok et al., 1997; Mahanta et al., 1998; Pandey et al., 1998, 2001, 2002a, b, 2009; Singh et al., 2000; Lee and Donaldson, 2001). Even catfishes have also been induced bred through the similar preparations/drugs (Ramaswamy and Sundararaj, 1956, 1957; Khan, 1972; Devaraj et al., 1972; Khan and Mukhopadhyay, 1975; Pathak et al., 1982; Zonneveld et al., 1988; Kohli, 1989, Kohli and Vidhayarthi, 1990; Rao and Janakiram, 1991; Alok et al., 1993, 1995, 1999; Tharakan and Joy, 1996; Goswami and Sarma, 1997; Kanungo et al., 1999;

Table 10.2: Induced Breeding and Seed Production of P. hypothalamus in the Three different Model Hatcheries at Raipur

Sl. No.	Date	Weight of Fish (kg)		Pituitary Dose (mg/kg)		Injection Time (hours)	Stripping Time (hours)	Total Eggs (in lakh)	Total Number of Good Eggs (in lakh)	Fertiliz-ation (per cent)	Hatch-lings (in lakh)	Percen-tage (per cent)	Fry	Finger-lings
		Male	Female	Male	Female									
								Circular Hatchery						
1.	24.8.2014	3.4	4.4	40	15, 20,	17.00	0.03	4.0	3.20	80	2.56	50	1,26,000	
2.	25.8.2014	3.0	5.0	50	15, 25	17.00	0.03	6.0	4.80	80	3.84	50	1,92,000	
3.	26.8.2014	4.2	6.5	64	20, 32	17.00	0.04	6.0	4.20	70	2.94	50	1,47,000	
4.	27.8.2014	4.5	6.0	60	22, 30	17.00	0.04	6.0	4.20	70	2.94	50	1,57,000	
5.	28.8.2010	4.0	5.50	40	20, 25	17.00	0.03	5.0	3.00	60	1.80	50	90,000	
								27.0	19.4		14,080		7,04,000	4,22,400
								Vertical Jar Hatchery						
6.	26.8.2014	2.9	3.5	30	10, 15	16.00	0.02	3.2	3.10	80	2.790	60	1,67,400	
7.	27.8.2014	3.0	4.9	40	15, 20	16.00	0.02	4.2	3.36	80	3.024	60	1,81,440	
8.	28.8.2014	3.6	4.7	50	18, 25	16.00	0.01	5.0	4.80	80	3.840	60	2,30,400	
9.	28.8.2014	3.0	5.2	50	17, 25	16.00	0.02	4.4	4.30	80	3.870	50	1,93,500	
10.	30.8.2010	3.0	5.2	50	15, 25	16.00	0.02	5.0	4.00	80	3.600	60	2,16,000	
								21.8	19.56		17.304		9,88,740	5,93,244
								Thailand Model Hatchery						
11.	09.08.2014	3.5	3.8	40	20, 25	8.00	9.00	2.0	1.00	50	0.50	30	15,000	
12.	10.08.2014	2.9	4.2	50	18, 30	7.30	8.30	4.0	1.20	30	0.60	30	18,000	
13.	11.08.2014	3.2	5.4	60	18, 40	8.30	9.30	3.5	1.05	30	0.525	25	15,750	
14.	12.08.2014	3.8	5.4	60	18, 40	8.30	9.30	3.0	1.20	40	0.60	30	18,000	
15.	13.08.2014	3.8	6.0	70	25, 40	9.00	10.00	4.8	1.90	40	0.95	30	28,500	
								17.3	6.35		3.175		95,250	65,550
						Total (fingerling production)								10,81,194

Nayak *et al.*, 2000; 2001; Lee and Donaldson, 2001; Singh *et al.*, 2002; Pandey and Koteeswaran, 2004; Sahoo *et al.*, 2005; Mishra *et al.*, 2011; Yadav *et al.*, 2011; Taslim and Ahemd, 2012; Chaturvedi *et al.*, 2012a, b, c, 2013). There exist report that the striped catfish has been bred successfully in Mekong Delta region of Vietnam by using high doses of human chorionic gonadotropin (HCG) (Bui *et al.*, 2010). We successfully induced bred *P. hypothalamus* through ovaprim administration under agro-climatic conditions of Raipur (Chhattisgarh) with better survival of fry and fingerlings (Table 10.2).

Table 10.3: Description of Three Hatcheries Used for Hatching of Eggs of *P. hypothalamus*

Sl.No.	Parameters	Circular Hatchery	Vertical Jar Hatchery	Thailand Hatchery	Specifications
1	Shape	Circular	Vertical Jar	Circular with Hatching trays	Eggs handling trays (45° (3x1x0')
2	Inlet	Horizontal base	Vertical bottom	Horizontal	Ground water
3	Water flow	120 l/m	4-6 l/m	6-10 l/m	Ground water
4	Egg loading	14,08,000	17,30,400	3,17,500	Ground water
5	Hatching (per cent)	60	50	30	Ground water
6	Spawn	7,040,000	9,88,740	65,500	Ground water
7	Water depth	3.0'	3.5'	3.0	Ground water

There exists a lot of scope for freshwater catfish farming in India for diversification of aquaculture and sustainable production (Dehadrai, 1978; Tripathi, 1990; Thakur, 1991; Nayak *et al.*, 2000). Since the culture of *P. hypothalamus* is more profitable among the catfishes, there exist more demand of this species for aquaculture in India and tropical regions of the America (Rahman *et al.*, 2006; Lakra and Singh, 2010; Mukai, 2011; Singh and Lakra, 2012; Hekimoglu *et al.*, 2014; McGee, 2015). Vietnam has shown the record production (1.0-1.5 million tonne per annum) of *P. hypothalamus* (Nguyen, 2007; Phuong and Onah, 2009; Phan *et al.*, 2009). Though the striped catfish is widely cultured in China, Vietnam, Thailand, Taiwan, Philippines, Cambodia, Indonesia, Lao People's Democratic Republic, Bangladesh, Nepal and India (Griffith *et al.*, 2010), this species has been declared Endangered in Vietnam due to over-exploitation, habitat degradation, changes in flow and water quality as well as over-harvesting of eggs, fry and juveniles for aquarium trade (Vidthayanon and Hogan, 2013; Anon, 2014). The success has been achieved earlier in induced spawning and seed production (10,50,000 fry and 6,30,000 fingerlings) of *P. hypothalamus* through ovaprim administration and hatchery development in Raipur (Chhattisgarh) (Chaturvedi *et al.*, 2014) for seed production of this species for conservation aquaculture (True *et al.*, 1996; Anders, 1998) which will reduce the pressure on collection of fry and juveniles from the wild natural habitats (Nguyen, 2009). In the present study, after rearing the fry in nursery ponds for 25-30 days, 5,93,244 fingerlings from Krundh-Liey Fish Farm, 4,22,400 fingerlings from State

Fisheries Farm, Raipur and 65,550 from Deepak Mandal Fish Farm (Thailand model hatchery) (total 10,81,144) were obtained paving the way for mass seed production of the commercially important catfish under agro-climatic conditions of Raipur (Chhattisgarh).

Acknowledgements

We are grateful to Sri. V.K. Shukla, Director, Chhattisgarh State Fisheries Department, Raipur for constant support and encouragement. Thanks are due to M/s Hemat Chaudrakar Fish Farm, Krundh-Liey Fish Farm and Mandal Fish Farm for providing facilities to carry out the work.

References

Ahmed, G.U., Chakma, A., Shamsuddin, M., Minar, M.H,; Islam, T. and Majumdar, M.Z., 2013. Growth performance of Thai Pangus (*Pangasianodon hypothalamus*) using prepared and commercial feed. *Int. J. Life Sci. Biotech. Pharm. Res.*, 2 (3): 92-102.

Ahmad, N. and Hasan, M.R., 2007. Sustainable livelihoods of pangus farming in rural Bangladesh. *Aqua.-Asia Magz.*, 12 (4): 5-10.

Alok, D., Krishnan, T,; Talwar, G.P. and Garg, L.C., 1993. Induced spawning of catfish, *Heteropneustes fossilis* (Bloch), using D-Lys^6 salmon gonadotropin-releasing hormone. *Aquaculture*, 115: 159-167.

Alok, D., Pillai, D., Talwar, G.P. and Garg, L.C., 1995. D-lys^6 salmon gonadotropin-releasing hormone analogue- domperidone induced ovulation in *Clarias batrachus* (L.). *Asian Fish. Sci.*, 8: 263-266.

Alok, D., Pillai, D. and Garg, L.C., 1997. Effect of D-lys^6 salmon sGNRH alone and in combination with domperidone on the spawning of common carp during the late spawning season. *Aquacult. Intern.*, 5: 369-374.

Alok, D., Talwar, G.P. and Garg, L.C., 1999. *In vivo* activity of salmon gonadotropin-releasing hormone (GnRH), its agonists with structural modifications at positions 6 and 9, mammalian GFnRH agonists and native cGNRH-II on the spawning of an Indian catfish. *Aquacult. Intern.*, 7: 383-392.

Anders, P.J., 1998 Conservation aquaculture and endangered species: can objective science prevail over risk anxiety? *Fisheries (Bethesda)*, 23 (11 : 28-31.

Anon, 2014. *Iridescent Shark (Pangasianodon hypothalamus)*. Wikipedia - The Free Encyclopaedia, pp. 1-4.

APHA, 1998. *Standard Methods for the Examination of Water and Wastewaters.* 20thEdn. APHA, AWWA and WPCF, Washington.

Bui, T.M., Phan, L.T., Ingram, B.A., Nguyen, T.T.T., Gooley, G.J., Nguyen, H.V., Nguyen, P.T. and DeSilva, S.S., 2010. Seed production practices of striped catfish, *Pangasainodon hypophthalmus* in the Mekong Delta region, Vitenam. *Aquaculture*, 306: 92-100.

Chaturvedi, C S., Raizada, S. and Pandey, A.K., 2012a. Breeding of *Clarias batrachus* in low-saline water under controlled condition in Rohtak (Haryana). *J. Exp. Zool. India*, 15: 379-382.

Chaturvedi, C.S., Singh, R.K. and Panday, A.K., 2012b. Successful induced breeding of endangered *Ompok pobda* (Hamilton-Buchanan) in Raipur, Chhattisgarh (India). *Biochem. Cell. Arch.*, 12: 321-325.

Chaturvedi, C.S., Somdutt and Panday, A.K., 2012c. Successful induced breeding and hatching of *Clarias batrachus* and *Heteropneustes fossilis* under agro-climatic conditions of Lucknow (Uttar Pradesh). *Natl. J. Life Sci.*, 9: 163-167.

Chaturvedi, C.S., Shukla, V. K., Singh, R.K. and Panday, A.K., 2013. Captive breeding and larval rearing of endangered *Ompok bimaculatus* under controlled condition at Raipur, Chhattisgarh (India). *Biochem. Cell. Arch.* 13: 133-136.

Chaturvedi, C.S.; Lakra W.S., Singh, R.K. and Pandey A.K., 2014. Successful induced breeding and hatchery development of *Pangasianodon hypophthalmus* (Sauvage, 1878) under controlled conditions of Raipur (Chhattisgarh), India. *J. Exp. Zool. India*, 17: 659-664.

Chaudhuri, H., 1960. Experiments on induced spawning of Indian major carps with pituitary injection. *Indian J. Fish.*, 7: 20-49

Chaudhuri, H. and Alikunhi, K.H., 1957. Observation on the spawning of Indian carps by hormone injection. *Curr. Sci.*, 2: 382-383.

Chaudhuri, H. and Singh, S.B., 1984. *Induced Breeding of Carps*. Indian Council of Agricultural Research, New Delhi.

Chaudhuri, H., Singh, S.B. and Sukumaran, K.K., 1966. Experiments on large-scale production of fish seed of the Chinese grass carp, *Ctenopharyngodon idellus* (C. and V.) and the silver carp, *Hypophthamichthys molitrix* (C and V) by induced breeding in ponds in India. *Proc. Indian Acad. Sci.*, 63B: 80-95.

Dehadrai, P. V., 1978. *A Brief Resumes of Research Information of Indian Catfishes*. Central Inland Fisheries Reasearch Institue (ICAR), Barrackpore, West Bengal (India).

Devaraj, K.V., Verghese, T.J. and Rao, G.S.P., 1972. Induced breeding of freshwater catfish, *Clarias batrachus* (Linn.), by using pituitary glands from marine catfish. *Curr. Sci.*, 41: 868-870

Dwivedi, S.N. and Ravindranathan, V., 1982. Carp hatchery model CIFE D 81- a new system to breed fish even when rain fails. *CIFE Bull.* No. 82. Central Institute of Fisheries Education, Mumbai, pp. 3-8.

Goswami, U.C. and Sarma, N.N., 1997. Pituitary dose optimization for induced ovulation, *in vitro* fertilization and production of normal fry of *Clarias batrachus* (Linn.). *Asian Fish. Sci.*, 10: 163-167.

Griffith, D.; van Khanh, P. and Trong, T. Q. (2010). *Cultured Aquatic Species Information Programme: Pangasius hypothalamus* (Sauvage, 1878). Fisheries and Aquaculture Department, FAO, Rome.

Kanungo, G., Sarkar, M., Singh, B.N.; Das, R.C. and Pandey, A.K., 1999. Advanced maturation of *Heteropneustes fossilis* by oral administration of human chorionic gonadotropin. *J. Adv. Zool.*, 20: 1-5.

Khan, H.A., 1972. Induced breeding of air-breathing fishes. *Indian Fmg.*, 22 (4): 44-45.

Khan, H.A. and Mukhopadhyay, S.K., 1975. Production of stocking material of some air- breathing fishes by hypophysation. *J. Inland Fish. Soc. India*, 7: 156-161.

Kohli, M.P.S., 1989. Natural breeding of air-breathing fishes in Andaman and Nicobar Islands. *J. Anadaman Sci. Assoc.*, 5: 96-97.

Kohli, M.P.S. and Vidhayarthi, S., 1990. Induced breeding, embryonic and larval development in *Heteropneustes fossilis* (Bloch) in the agro-climatic condition of Maharashtra. *J. Indian Fish Assoc.*, 20: 15-19.

Kumar, K., Prasad K.P., Raman, R.P., Kumar, S. and Purushothaman, C.S., 2013. Association of *Enterbacter cloacae* in the mortality of *Pangasainodon hypophthalmus* (Sauvage, 1878) reared in culture pond in Bhimavaram, Andhra Pradesh, India. *Indian J. Fish.* 60: 147-149.

Lakra, W.S. and Singh, A.K., 2010. Risk analysis and sustainability of *Pangasianodon hypothalamus* culture in India. *Aqua.-Asia Magz.*, 15 (1): 34-37.

Lakra, W.S., Mishra, A., Dayal, R. and Pandey, A.K., 1996. Breeding of Indian major carps with the synthetic hormone drug ovaprim in Uttar Pradesh. *J. Adv. Zool.*, 17: 105-109.

Laxmappa, B., 2004. Status of murrel farming in Andhra pradesh. *Fishing Chimes*, 23 (12): 60-61.

Lee, C.-S. and Donaldson, E.M., 2001. *Reproductive Biotechnology in Fish Aquaculture.* Elsevier, Amsterdam.

Mahanta, P.C., Rao, K.G., Pandey, G.C. and Pandey, A. K., 1998. Induced double spawning of an Indian major carp, *Labeo rohita*, in the same breeding season under the agro-climatic conditions of Assam. *J. Adv. Zool.*, 19: 99-104.

McGee, M.V., 2015. *Pangasius hypothalamus*, a potential aquaculture species for tropical regions of the Americas (unpublished)

Mishra, R.K., Yadav, A.K., Varshney, P.K.; Pandey, A.K. and Lakra, W.S., 2011. Comparative effects of different hormonal preparations in induced spawning of the freshwater catfish, *Heteropneustes fossilis. J. Appl. Biosci.*, 37: 27-30.

Mukai, Y., 2011. High survival rates of sutchi catfish, *Pangasianodon hypothalamus*, larvae reared under dark conditions. *J. Fish. Aquat. Sci.*, 6: 285-290.

Nandeesha, M.C., Das, S.K., Nathaniel, E., Vargehese, T.J. and Shetty, H.P.C., 1989. Ovaprim - a new drug for induced breeding of carps. *Fishing Chimes*, 9 (4): 13-15.

Nandeesha, M.C., Rao, K.G., Jayanna, R., Parker, N.C., Verghese, T.J., Keshavanath, P. and Shetty, H.P.C., 1990. Induced spawning of Indian major carps through single application of ovaprim. In: *Proceedings of the Second Asian Fisheries Forum* (Eds.) Hirano, R. and Hanyu, I. Asian Fisheries Society, Manila, Philippines, pp. 581-585.

Nayak, P.K., Pandey, A.K., Singh, B.N., Mishra, J., Das, R.C. and Ayyappan, S., 2000. *Breeding, Larval Rearing and Seed Production of the Asian Catfish, Heteropneustes fossilis* (Bloch). Central Institute of Freshwater Aquaculture, Bhubabeswar, 68 p.

Nayak, P.K., Mishra, T.K., Singh, B.N., Pandey, A.K. and Das, R.C., 2001. Induced maturation and ovulation in *Heteropneustes fossilis* by using LHRHa, pimozide and ovaprim for production of quality eggs and larvae. *Indian J. Fish.*, 48: 269-275.

Nguyen, T.T.T., 2009. Patterns of use and exchange of genetic resources of the striped catfish, *Pangasainodon hypophthalmus* (Sauvage, 1878). *Rev. Aquacult.*, 1: 224-231.

Nguyen, V.H., 2007. Vitenam catfish and marine shrimp production: an example of growth and sustainability issues. *Aqua. Asia-Pacific*, 3: 36-39.

Pandey, A.K. and Koteeswaran, R., 2004. Ovatide induced breeding of the Indian catfish, *Heteropneustes fossilis* (Bloch). *Proc. Zool. Soc. (Calcutta)*, 57: 35-38.

Pandey, A.K., Patiyal, R.S., Upadhyay, J.C., Tyagi, M. and Mahanta, P.C., 1998. Induced spawning of endangered golden mahseer (*Tor putitora*) with ovaprim at State Fish Farm near Dehradun. *Indian J. Fish.*, 45: 457-459.

Pandey, A.K., Mahapatra, C.T., Sarkar, M., Kanungo, G., Sahoo, G.C. and Singh, B.N., 2001. Ovatide induced spawning in Indian major carp, *Catla catla* (Hamilton-Buchanan), for mass scale seed production. *J. Adv. Zool.* 22: 70-73.

Pandey, A.K., Mahapatra, C.T., Sarkar, M., Kanungo, G. and Singh, B.N., 2002a. Ovatide induced spawning in Indian major carp, *Cirrhinus mrigala*, for mass scale seed production. *J. Exp. Zool. India*, 5: 81-85.

Pandey, A.K., Mahapatra, C.T., Sarkar, M., Kanungo, G. and Singh, B.N., 2002b. Ovatide induced spawning in Indian major carp, *Labeo rohita* (Hamilton-Buchanan). *Aquacult,* 3: 1-5.

Pandey, A.K., Mahapatra, C.T., Kanungo, G. and Singh, B.N., 2009. Induced breeding of the Indian major carp, *Labeo rohita*, with synthetic hormone drug WOVA-FH. In: *Recent Advances in Hormonal Physiology of Fish and Shellfish Reproduction* (Eds.) Singh, B.N. and Pandey, A.K. Narendra Pub. House, Delhi, pp. 257-260.

Pathak, S.C., Kohli, M.P.S. and Munset, S.K., 1982. *Proceedings of the Sixth Workshop on All India Co-ordinated Research Project on Air-breathing Fish Culutre.* Central Inland Fisheries Reseatrch Instutute, Barrackpore, West Bengal (India).

Peter, R.E., Lin, H.R. and van der Kraak, G., 1988. Induced ovulation and spawning of cultured freshwater fish in China: advances in application of GnRH analogues and dopamine antagonists. *Aquaculture*, 74: 1-10.

Peter, R.E., Lin, H.R., van der Kraak, G. and Little, M., 1993. Releasing hormones, dopamine antagonists and induced spawning. In : *Recent Advances in Aquaculture. IV* (Eds.) Muir, J.F. and Roberts, R.J. Blackwell Science Publications, London, pp. 25-30.

Phan, L.T., Bui, T.M., Nguyen, T.T.T., Gooley, G.J., Ingram, B.A., Nguyen, H.V., Nguyen, P.T. and De Silva, S.S., 2009.1 Current staus of farming practices of striped catfish, *Pangasainodon hypophthalmus* in the Mekong Delta, Vitenam. *Aquaculture*, **296** : 227-236.

Phuong N T and Oanh D T H (2009) Striped catfish (*Pangasianodon hypothalamus*) aquaculture in Vietnam: an unprecedented development within a decade. In: *Success Stories in Asian Aquaculture* (Eds.) De Silva, S.S. and Davy, F.B. NACA and IDRC, Bangkok and Ottawa, pp133-149.

Rahman, M.M., Islam, M.S., Haldar, G.C. and Tanaka, M., 2006. Cage culture of sutchi catfish, *Pangasius sutchi* (Flower, 1937): effects of stocking density on growth, survival, yield and farm profitability. *Aqua. Res.*, 37: 33-39.

Ramaswamy, L.S. and Sundararaj, B.I., 1956. Induced spawning in the Indian catfish. *Science*, 123: 1080.

Ramaswamy, L.S. and Sundararaj, B.I., 1957. Induced spawning in the catfish, *Clarias*. *Naturewissen*, **44**: 344.

Rao, G.R.M. and Janakiram, K., 1991. An effective dose of pituitary for breeding *Clarias batrachus*. *J. Aquacult. Trop.*, 6: 207-210.

Sahoo, S.K., Giri, S.S. and Sahu, A.K., 2005. Induced spawning of Asian catfish, *Clarias batrachus* (Linn.): effect of various latency periods and sGnRHa and domperidone doses on spawning performance and egg quality. *Aquacult. Res.*, 36: 1273-1278.

Singh, A.K. and Lakra, W.S., 2012. Culture of *Pangasianodon hypothalamus* into India: impacts and present scenario. *Pak. J. Biol. Sci.*, 15: 19-26.

Singh, B.N., Das, R.C., Sahu, A., Kanungo, G., Sarkar, M., Sahoo, G.C., Nayak, P.K. and Pandey, A.K., 2000. Balanced diet for broodstocks of *Catla catla* and *Labeo rohita* and induced breeding using ovaprim. *J. Adv. Zool.*, 21: 92-97.

Singh, B.N., Das, R.C., Sarkar, M., Kanungo, G., Sahoo, G.C. and Pandey, A.K., 2002. Balanced diet for rearing of singhi, *Heteropneustes fossilis* (Bloch), brood fish and induced breeding using ovaprim. *J. Ecophysiol. Occup. Hlth.*, 2: 57-64.

Taslim, K. and Ahemd, F., 2012. Study on seed produciton technique of indigenous magur, *Clarias batrachus*, singhi (*Heteroprieustes fossilis*) and pabda (*Ompok pabda*) through induced breeding. *Bull. Eniviron. Pharma. Life Sci.*, 1 (4): 16-32.

Thakur, N.K., 1991. Possibilities and problems of catfish culture in India. *J. Inland Fish. Soc. India*, 23 (2): 80-90.

Tharakan, B. and Joy, K.P., 1996. Effects of mammalian gonadrotropin-releasing hormone analogue, pimozide, and the combination on plasma gonadotropin levels in different seasons and induction of ovulation in female catfish. *J. Fish Biol.*, 48: 623-632.

Tripathi, S.D., 1990. Present status of breeding and culture of catfish in South Asia, *Aquatic Living Resour.*, 9: 219-228.

True, C.D., Silva-Lora, A. and Castro-Castro, M., 1996. Is aquaculture the answer for the endangered totoaba? *World Aquacult.*, 27 (4): 38-43.

Verghese, T.J., Rao, G.P.S., Devaraj, K.V. and Chandrashekar, B., 1975. Preliminary observations on the use of marine catfish pituitary glands for induced spawning of Indian major carps. *Curr. Sci.*, **44**: 75-78.

Yadav, A.K., Mishra, R.K., Singh, S.K., Varshney, P.K., Pandey, A.K. and Lakra, W.S., 2011. Induced spawning of Asian catfish, *Clarias batrachus*, with different doses of sGnRH-based drugs. *J. Exp. Zool. India*, 14: 199-202.

Vidthayanon, C. and Hogan, Z., 2013. *Pangasianodon hypothalamus: The IUCN Red List of Threatened Species*. IUCN, Gland, Switzerland.

Zonneveld, N., Rustidja, W.J.R.V. and Mundane, W., 1988. Induced spawning and egg incubation of Asian catfish, *Clarias batrachus*. *Aquaculture*, 74: 41-47.

Chapter 11

Allelopathic Interactions in Freshwater Ecosystems with Special Reference to Zooplankton

S.S.S. Sarma and S. Nandini

ABSTRACT

Allelopathic interactions are well known in the terrestrial plant communities. However, in aquatic ecosystems, allelopathic studies have received much less attention. In this review, we summarize the major allelopathic interactions among the various groups of aquatic organisms including the plankton. A brief overview of different terms related to infochemicals such as allelochemicals, kairomones, pheromones, allomones, synomones, conditioned-medium and alarm substances widely used in aquatic allelopathic research are explained. The role of allelopathy in aquatic ecosystems largely is exemplified by the competitive interactions and predator-prey systems. Other interactions including host-epizoic relationships are rarely investigated. Finally, the terrestrial allelopathy has been successfully applied in the field (as in agroecology). However, aquatic allelopathy is rarely used for field conditions, although there is some potential (e.g., biomanipulation of water bodies using allelochemicals from macrophytes). Our review emphasizes the need to do more research for the commercial application of aquatic allelopathy in regulating the structure of freshwater bodies.

Keywords: *Zooplankton, Biomanipulation, Allelopathic studies, Macrophytes.*

Introduction

Allelopathic interactions among organisms are well studied in terrestrial systems because unlike animals, plants lack movement and therefore must develop strategies to fend off the invading herbivores and competitors (Cheema *et al.*, 2013). The development of the defence mechanisms of terrestrial plants include morphological features (*e.g.*, thorns), life history strategies (*e.g.*, reducing the number of seeds: Agrawal 1998) or production of chemicals (*e.g.*, toxins: Kruse *et al.*, 2000), though aspects related to behaviour (*e.g.*, gravitropism: Lupini *et al.*, 2013) are rare. Production of penicillin from the fungus *Penicillium* inhibiting the growth of bacteria, first observed by Alexander Fleming in 1928, is considered as a classic example of allelopathic effects (Fleming 1929, Rice 1974).

Terrestrial plants have developed different mechanisms including through the production of secondary metabolites to inhibit the growth of competing species. However, it is important to consider that not all plant allelochemicals have the adverse effect and in some cases they actually stimulate the growth of other plants or other organisms (Anaya 1999). Although it is known that chemical defence methods are not restricted exclusively to the terrestrial systems, efforts to document them in aquatic ecosystems were scarce until the last century (Brönmark and Hansson 2012). Certain ecological interactions among the aquatic organisms cannot be explained by conventional mechanisms of competition or predation. For example, for a successful predation, predators have to be in physical contact with their prey. However, if the prey are far away from the predator, they still can be paralyzed or killed through the secretion of secondary metabolites (*e.g.*, toxins: Barbosa and Castellanos 2005). When chemicals secreted by a predator are perceived by a prey species, it responds by changing its behaviour, morphology and/or life history variables (Brönmark and Hansson 2012). Thus, one of the most common modes of defence in sessile aquatic organisms is the production of chemicals known as allelochemicals vast majority of which is made up of secondary metabolites.

In the freshwater bodies, depending on water depth and the quantity of nutrients, aquatic vegetation can be of different types (*e.g.*, heliophytes, Hygrophytes, rooted-macrophytes, floating macrophytes, submerged macrophytes and phytoplankton) (Murillo *et al.*, 2009). In the macrophyte-dominated aquatic ecosystems, interactions among the species can be different; however, it is also common to find allelopathic effects in the littoral zones. In such waterbodies, the allelopathic interactions occur among different groups of organisms: within the species of macrophytes, macrophytes and plankton, between phytoplankton and zooplankton, within the plankton and between fish and invertebrates (Moss 2010). Many of these interactions occur through infochemicals that are not covered in the classic definition of allelopathy. For example, the International Society of Allelopathy defines allelopathy as a process involving plant-produced secondary metabolites, algae, bacteria and fungi which influence the growth and development of any biological system (Olofsdotter 1998). Even with this broad definition, the chemically-mediated interactions are still far from complete, especially when applied to aquatic ecosystems. For example, many conventional definitions of infochemicals including pheromones do not also take into account of self-regulation or autotoxicity of a population through its own secondary metabolites; examples of this type are common in zooplankton especially in rotifers (Snell 1998), cladocerans (Matveev 1993) and in some turbellarians (Dumont *et al.*, 2014).

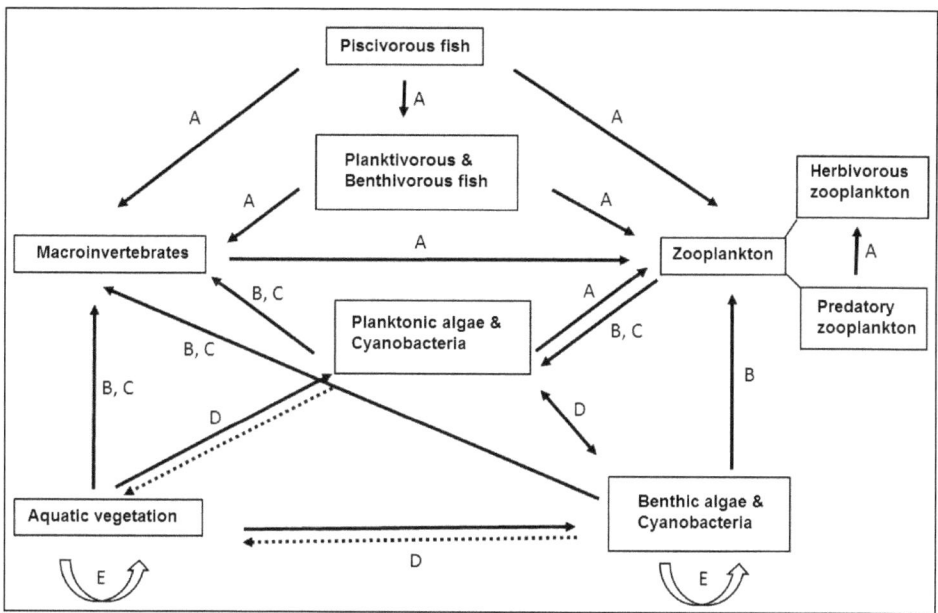

Figure 11.1: Possible Forms of Allelopathic Interactions in Aquatic Ecosystems.
A: Kairomones (avoiding enemy); B: Kairomones (foraging); C: feeding inhibitors; D: allelopathic activity; E: autotoxicity. Solid arrows: strong evidence; dashed arrows: weak evidence (slightly modified from Gross (2003)).

In the literature many terms (infochemicals, allelochemicals, kairomones, *etc.*) are used to explain the chemical interactions among organisms. In this review, we follow the scheme provided by Dicke and Sabelis (1988), where allelochemicals are treated in a broad sense (depending on the effect that occurs due to the chemical signal in the transmitter or in the receiver) to understand the ecological interactions in the aquatic ecosystems (Table 11.1). While studying the allelopathic effects in waterbodies natural chemicals released by the transmitter are considered. It is also possible that under laboratory conditions, many pure organic chemicals such as phenols may have similar adverse effects on the target species. Such pure compounds are not considered here as allelochemicals.

Various natural substances may have allelopathic effects on aquatic organisms. These include glucosinolates, phenolic compounds, terpenoids, alkaloids, and many other natural toxins such as microcystins. Though many of them have negative allelopathic reactions (*e.g.*, inhibiting growth of other species), some of them may have other functions too (Haig 2008). There are many factors that determine the production of allelochemicals by organisms. Certain substances such as microcystins are produced in response to both abiotic and biotic factors. For example, microcystins produced by the cyanobacterium *Microcystis* is due to the exposure to light (Wiedner *et al.*, 2003) and changes in the nutrient levels (Oh *et al.*, 2000) and the presence of herbivores such as daphniids and rotifers (Jang *et al.*, 2007, Pérez-Morales *et al.*, 2015). Studies from the marine systems indicate that the diatom *Pseudo-nitzschia*

Table 11.1: Definitions of Chemical Substances of Biological Origin that Mediate Interactions between Individuals in Aquatic Ecosystems (Dicke and Sabelis 1988, Sbarbati and Osculati 2006, Brönmark and Hansson 2012)

- ☆ *Infochemical*: Chemical substances of biological origin involved in the interaction between two individuals of the same or different species.
- ☆ *Pheromones*: infochemicals that mediate interactions between two individuals of the same species (*e.g.*, male and female).
- ☆ *Allelochemicals*: infochemicals that mediate stimulatory or inhibitory interactions between two individuals of different species.
- ☆ *Synomones*: Allelochemicals released by a species, which, when in contact with another species, the receiver exhibits a physiological or a behavioural response that is adaptively favourable for both species.
- ☆ *Kairomones*: Allelochemicals released by a species, which, when in contact with another species the receiver exhibits a physiological or a behavioural response that is adaptively favourable for the receiver but not for the transmitter.
- ☆ *Allomones*: Allelochemicals released by a species, which, when in contact with another species the receiver exhibits a physiological or a behavioural response that is adaptively favourable for the transmitter, but not to the receiver.
- ☆ *Alarm-substance*: infochemicals released from a partly bitten prey (by a predator) which, when in contact with other individuals of the same population, the receiver exhibits a physiological or a behavioural response that is adaptively favourable to the receiver.
- ☆ *Conditioned-medium*: culture filtrate, containing metabolic products (exudates, secretions, waste products, infochemicals, among others) from an aquatic species previously maintained in the medium for a specified period of time.

produces domoic acid (Rao *et al.*, 1988) and its concentration increases in the presence of herbivores such as calanoid copepods (Tammilehto *et al.*, 2015). Thus, for some phytoplankton species, toxins offer a permanent defence against herbivores that guarantees the survival of the species, even if its production involves some energy costs. Further, phytoplankton species are adapted to adjust the energy allocation between toxin production and for growth and reproduction (Wolfe 2000).

The effect of allelochemicals released by a species is not always directed to a single species. For example, microcystins are produced specifically to reduce grazing by zooplankton, but they can also kill vertebrate predators such as fish, which do not feed on phytoplankton (Lindholm *et al.*, 1999). Examples that contradict this situation are also documented in literature. For example, allelochemicals released by the predatory rotifer *Asplanchna* are specific to their prey (rotifers, mainly from the family Brachionidae and some from the family Lecanidae) and induce morphological changes; but it is apparently ineffective against cladocerans, although some of them (Chydoridae) are the prey for the members of Asplanchnidae (Nandini and Sarma 2005). The kairomones may also cause opposing effects depending on receiving prey species. For example, fish kairomones tend to reduce the body size of prey rotifer species (Duncan 1983), but in the cladoceran prey (*e.g.*, *Daphnia*) they induce elongation of posterior spine (Spaak and Boersma 1997).

The stability of allelochemicals varies considerably. The asplachnin, a kairomone from *Asplanchna*, is heat stable and can withstand a temperature of 100°C up to one hour (Snell 1998), while others are easily degraded (Haig 2008). A stored conditioned-medium (see Table 11.1) generally loses its allelopathic properties and therefore, the transmitter must produce such substances continuously to ensure their effectiveness. This suggests a constant energy cost by the producers during the period when the threat is present (Lass and Spaak 2003). Kairomones are produced by a predator and prey species perceive them so as to modify their morphology, behaviour and life history variables in order to minimize the risk of, or the escape from the predators. This suggests that if a predator is capable of inducing a defence on its prey, its consumption can be expected to be lower; this has been considered as a kind of self-regulation of the predator's population. If this occurs, the predator and prey can coexist (Gilbert 2014). The first anecdotal evidence of this came from De Beauchamp (1952), who observed that in the water bodies where *Asplanchna* occurred, its prey *Brachionus calyciflorus* was well armed with strong postero-lateral spines. Later, Gilbert (1966) experimentally induced formation of spines in the same brachionid species using *Asplanchna*-conditioned medium. Thus, the substances released into the medium by *Asplanchna* are recognized as kairomones. It has been suggested that induction of spines by a predator is an effective strategy that ensures the survival of the predator itself, as prey with spines allow the availability of prey for a longer period of time during which the predator can at least maintain a population. If the prey species are unable to develop this type of morphologically induced anti-predator defences, they would disappear from the system (Gilbert and Waage 1967). This is sometimes described as the intelligent predator hypothesis. In addition, if the rate of predation is not offset by the recruitment of prey density, and no other alternative prey is available, then the predator population may disappear (Dumont et al., 1990).

Allelopathic Interactions between Macrophytes and Plankton

It is known that macrophytes, like land plants, compete with other aquatic photosynthetic species including phytoplankton for nutrients, light, space, *etc*. In this type of competitive interactions, the end result would be influenced not only by the capacity to utilize the limited resources, but also the ability to produce allelopathic substances that inhibit the growth of the competing species (Hilt and Gross 2008). It has been reported that macrophytes have a strong antagonistic impact on phytoplankton (Wium-Andersen *et al.*, 1982, Körner and Nicklisch 2002).

Several reviews, especially those based on studies in aquatic ecosystems of temperate regions, document the allelopathic effects among the different species of macrophytes and between macrophytes and phytoplankton (Hanson and Butler 1994, Gross *et al.*, 2007, Hilt and Gross 2008). Field and laboratory data show that the allelochemicals exuded by *Hydrilla* and *Ceratophyllum* inhibit the growth of other species of macrophytes (Gopal and Goel 1993). Simple field experiments allow us to study the role of allelochemicals from macrophytes on phytoplankton. These tests can be performed using shallow mesotrophic or eutrophic waterbodies, since waves and other physical factors are held constant (Moss 2010). When living macrophytes and plastic models resembling the macrophytes are placed in waterbodies dominated by phytoplankton, within a few weeks phytoplankton-free zones appear around the living macrophytes but without such clear areas around the plastic plants implying the adverse impact allelochemicals from macrophytes on phytoplankton (van Donk and van de Bund 2002). Though such cause and effect relations are easy to understand in aquatic systems, it is necessary to use molecular tools to explore the mechanisms involved in such processes (Addisie and Medellin 2012).

Allelopathic properties of certain macrophytes can be used to control toxic cyanobacterial blooms in different water bodies (Moss 2010). Controlling excessive growth of phytoplankton (especially cyanobacterial blooms) through biological (*e.g.*, biomanipulation), physical (*e.g.*, ultrasonication: Ahn *et al.*, 2003) or chemical (*e.g.*, van Oosterhout and Lürling, 2011) means is a challenge in water management practices. Macrophytes are used as a refuge for cladocerans against fish predation. This increases foraging pressure on phytoplankton by the cladocerans and this in turn would improve water clarity or transparency (Gulati *et al.*, 1990). Macrophytes also reduce the availability of nutrients for phytoplankton and thus suppress their growth (Moss 2010). The introduction of exotic macrophytes to control undesirable phytoplankton blooms for improving water quality requires additional considerations. For example, the exotic macrophyte *Eichhornia crassipes* was introduced into Mexican waters for its aesthetic importance and soon became invasive in a large number of inland water bodies (Gutiérrez *et al.*, 1996). The introduction of non-native aquatic macrophytes to reduce the cyanobacterial blooms could be possible, but this should be first tried using native macrophytes, which are generally non-aggressive to the local environment.

Production of allelochemicals by the macrophytes appears to be seasonal (Gross 2000). In some months, the quantity and type of allelochemicals may be different from those in other months. Besides the seasonal influence, the age of the macrophytes also has a role in the production of allelochemicals (Mulderij *et al.*, 2003). Usually metabolically active (young) plants have higher potential to produce secondary metabolites than old or dying plants. Although all structures macrophytes are capable of producing allelochemicals, they release most of these metabolites through their root systems (Gross 2003). Therefore, for quantitative studies on the production of allelochemicals by macrophytes weight of the root may be used as an index of secondary metabolites.

Most studies dealing with the cause-effect relationships use the conditioned-medium from macrophytes and evaluate their effects on a particular species of phytoplankton. Only a few studies have considered the isolation of the active chemical compounds from mixtures of allelopathic substances (Figure 11.2). A well-known example is the study by Wium-Andersen *et al.* (1982) who identified that two sulphur compounds (dithiane and trithiane) from Chara are involved in inhibiting phytoplankton growth. In another study, Nakai *et al.* (1999) evaluated the allelopathic effect of nine species of macrophytes on three species of cyanobacteria. They found that only *Cabomba caroliniana* and *Myriophyllum spicatum* inhibited the growth of the cyanobacteria, with the effect of the latter being much stronger. Their results also suggest that the inhibitory effect of allelochemicals of macrophytes to phytoplankton is species-specific. Thus, it can be expected that only certain species of cyanobacteria are affected by the allelochemicals from a given macrophyte species. Furthermore, it is also known that the macrophyte-conditioned medium is prone to

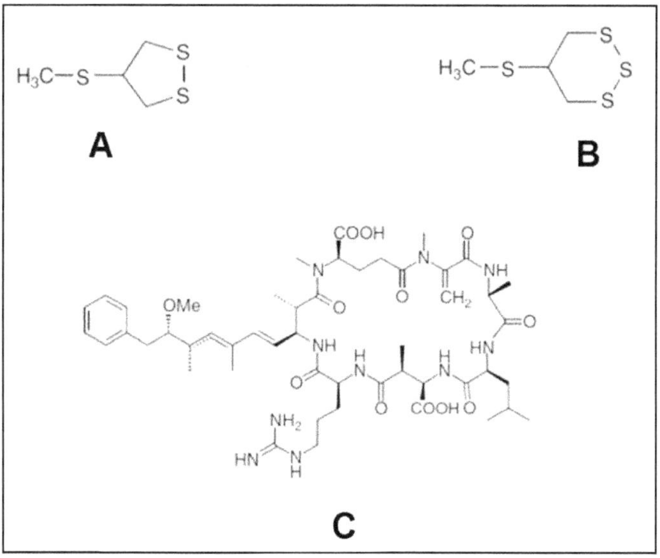

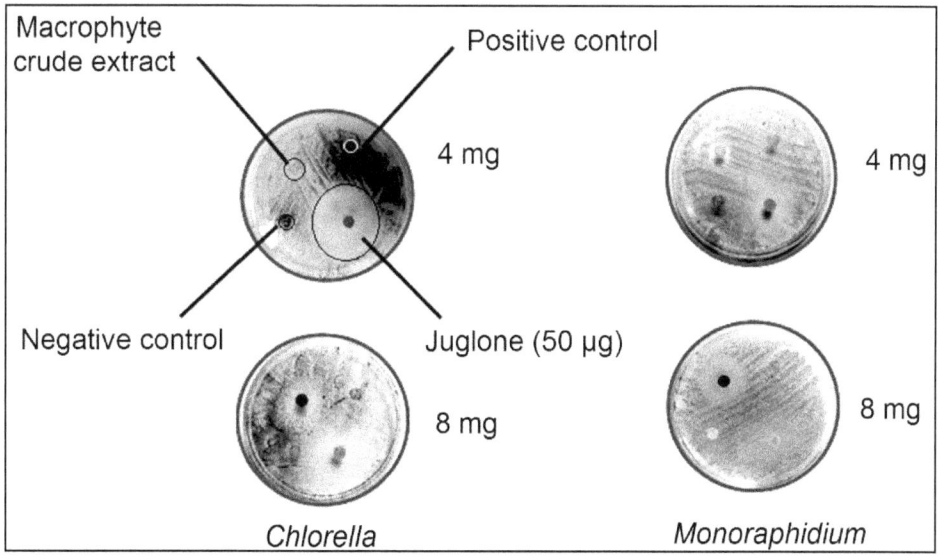

Figure 11.2: Some Active Chemical Compounds of Allelopathic Substances.
A: Dithiane, B: Trithiane, C: Microcystin LR (Gross 2003).

degradation by bacteria and therefore the inhibitory effects may vary considerably depending on the renewal of the medium. Therefore, the conditioned medium should be supplied continuously in order to avoid loss of allelopathic effects.

Figure 11.3: Agar Diffusion Bioassays Using Sensidiscs.
The effect of the macrophyte extract is measured by the diameter of inhibition zone (clear area without algae). Juglone is used as reference chemical in phytotoxicity tests. The negative control contains only the solvent (ethyl acetate) and while the positive control group is with distilled water (Espinosa-Rodríguez et al., 2016a).

Chemical components extracted from macrophytes are often quantified by agar diffusion assays using sensidiscs (Figure 11.3). Using this approach, Espinosa-Rodríguez et al. (2016a) have quantified the inhibitory effects of different concentrations of crude extracts of ethyl acetate and methanol from the macrophyte *Egeria densa* on the growth of four algal taxa (*Scenedesmus acutus*, *Chlorella vulgaris*, *Monoraphidium* sp. and *Nitzschia palea*). They have observed that the size of the algal inhibition halos (of up to 20 mm in diameter) were dependent on the concentration of the extract, the solvent type used and the tested phytoplankton species, but in general *S. acutus* was most sensitive to ethyl acetate extract while *N. palea* was least sensitive to this. They further noted that this macrophyte has more than one chemical substance with allelopathic activity (Figure 11.4). It has been observed that macrophytes through their allelochemicals differentially affect each phytoplankton group, cyanobacteria being the most affected followed by diatoms; green algae such *Scenedesmus* as may show some positive effects through other mechanisms, e.g., hormesis. The hermetic reaction exhibited by a species is a dose response phenomenon where low dose causes stimulation and high dose leads to inhibition (Hilt and Gross 2008).

Species of zooplankton are also affected by the allelochemicals from macrophytes. Though zooplankton species do not compete with macrophytes

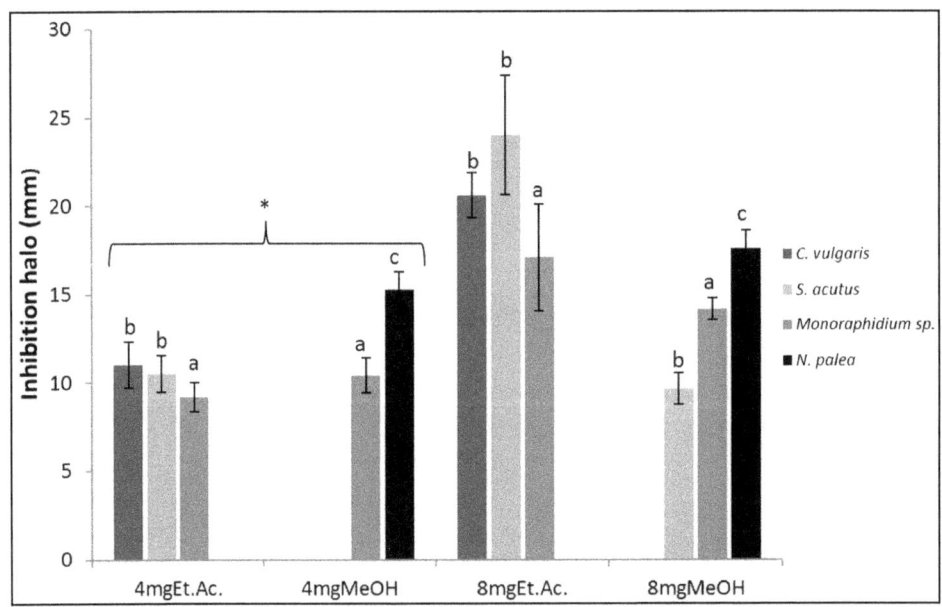

Figure 11.4: Inhibition Halo Produced by of 4 and 8 mg per Sensidisk of Ethyl Acetate and Methanolic Crude Extracts of *Egeria densa* on *S. acutus*, *C. vulgaris*, *Monoraphidium* sp. and *N. palea* (Treatments without effect were discarded).

Data bars carrying lowercase letters indicate a statistically significant difference with two-way ANOVA and the post-hoc Tukey test (p<0.05). The asterisk (*) indicates significant differences between the concentrations of crude extracts (after Espinosa-Rodríguez *et al.*, 2016a).

for nutrients, space or sunlight, they may cause some damage to the tender vegetation through the plant surface scraping or feeding. Compared to macrophyte-phytoplankton interactions, studies on macrophyte-zooplankton relationships are far less. Recent studies suggest that the macrophyte-conditioned medium does affect both the lifespan and reproductive output of cladocerans. For example, Espinosa-Rodríguez et al. (2016b) have observed that three cladoceran species of *Simocephalus*, *S. exspinosus*, *S. serrulatus* and *S. mixtus*, when fed algae grown on macrophyte-conditioned medium, had longer lifespan and higher reproductive output than those in controls. In addition, when these cladocerans were grown on macrophyte-conditioned medium but fed normal algae, their survival and reproduction were also enhanced (Figures 11.5 and 11.6). The enhanced survival and reproductive rates of cladocerans when exposed to macrophyte-allelochemicals are possibly a

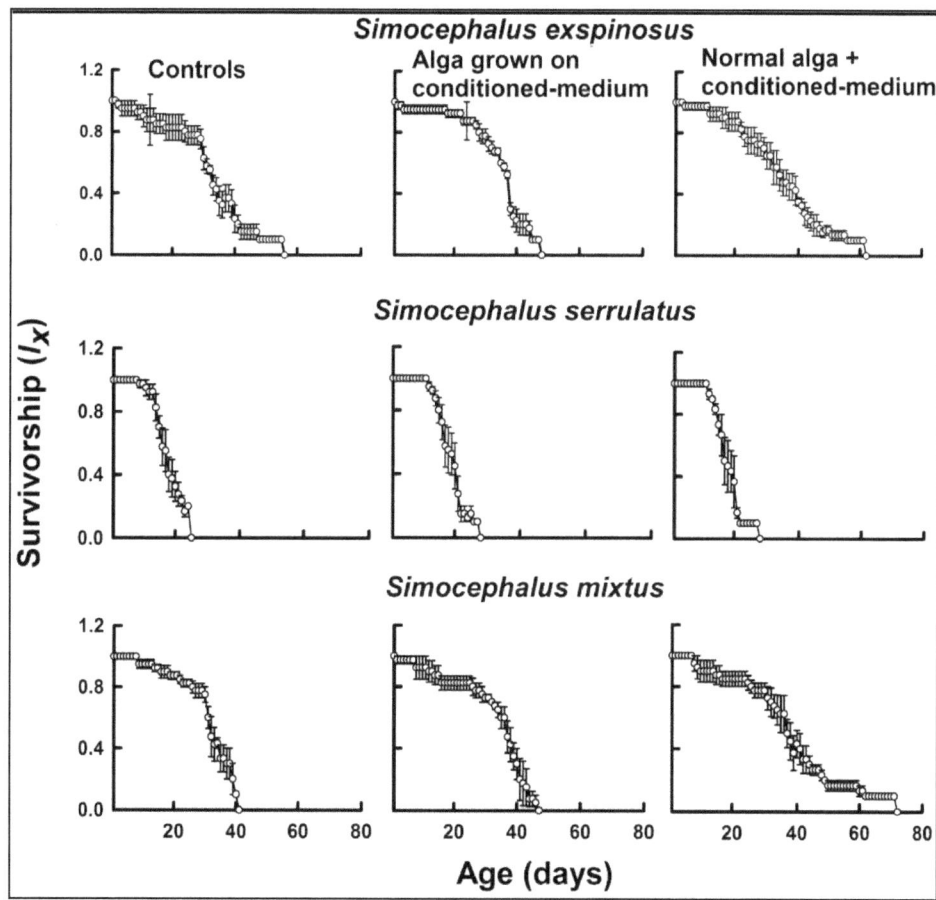

Figure 11.5: Age-specific Survivorship Curves of Three Species of Cladocerans in Controls (First column), in Treatment Containing Alga Cultured on Macrophyte-Conditioned Medium (Middle column) and in Treatment Containing Normal Alga but with Macrophyte-Conditioned Medium (Third column) (from Espinosa-Rodríguez et al., 2016b).

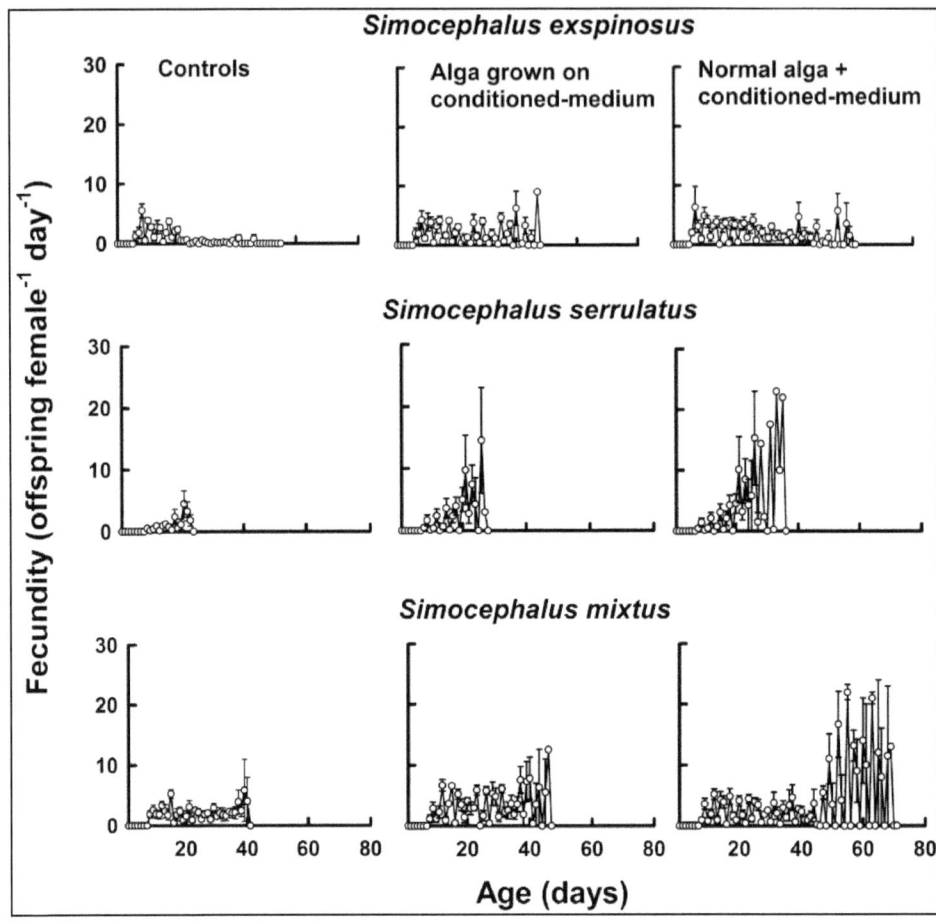

Figure 11.6: Age-specific Fecundity Curves of Three Species of Cladocerans in Controls (First column), in Treatment Containing Alga Cultured on Macrophyte-Conditioned Medium (Middle column) and in Treatment Containing Normal Alga but with Macrophyte-Conditioned Medium (Third column)
(from Espinosa-Rodríguez et al., 2016b).

hormetic response. Perhaps, in order to avoid exposure to plant allelochemicals, daphniids that seek shelter from macrophytes against fish predation, keep a short distance away from these plants (Brönmark and Hansson, 2012).

Allelopathic Interactions within Zooplankton Species

There is some information on the interactions between zooplankton (cladocerans and rotifers) and their vertebrate predators (mainly fish) and between zooplankton and invertebrate predators such as dipteran larvae (*Chaoborus*), cyclopoids, flatworms and *Asplanchna*, among others (Kerfoot and Sih 1987). However, information on typical intra-zooplankton interactions through allelochemicals is limited (Dumont et al., 1990). Free-living ciliates and rotifers feed on similar resources such as bacteria, algae and particulate organic matter, such as detritus. In rotifer mass culture tanks ciliates often invade as contamination (Sarma 1991). Rothbard (1975) noted that the ciliate *Euplotes* produces secondary metabolites, which have an inhibitory effect on the reproduction of rotifers; this in turn leads to the decrease of rotifer populations and therefore the ciliates gain a competitive advantage. In addition, in rotifer culture tanks, planktonic ciliates when present cause an aggregation of the algae such as *Chlorella*. Such algal aggregations are usually settled at the bottom of culture tanks and remain inaccessible for planktonic rotifers, while ciliates are able to feed on these algal aggregates by browsing through the inter-cellular spaces. Forming aggregations of algae is probably due to the secondary metabolites from ciliates. This is similar to allelochemicals released by daphniids which generate colony formation in some algal genera such as *Scenedesmus* (Lampert *et al.*, 1994). Sánchez Rodríguez *et al.* (2010) have quantified the effect of allelochemicals produced by *Paramecium caudatum* on the population growth of rotifers, *Brachionus calyciflorus* and *Plationus patulus*. They have observed that, compared with controls, the population density of rotifers decreased due to allelochemicals from ciliates (Figure 11.7). Thus, the role of allelochemicals from such interactions results in a competitive advantage for ciliates as both ciliates and rotifers share the same limiting resources.

Allelochemicals from predators fall in the category of kairomones since the prey perceives them and gets ecological benefits (Tollrian and Harvell, 1999). There are several ways by which a prey species reacts to the presence of kairomones. The most common responses are changes in prey morphology. Visual predators such as fish usually prefer the largest prey item that they can handle. Hence, in the

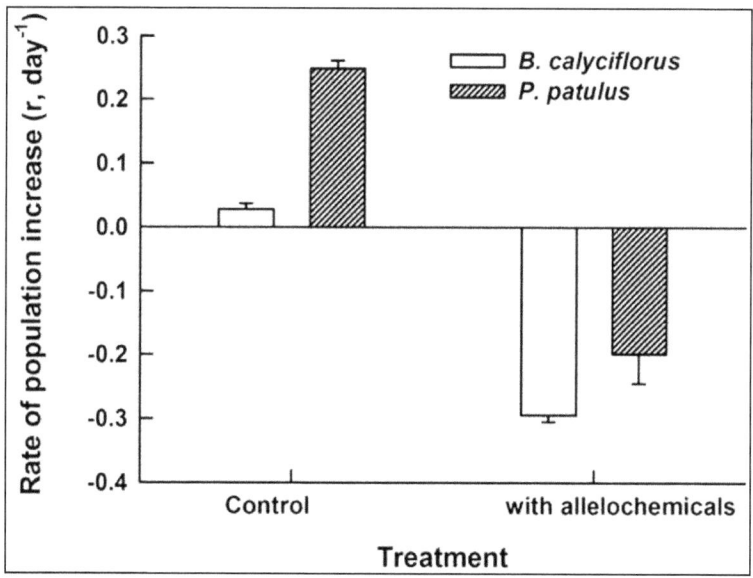

Figure 11.7: Rate of Population Increase (*r*) per day of *B. calyciflorus* and *P. patulus* in Controls and in the Presence of Allelochemicals from Ciliate (*Paramecium caudatum*) Culture Filtrate. Shown are the mean±standard errors based on four replicates (Sánchez Rodríguez *et al.*, 2010).

presence of fish kairomones rotifers reduce their body size so that fish avoid them because feeding on such smaller items is not energetically profitable (Zaret 1980). On the other hand, cladocerans such as *Daphnia* develop tail spines which become larger than the gape size so that the larval fish cannot handle them and in this way both rotifer and cladocerans adapt different strategies to minimize fish predation (Lass and Spaak 2003). Kairomones from invertebrate predators such as *Asplanchna* increase the body size of their prey species. For example, many laboratory studies have shown that in the presence of *Asplanchna*, brachionids increase their body size by about 20 per cent and the spine lengths to a much higher degree (>120 per cent) (Gilbert 2014). When a prey is permanently well armed with defensive structures such as spines, the presence of kairomones has a very little effect on the elongation of these structures. For example, Sarma *et al.* (2011) cultured unspined *Brachionus calyciflorus* and permanently spines *Plationus macracanthus* in the presence of kairomones from *Asplanchna brightwellii* and documented that the postero-lateral spines of *P. macracanthus* increased by about 15 per cent, while in *B. calyciflorus* the increase was much more (150 per cent).

The changes in the morphology of prey species in the presence of kairomones appear to depend on the concentration of these allelochemicals. For example, Sarma *et al.* (2011) have shown that an increase in the predators' density has a far greater effect on the prey morphology. When the conditioned-medium derived from extremely high density of predators is used for prey behavioural studies, it is difficult to distinguish the possible role of waste products (*e.g.*, ammonia) from the kairomones (Gilbert 2013). The ability to induce morphological changes in the prey items apparently is dependent on the voracity of the predator too. At a given predator density, more voracious species are likely to induce larger morphological changes than the weak predators. This has been shown in different species of *Asplanchna*, where *A. sieboldii*, one the most voracious predators (Wallace *et al.*, 2006), caused the longest spine induction in *B. calyciflorus* (Gama-Flores *et al.*, 2011). Similar results are reported by Nandini *et al.* (2011) on predators from different phyla such as Platyhelminthes and Arthropoda.

Kairomone-induced life history changes in the prey population are of great ecological importance in zooplankton studies (Brönmark and Hansson 2012). Due to pressure from predators, certain prey species may switch to cyst production from the regular parthenogenetic mode of reproduction. Since cysts are difficult to digest or even pass through unharmed from the predators' gut, formation of cysts is yet another life history strategy to offset predation pressure (Radzikowski 2013).

Mechanisms of prey alert in the presence of infochemicals from the predators have received considerably less attention. Partly bitten prey induces some behaviourial changes in its population (*e.g.*, higher swimming speed, patchy distribution away from the predators *etc.*) so that the individuals can avoid predation (Pestana *et al.*, 2013). For experiments designed to study the effect of kairomones, it is necessary keep the predators in an indirect contact (*e.g.*, separated by a mesh) with the prey so that there is a constant release of allelochemicals and at the same time predators do have access to prey. However, in order to keep predators hunger-free, a certain quantity of prey items are offered to them. The prey species used to feed the predators in test jars is different from the species that is used to study the impact of kairomones (Gama-Flores *et al.*, 2011) and this would also prevent the partially damaged prey to influence the population of test species through the release of alarm or alert chemicals (Brönmark and Hansson, 2012).

Equally important is the role of infochemicals within the population of a given species (pheromones). Density-dependent population regulation in zooplankton is long known but the precise mechanism is not often documented (Pourriot and Snell 1983). It is now known that the metabolic products including infochemicals released, when the population density of a given rotifer density is high, have a potential role in reducing population growth rates and egg ratios. It is known that overcrowding is one of the most common causes of sexual reproduction in cyclic parthenogens (Sarma *et al.*, 2005).

Generally, pheromones are not dependent on population density because a single individual is able to produce it. It has been observed that overcrowding of a given species of rotifer, causes the production of infochemicals that regulate the mode of reproduction. Thus these infochemicals these are dependent on population density and induce a change in the reproduction, from parthenogenesis to sexual mode. Thus, if the population density of a zooplankton species is below threshold levels of these infochemicals, then there appears to be no change in the mode of reproduction and the population continues to reproduce asexually (Mitchell and Carvalho 2002). This is well documented for rotifer species of the genus *Asplanchna*. For example, when the population density of *Asplanchna girodi* increases beyond a threshold level through parthenogenesis, then males begin to appear to fertilize the mictic (sexual) females leading to the production of resting eggs or cysts (Serra and King 1999). Thus, parthenogenetic females avoid intraspecific competition for limiting resources. It is possible that such a cue to shift from parthenogenesis to sexual mode of reproduction is due to both, accumulation of waste products and infochemicals. However, it is more than likely that the latter is mainly responsible for the production of resting eggs, based on the data of Hagiwara *et al.* (1994) who sonicated population of *Brachionus plicatilis* and added this extract to live rotifer cultures. The population that received conspecific extract had higher rates of resting egg production. Since the extracts lacked the waste metabolic products, it appears that conspecific infochemicals are possibly responsible for induced sexual reproduction in rotifers. Studies carried out for cladocerans also show the density-dependent ephippial production (Lürling *et al.*, 2003). In certain turbellarians too adults induce the neonates to switch to sexual reproduction using chemical cues (Fiore 1971).

Allelopathic Interactions between Littoral Invertebrate Predators and Zooplankton Prey

Littoral predatory invertebrates that are in contact with zooplankton in shallow waters exert some allelopathic pressure on the survival and reproduction of rotifers and cladocerans. *Hydra* is a littoral cnidarian that often feeds on rotifers and cladocerans and other smaller crustaceans (Quinn *et al.*, 2012). However, zooplankton within the vicinity of *Hydra* are also indirectly affected by the toxins and/or allelochemicals secreted by the cnidarians. Rivera-De la Parra et al. (2016) have tested the effect on *Hydra*-conditioned medium on the lifetable demography of the cladoceran *Daphnia* cf. *mendotae*. They observed that the reproductive rates and the rate of population increase (*r*, per day) of *D*. cf. *mendotae* decreased due to *Hydra*-allelochemicals. In addition, reproduction-related variables of the daphniids varied significantly in relation to the density of *Hydra* in the test jars (Figure 11.8).

Turbellarians (Phylum Platyhelminthes) are yet another group of littoral invertebrates influencing the abundance and diversity of rotifers and cladocerans

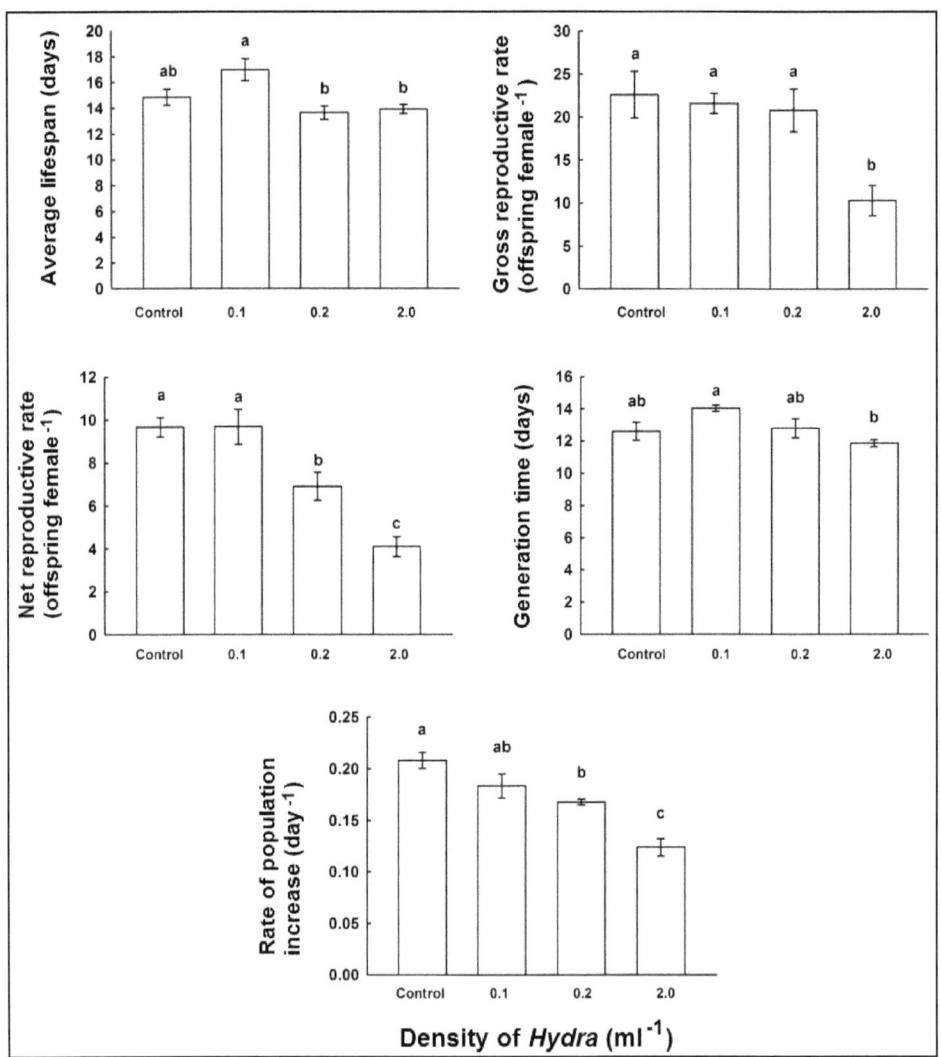

Figure 11.8: Demographic Variables (Average lifespan, gross reproductive rate, net reproductive rate, generation time and population growth rate) of *Daphnia* cf. *mendotae* Cultured in Presence of Allelochemicals from *Hydra* under Four Densities (ind. ml^{-1}). Shown are mean ± standard error for each treatment (Rivera-De la Parra et al., 2016).

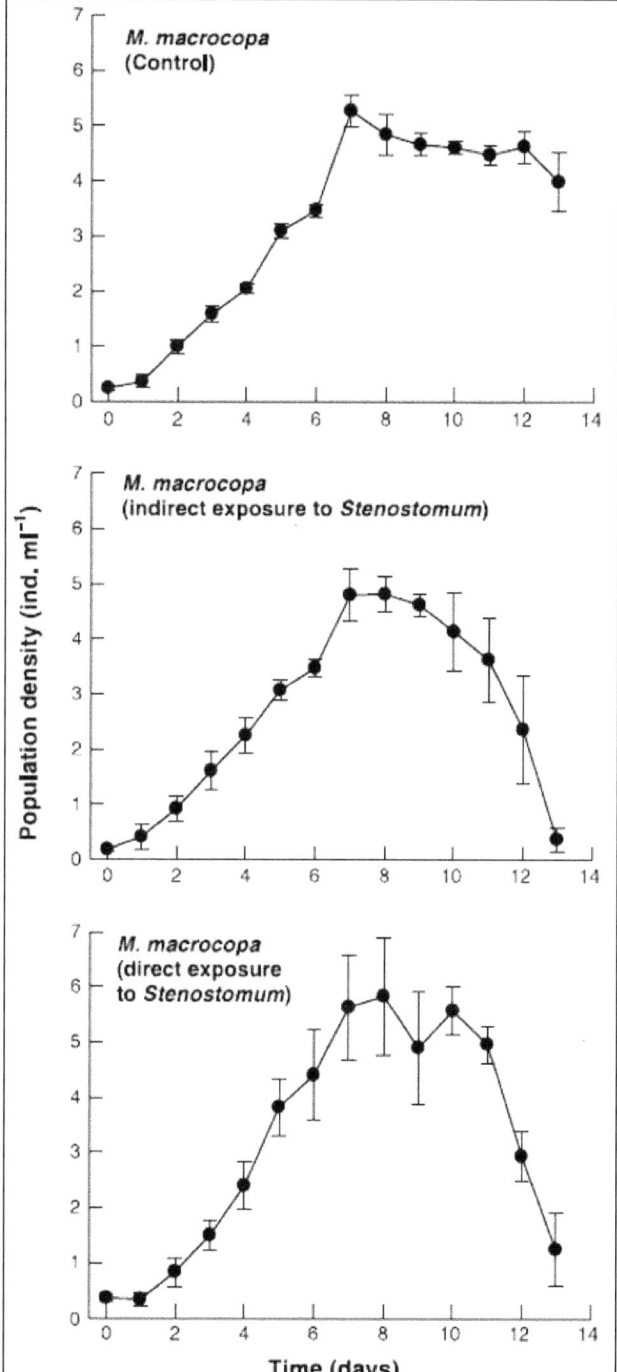

Figure 11.9: Population Growth Curves of *Moina macrocopa* in the Absence (Controls), Direct Presence of and Indirect Contact with *Stenostomum* sp. Shown are mean ± SE values based on four replicate recordings (Nandini *et al.*, 2011).

through their allelochemicals. Dumont *et al.* (2014) reviewed most recent information on some aspects of turbellarians including their ecology. Toxin production in the turbellarian *Rhynchomesostoma* is detrimental to its cladoceran prey including *Daphnia* and *Ceriodaphnia*. Excessive toxins produced by the high population density of this turbellarian is toxic to itself leading to the decline of the species abundance (Dumont *et al.*, 2014). A recent study on effects of allelochemicals from *Stenostomum* cf. *leucops* to the cladoceran *Moina macrocopa* reveals that the population growth rates of cladocerans declined due to the toxins from the turbellarians (Nandini *et al.*, 2011) (Figure 11.9). Interestingly, compared to controls, regardless of the mode of exposure to the turbellarians, *i.e.*, direct contact or via allelochemicals, the population density of *Moina* declined after about 10 days suggesting that the allelochemicals from the medium possibly accumulated in *Moina*. Accumulation of allelochemicals from other groups (*e.g.*, cyanobacteria) to higher trophic level (*e.g.*, herbivores, cladocerans) has also been reported (Ferrão-Filho and Kozlowsky-Suzuki 2011).

Cyanobacterial Allelochemicals and Zooplankton

Considerabale data on the role of toxic cyanobacteria such as *Microcystis* to zooplankton, particularly rotifers and cladocerans are available in literature. Most studies are concerned with the isolation and characterization of cyanotoxins (Whitton 2012) and the direct effect of feeding zooplankton with toxic cyanobacteria (Pérez-Morales *et al.*, 2014). Only few studies have tested the effect of cyanotoxins on growth, survival and reproduction of zooplankton (Lürling *et al.*, 2011). Using crude extracts of cyanotoxins, Zamora Barrios *et al.* (2015) have shown that the reproductive output of the rotifer *Plationus patulus* was lower at higher microcystin concentrations as compared to the controls. For the cladoceran *Ceriodaphnia cornuta* too, reproduction was restricted towards the end of the lifespan in the presence of cyanotoxins but in controls the age-specific fecundity was high throughout the lifespan (Figure 11.10). Toxins from cyanobacteria are endotoxins and hence are liberated once the cells are ingested by the herbivores. Due to natural cell lysis and application of chemical treatments such as algicides, these cyanotoxins are also freely liberated into the water (van Apeldoorn *et al.*, 2007). Though the FAO describes the safety limit of cyanotoxins as 1 µg/L for drinking water, in some reservoirs this limit is often exceeded. For example, in the drinking water reservoir, Valle de Bravo, Mexico, occasionally the quantity of cyanobacteria exceeds 10 µg/L (Alillo-Sánchez *et al.*, 2014). It is not fully known the consequences of high cyanotoxins on the zooplankton community structure in drinking water reservoirs.

The production of cyanotoxins by *Microcystis aerugina* and possibly by many other cyanobacteria is seasonal (Nandini *et al.*, 2016). Usually summer months are ideal for both cyanobacterial bloom formation and higher levels of cyanotoxins per cell (Znachor *et al.*, 2006). Since these cyanotoxins cause adverse effect to the consumers, they are treated as allelochemicals. Structurally microcystins contain three D-amino acids: alanine, methylaspartic acid and glutamic acid together with two uncommon N-methyldehydroalanine and 3-amino-9-methoxy-2,6,8,-trimethyl-10-phenyldeca-4,6-dienoic acid and the last one is associated with toxicity to organisms (Mazur and Pliński 2001). Cyanobacteria such as *Spirulina platensis*, are also known to produce antioxidants (Ismaeil *et al.*, 2014) which often has a positive

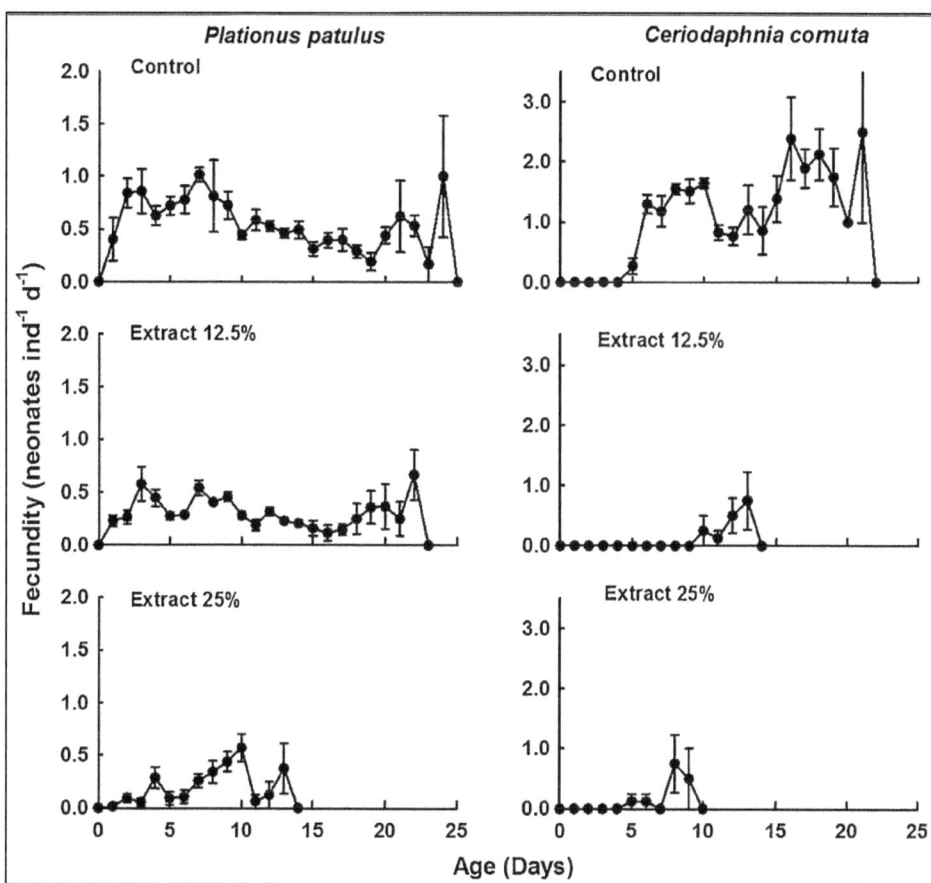

Figure 11.10: Fecundity Curves of *P. patulus* and *C. cornuta* in Controls (Absence of crude extracts) and different Proportions of the *Anabaena* Crude Extracts. Mean ± SD based on four replicates (Zamora Barrios *et al.*, 2015).

effect on the growth of rotifers (Snell *et al.*, 2012). Microcystins are extremely stable against hydrolysis, oxidation *etc.* and can even withstand boiling temperatures (Harada *et al.*, 1996).

Laboratory tests show that some species of cladocerans and rotifers discriminate the single cell form of *Microcystis* from edible algae of comparable size (Pérez-Morales *et al.*, 2014). In addition, when toxic *Microcystis* is mixed with edible algae, especially in higher proportion of the latter, zooplankton species show positive growth rates (Figure 11.11) suggesting that some natural resistance is present in zooplankton against the toxins (Alva-Martínez *et al.*, 2009). Fernández *et al.* (2014) studied the demographic responses of the cladoceran *Simocephalus vetulus* in the presence of ostracods (*Heterocypris incongruens*), fish (*Oreochromis*) kairomones and fed one of the three diets (green alga *Scenedesmus*, toxic cyanobacteria *Microcystis* and *Limnothrix*). Their results suggest that coexistence with ostracods was beneficial for cladocerans when food quality is poor. The ecological significance of such studies

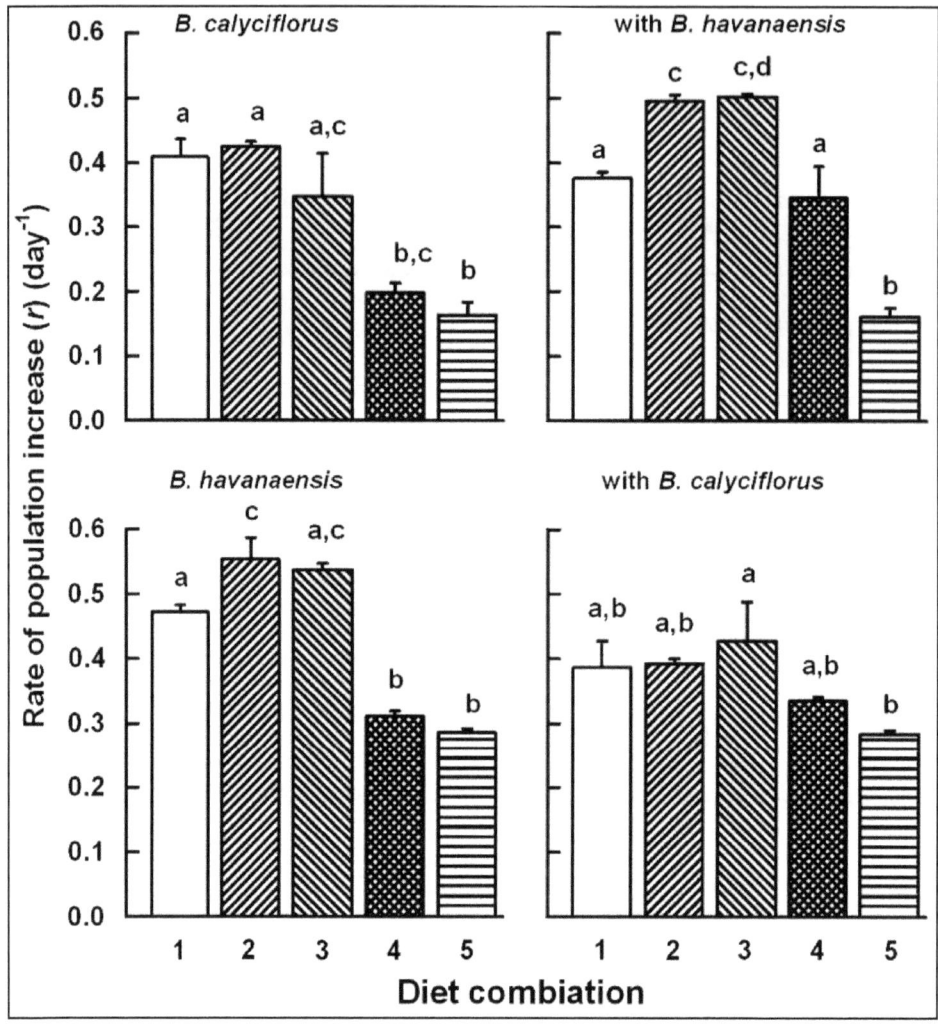

Figure 11.11: Rate of Population Increase (r, day^{-1}) of *B. calyciflorus* and *B. havanaensis* Fed different Diets (*Chlorella*, *Microcystis* or their proportions) Grown Separately or in Competition.

Shown are the mean±standard error based on four replicates. (1) *C. vulgaris* 100 per cent +*M. aeruginosa* 0 per cent; (2) *C. vulgaris* 75 per cent +*M aeruginosa* 25 per cent; (3) *C. vulgaris* 50 per cent +*M. aeruginosa* 50 per cent; (4) *C. vulgaris* 25 per cent +*M. aeruginosa* 75 per cent; (5) *C. vulgaris* 0 per cent +*M. aeruginosa* 100 per cent. For each treatment, bars carrying identical letters are not statistically significant ($p<0.05$, Tukey test) (Alva-Martínez et al., 2009).

seems to be that a resistant grazer can partly neutralize the toxins and hence the allelopathic effects no longer reach a sensitive co-existing zooplankton species.

Allelochemicals from Vertebrates and their Impact on Zooplankton

Exhaustive information is available about the fish kairomone effects on zooplankton, especially the temperate cladoceran *Daphnia* (Tollrian and Harvell 1999). Fish kairomones cause vertical migration and many other behavioural modifications and morphological changes in zooplankton (Gliwicz 2003). Kairomones are largely responsible for changes in certain life history traits of zooplankton such as the age at first reproduction, size at maturity and rates of offspring production. It is known that daphniids exposed to fish kairomones undergo phenotypic changes. Most daphniids when exposed to fish kairomones have the reduced age at maturity and size at maturity and enhanced neonate production in the first clutch (Lass and Spaak 2003). There is also some indication that density of fish increases the kairomone levels in the medium; and daphniids' response is generally elevated with increasing levels of fish allelochemicals (Reede 1995). Information about the minimum time required for the predator to release allelochemicals into medium is unavailable. In the laboratory experiments, the duration to obtain the conditioned medium is variable from 6 to 24h, although the most common is the latter (Gama-Flores *et al.*, 2013).

The response of rotifers to fish kairomones is equally interesting. The phenotypic responses, such as reduced body size, of rotifers exposed to fish kairomones are known. The induction of spines in the prey rotifers due to fish kairomones is not well-reported (Gilbert 2013). One of the common rotifers, *Brachionus rubens* lives in an epizoic mode on the carapace of cladocerans. This epizoic mode helps the rotifer to escape predation from invertebrates such as *Asplanchna* (Iyer and Rao 1995) and avoids interference competition from cladocerans (Iyer and Rao 1993). Peña-Aguada *et al.* (2008) studied the effect of kairomones from copepods, *Asplanchna* and fish on the epizoic tendency of *B. rubens*. Their results indicate that this epizoic nature of *B. rubens* was not an induced mechanism to avoid predation since there was no significant difference in the incidence of epizoic attachment to the carapace of daphniids with respect to the infochemicals of vertebrate or invertebrate origin.

Compared to fish, data on the effect of allelochemicals derived from other vertebrates to zooplankton are much limited. García *et al.* (2007) evaluated the morphometry and life-history characteristics of the rotifer *Brachionus havanaensis* in the presence of kairomones from salamander and copepod predators using population growth and life-table demography approach. Morphometric data on the body size of *B. havanaensis* showed that the total lorica length, anterior, and posterior spine lengths of *B. havanaensis* were significantly longer in the presence of kairomones from both vertebrate and invertebrate predators than in controls (Figure 11.12). Rotifers cultured using salamander-conditioned medium showed higher population growth than that using copepod kairomones. Shortest lifespan of rotifers was observed in treatments containing salamander kairomones. In addition, the reproductive rates of rotifers were significantly higher in the salamander-conditioned medium than in the other treatments including controls. These results suggest that rotifers show differential life history strategies to kairomones from invertebrate and vertebrate predators.

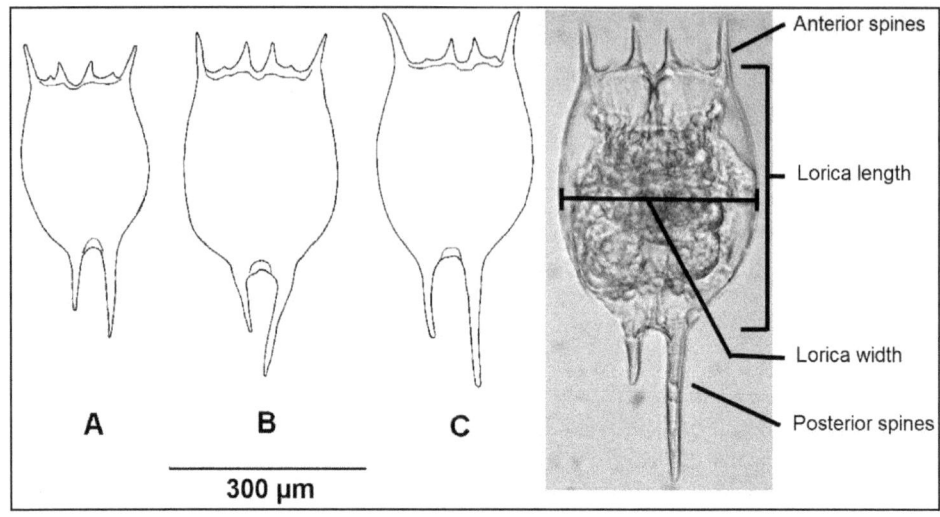

Figure 11.12: Lorica Morphology of Adult Parthenogenetic Female *Brachionus havanaensis* Cultured in Controls (A) and in the Presence of Kairomones from Copepods or Salamanders. All figures were drawn on the same scale (García *et al.*, 2007).

Gama-Flores *et al.* (2013) studied the survivorship and reproductive variables of *Moina macrocopa*, exposed to allelochemicals from fish, axolotl and frog tadpoles (Table 11.2). They found that survivorship variables were less affected due to vertebrate allelochemicals than those of reproduction of cladocerans. Compared with controls, there was a significant increase in the reproductive output exposed to vertebrate conditioned medium; the highest reproduction was observed in the tadpole conditioned medium while the lowest was with *Ambystoma*-conditioned medium. The three conditioned-media from the predators induced differential responses in life history traits of *M. macrocopa*.

Table 11.2: Demographic Variables of *Moina macrocopa* in Controls and Treatments Exposed to Conditioned-Medium from Axolotl, Fish and Tadpoles

Variable	Control	Ambystoma	Fish	Tadpole
ALS	9.49±0.11[a,c]	7.78±0.15[b]	10.61±0.10[c]	9.62±0.19[d,e]
GRR	35.67±1.36[a]	50.32±1.40[a,d]	92.18±3.04[c,d]	89.9±7.56[b,d]
NRR	16.41±0.29[a]	22.10±0.61[b]	37.43±0.71[c]	33.06±0.55[d]
GT	6.42±0.08[a,e]	6.15±0.16[b,e]	9.32 ±0.21[c]	8.08±0.17[d]
r	0.75±0.02[a]	0.62±0.01[b,e]	0.18±0.02[c]	0.62±0.03[d,e]

ALS: Average lifespan (days), GRR: Gross reproductive rate (offspring/female), NRR: Survival-weighted offspring/female), GT: Generation time (days) and *r*: Rate of population increase per day. Data show the mean ± standard error based on 5 replicates. For each variable, data carrying similar alphabets are not statistically significant (p>0.05, Tukey test) (Gama-Flores *et al.*, 2013).

Conclusions

Allelopathic interactions in aquatic ecosystems are more complex than previously thought. Different groups of organisms, from microorganisms to vascular plants and lower invertebrates to vertebrates produce secondary metabolites that have allelopathic properties, mainly against target species. In spite of considerable advances made in this field, the very definitions of various components of infochemicals of which allelopathy is a part are still in vague. The infochemical effects in aquatic species include changes in the morphology and behaviour of target species and even self-destruction via toxic metabolites. Literature contains abundant data about the allelopathic effects of secondary metabolites from macrophytes to phytoplankton and fish kairomones to zooplankton. However, for several sedentary groups of invertebrates such as bryozoans, anemones and polychaetes, data on the allelopathic effects are scanty. Compared to terrestrial systems (such as agroecology), commercial exploitation of allelopathic knowledge in aquatic ecosystems is still not at a reachable distance. Hence, further research in this area is urgently needed.

Acknowledgements

We thank Bansi Lal Kaul, Ana Luisa Anaya Lang and Cristian Alberto Espinosa Rodríguez for suggestions during the preparation of this manuscript. Funding from PAPIIT- DGAPA - IN213513, UNAM is acknowledged.

References

Addisie, Y. and Medellin, A.C., 2012. Allelopathy in aquatic macrophytes: Effects on growth and physiology of phytoplanktons. *African Journal of Plant Science* 6: 270-276.

Agrawal, A.A., 1998. Induced responses to herbivory and increased plant performance. *Science,* 279: 1201-1201.

Alillo-Sánchez, J.L., Gaytán-Herrera, M.L., Martínez-Almeida, V.M. and Ramírez-García, P., 2014. Microcystin-LR equivalents and their correlation with *Anabaena* spp. in the main reservoir of a hydraulic system of Central Mexico. *Inland Waters,* 4: 327-336.

Alva-Martínez, A.F., Sarma, S.S.S. and Nandini, S., 2007. Effect of mixed diets (cyanobacteria and green algae) on the population growth of the cladocerans *Ceriodaphnia dubia* and *Moina macrocopa*. *Aquatic Ecology,* 41: 579-585.

Alva-Martínez, A.F., Fernández, R., Sarma, S.S.S. and Nandini, S., 2009. Effect of mixed toxic diets (*Microcystis* and *Chlorella*) on the rotifers *Brachionus calyciflorus* and *Brachionus havanaensis* cultured alone and together. *Limnologica*, 39: 302-305.

Anaya, A.L., 1999. Allelopathy as a tool in the management of biotic resources in agroecosystems. *Critical Reviews in Plant Sciences,* 18: 697-739.

Ahn, C-Y., Park, M-H., Joung, S-H., Kim, H-S., Jang, K-Y. and Oh, H-M., 2003. Growth inhibition of cyanobacteria by ultrasonic radiation: Laboratory and enclosure studies. *Environ. Sci. Technol.,* 37: 3031-3037.

Barbosa, P. and Castellanos, I. (Eds), 2005. *Ecology of Predator-Prey Interactions.* Oxford University Press, London.

Brönmark, C. and Hansson, L-A. (Eds), 2012. *Chemical Ecology in Aquatic Systems.* Oxford University Press, London.

Cheema, Z.A., Farooq, M. and Wahid, A., 2013. *Allelopathy. Current Trends and Future Applications.* Springer, New York.

Dicke, M. and Sabelis, M.W., 1988. Terminology of chemicals involved in interactions between individual organisms: should it be based on cost-benefit analysis rather than on origin of compounds? *Functional Ecology,* 2: 131-139.

De Beauchamp, P., 1952. Un facteur de la variabilité chez les rotifères du genre *Brachionus.* Comptes rendus de l'Académie des sciences, Paris 234: 573-575.

Dumont, H.J., Tundisi, J.G. and Roche, K. (Eds), 1990. Intrazooplankton predation. *Hydrobiologia,* 198: 1-242.

Dumont, H.J., Rietzler, A.C. and Bo-Ping, H., 2014. A review of typhloplanid flatworm ecology, with emphasis on pelagic species. *Inland Waters,* 4: 257-270.

Duncan, A., 1983. The composition, density and distribution of the zooplankton of Parakrama Samudra. In: *Limnology of Parakrama Samudra,* Sri Lanka, ed. Schiemer F pp. 85-94. Junk, The Hague.

Espinosa-Rodríguez, C.A,. Valencia-Del Toro, G., Sarma, S.S.S. and Nandini, S., 2015a. Allelopathic activity and chemical analysis of crude extracts from the macrophyte *Egeria densa* on selected phytoplankton species. *Allelopathy Journal,* 37: 147-160.

Espinosa-Rodríguez, C.A., Rivera-De la Parra, L., Martínez-Téllez, A., Gómez-Cabral, G.C., Sarma, S.S.S. and Nandini, S., 2016b. Allelopathic interactions between the macrophyte *Egeria densa* and plankton (alga, *Scenedesmus acutus* and cladocerans, *Simocephalus* spp.): a laboratory study. *Journal of Limnology,* 75(s1): 151-160.

Fernández, R., Nandini, S., Sarma, S.S.S. and Castellanos-Paez, M.E., 2014. Effects of cyanobacteria, fish kairomones, and the presence of ostracods on the demography of *Simocephalus vetulus* (Cladocera). *Invertebrate Biology,* 133: 371-380.

Ferrão-Filho, A.S. and Kozlowsky-Suzuki, B., 2011. Cyanotoxins: Bioaccumulation and effects on aquatic animals. *Mar Drugs,* 9: 2729-2772.

Fiore, L., 1971. A mechanism for self-inhibition of population growth in the flatworm *Mesostoma ehrenbergii* (Focke). *Oecologia,* 7: 356-360.

Fleming, A., 1929. On the antibacterial action of cultures of a penicillium with special reference to their use in the isolation of *B. influenzae. British Journal of Experimental Pathology,* 10: 226-236.

Gama-Flores, J.L., Huidobro-Salas, M.E., Sarma, S.S.S. and Nandini, S., 2011. Effects of predator (*Asplanchna*) type and density on morphometric responses of *Brachionus* calyciflorus (Rotifera). *Allelopathy Journal,* 27: 289-300.

Gama-Flores, J.L., Huidobro-Salas, M.E., Sarma, S.S.S. and Nandini, S., 2013. A laboratory study on the effects of allelochemicals released by vertebrates (fish, salamander and tadpole) on the life history parameters of *Moina macrocopa* (Cladocera). *Allelopathy Journal,* 31: 415-425.

García, C.E., Chaparro-Herrera, D.J., Nandini, S. and Sarma, S.S.S., 2007. Life history strategies of *Brachionus havanaensis* subject to kairomones of vertebrate and invertebrate predators. *Chemistry and Ecology,* 23: 303-313.

Gilbert, J.J., 1966. Rotifer ecology and embryological induction. *Science,* 151: 1234-1237.

Gilbert, J.J. and Waage, J.K., 1967. *Asplanchna, Asplanchna*-substance, and posterolateral spine length variation of the rotifer *Brachionus calyciflorus* in a natural environment. *Ecology* 48, 1027-1031.

Gilbert, J.J., 2013. The cost of predator-induced morphological defense in rotifers: experimental studies and synthesis. *Journal of Plankton Research,* 35: 461-472.

Gilbert, J.J. 2014. Morphological and behavioral responses of a rotifer to the predator Asplanchna. *Journal of Plankton Research,* 36: 1576-1584.

Gliwicz, Z.M., 2003. Between hazards of starvation and risk of predation: The ecology of offshore animals. *Excellence in Ecology* No. 12, Oldendorf/Luhe, Germany.

Gopal, B. and Goel, U., 1993. Competition and allelopathy in aquatic plant communities. *Botanical Review,* 59: 155-210.

Gross, E.M., 2003. Allelopathy of aquatic autotrophs. *Critical Reviews in Plant Sciences,* 22: 313-339.

Gross, E.M., 2000. Seasonal and spatial dynamics of allelochemicals in the submersed macrophyte *Myriophyllum spicatum*. Verhandlungen Internationale Vereinigung für theoretische und angewandte. *Limnologie,* 27: 2116-2119.

Gross, E.M., Hilt, S., Lombardo, P. and Mulderij, G., 2007. Searching for allelopathic effects of submerged macrophytes on phytoplankton - state of the art and open questions. *Hydrobiologia,* 584: 77-88.

Gulati, R.D., Lammens, E.H.R.R., Meijer, M-L. and van Donk, E. (Eds), 1990. Biomanipulation - tool for water management. *Hydrobiologia,* 200/201, 1-628.

Gutiérrez, E., Huerto, R., Saldaña, P. and Arreguín, F., 1996. Strategies for water-hyacinth (*Eichhornia crassipes*) control in Mexico. *Hydrobiologia,* 340: 181-185.

Hagiwara, A., Hamada, K., Hori, S. and Hirayama, K., 1994. Increased sexual reproduction in *Brachionus plicatilis* (Rotifera) with the addition of bacteria and rotifer extracts. *Journal of Experimental Marine Biology and Ecology,* 181: 1-8.

Haig, T., 2008. Chapter 4. Allelochemicals in plants. In: *Allelopathy in Sustainable Agriculture and Forestry,* eds. Zeng RS, Mallik AU and Luo SM, pp. 63-104. Springer, New York.

Hanson, M.A. and Butler, M.G., 1994. Responses of plankton, turbidity, and macrophytes to biomanipulation in a shallow Prairie Lake. *Canadian Journal of Fisheries and Aquatic Sciences,* 51: 1180-1188.

Harada K-I, Tsuji K, Watanabe MF and Kondo F (1996). Stability of microcystins from cyanobacteria-III. Effect of pH and temperature. *Phycologia*, 35: 83-88.

Hilt, S. and Gross, E.M., 2008. Can allelopathically active submerged macrophytes stabilise clear-water states in shallow lakes? *Basic and Applied Ecology*, 9: 422-432.

Ismaiel, M.M.S., El-Ayouty, Y.M. and Piercey-Normore, M.D., 2014. Antioxidants characterization in selected cyanobacteria. *Annals of Microbiology*, 64: 1223-1230.

Iyer, N. and Rao,T.R., 1993. Effect of the epizoic rotifer *Brachionus rubens* on the population growth of three cladoceran species. *Hydrobiologia*, 255/256: 325-332.

Iyer, N. and Rao, T.R., 1995. The epizoic mode of life in *Brachionus rubens* Ehrenberg as a deterrent against predation by *Asplanchna* intermedia Hudson. *Hydrobiologia*, 313/314: 377-380.

Jang, M-H., Jung, J-M. and Takamura, N., 2007. Changes in microcystin production in cyanobacteria exposed to zooplankton at different population densities and infochemical concentrations. *Limnology and Oceanography*, 52: 1454-1466.

Kerfoot, W.C. and Sih, A. (Eds.), 1987. *Predation: Direct and Indirect Impacts on Aquatic Communities*. University Press of New England, Hanover.

Körner, S. and Nicklisch, A., 2002. Allelopathic growth inhibition of selected phytoplankton species by submerged macrophytes. *Journal of Phycology*, 38: 862-871.

Kruse, M., Strandberg, M. and Strandberg, B., 2000. Ecological Effects of Allelopathic Plants – a Review. NERI Technical Report No. 315, National Environmental Research Institute, Silkeborg, Denmark.

Lampert, W., Rothhaupt, K.O. and von Elert, E., 1994. Chemical induction of colony formation in the green alga (*Scenedesmus acutus*) by grazers (*Daphnia*). *Limnology and Oceanography*, 39: 1543-1550.

Lass, S. and Spaak, P., 2003. Chemically induced anti-predator defences in plankton: a review. *Hydrobiologia*, 491: 221-239.

Lindholm, T., Öhman, P., Kurki-Helasmo, K., Kincaid, B. and Meriluoto, J., 1999. Toxic algae and fish mortality in a brackishwater lake in Åland, SW Finland. *Hydrobiologia*, 397: 109-120.

Lupini, A., Araniti, F., Sunseri, F. and Abenavoli, M.R., 2013. Gravitropic response induced by coumarin: Evidences of ROS distribution involvement. *Plant Signaling and Behavior*, 8(2): e23156.

Lürling, M., Roozen, F., van Donk, E. and Goser, B., 2003. Response of *Daphnia* to substances released from crowded congeners and conspecifics. *Journal of Plankton Research*, 25: 967-978.

Lürling, M., Faassen, E.J. and van Eenennaam, J.S., 2011. Effects of the cyanobacterial neurotoxin β-N-methylamino-L-alanine (BMAA) on the survival, mobility and reproduction of *Daphnia magna*. *Journal of Plankton Research*, 33: 333-342.

Matveev, V., 1993. An investigation of allelopathic effects of *Daphnia*. *Freshwater Biology*, 29: 99-105.

Mazur, H. and Pliñski, M., 2001. Stability of cyanotoxins, microcystin-LR, microcystin-RR and nodularin in seawater and BG-11 medium of different salinity. *Oceanologia,* 43: 329-339.

Mitchell, S.E. and Carvalho, G.R., 2002. Comparative demographic impacts of 'info-chemicals' and exploitative competition: an empirical test using *Daphnia magna. Freshwater Biology,* 47: 459-471.

Moss, B., 2010. *Ecology of Freshwaters: A View for the Twenty-first Century,* 4th edn. Wiley-Blackwell, London.

Mulderij, G,, van Donk, E. and Roelofs, J.G.M., 2003. Differential sensitivity of green algae to allelopathic substances from Chara. *Hydrobiologia,* 491: 261-271.

Murillo, P.G., Zamudio, R.F. and Bracamonte, S.C., 2009. Habitantes del agua. Macrófitos. Agencia Andaluza del Agua, España.

Nakai, S., Inoue, Y., Hosomi, M. and Murakami, A., 1999. Growth inhibition of blue-green algae by allelopathic effects of macrophytes. *Water and Science Technology,* 8: 47-53.

Nandini, S. and Sarma, S.S.S., 2005. Life history characteristics of *Asplanchnopus multiceps* (Rotifera) fed rotifer and cladoceran prey. *Hydrobiologia,* 546: 491-501.

Nandini, S., Sarma, S.S.S. and Dumont, H.J., 2011. Predatory and toxic effects of the turbellerian (*Stenostomum* cf. *leucops*) on the population dynamics of *Euchlanis dilatata, Plationus patulus* (Rotifera) and *Moina macrocopa* (Cladocera). *Hydrobiologia,* 662: 171-177.

Nandini, S., Miracle, M.R., Vicente, E., Sarma, S.S.S. and Gulati, R.D., 2016. *Microcystis* extracts and single cells have differential impacts on the demography of cladocerans: a case study on *Moina* cf. *micrura* isolated from the Mediterranean coastal shallow lake (L'Albufera, Spain). Hydrobiologia (in press).

Oh, H-M., Lee, S.J., Min-Ho, Jang M-H. and Yoon, B-D., 2000. Microcystin production by *Microcystis aeruginosa* in a phosphorus-limited chemostat. *Applied and Environmental Microbiology,* 66: 176-179.

Olofsdotter, M. (Ed.), 1998. *Allelopathy in Rice.* International Rice Research Institute, Manila.

Pérez-Morales, A., Sarma, S.S.S. and Nandini, S., 2014. Feeding and filtration rates of zooplankton (rotifers and cladocerans) fed toxic cyanobacterium (*Microcystis aeruginosa*). *Journal of Environmental Biology,* 35: 1013-1020.

Pérez-Morales, A., Sarma, S.S.S. and Nandini, S., 2015. Zooplankton-induced microcystins production in Microcystis aeruginosa (Kützing). *Hidrobiológica,* 25(3): 411-415.

Pestana, J.L.T., Baird, D.J. and Soares, A.M.V.M., 2013. Predator threat assessment in *Daphnia magna*: the role of kairomones versus conspecific alarm cues. *Marine and Freshwater Research,* 64: 679-686.

Peña-Aguado, F., Morales-Ventura, J., Nandini, S. and Sarma, S.S.S., 2008. Influence of vertebrate and invertebrate infochemicals on the population growth and epizoic tendency of *Brachionus rubens* (Ehrenberg) (Rotifera: Brachionidae). *Allelopathy Journal*, 22: 123-130.

Pourriot, R. and Snell, T.W., 1983. Resting eggs in rotifers. *Hydrobiologia*, 104: 213-224.

Quinn, B., Gagné, F. and Blaise, C., 2012. *Hydra*, a model system for environmental studies. *International Journal of Developmental Biology*, 56: 613-625.

Radzikowsk, J., 2013. Resistance of dormant stages of planktonic invertebrates to adverse environmental conditions. *Journal of Plankton Research*, 35: 707-723.

Rao, D.V.S., Quilliam, M.A. and Pocklington, R., 1988. Domoic acid: A neurotoxic amino acid produced by the marine diatom *Nitzschia pungens* in culture. *Canadian Journal Fisheries and Aquatic Sciences*, 45: 2076-2079.

Reede, T., 1995. Life history shifts in response to different levels of fish kairomones in *Daphnia*. *Journal of Plankton Research*, 17: 1661-1667.

Rice, E.L., 1974. *Allelopathy*. Academic Press, New York

Rivera-De la Parra, L., Sarma SSS and Nandini S. (2016). Direct and Indirect effects of predation by Hydra (Cnidaria) on cladocerans (Cladocera). *Journal of Limnology*, 75(s1): 39-47.

Rothbard, S., 1975. Control of *Euplotes* sp. by formalin in growth tanks of *Chlorella* sp. used as medium for the rotifer *Brachionus plicatilis*, which serves as food for hatchlings. *Bamidgeh*, 27: 101-109.

Sarma, S.S.S., 1991. Rotifera. In: *Manual on Culture of Live Food Organisms*, (Eds.) T.J. Pandian and M.P. Marian. Marine Products Export Development Authority, Government of India, Cochin, India: 47-61.

Sarma, S.S.S., Resendiz, R.A.L. and Nandini, S., 2011. Morphometric and demographic responses of brachionid prey (*Brachionus calyciflorus* Pallas and *Plationus macracanthus* (Daday)) in the presence of different densities of the predator *Asplanchna brightwellii* (Rotifera: Asplanchnidae). *Hydrobiologia*, 662: 179-187.

Sarma, S.S.S., Gulati, R.D. and Nandini, S., 2005. Factors affecting egg-ratio in planktonic rotifers. *Hydrobiologia*, 546: 361-373.

Sánchez Rodríguez, M.R., Avila, L.A.N., Sarma, S.S.S., Nandini, S. and Vásquez, A.L., 2010. Allelopathic effects of ciliate (*Paramecium caudatum*) (Ciliophora) culture filtrate on the population growth of brachionid rotifers (Rotifera: Brachionidae). *Allelopathy Journal*, 26: 123-130.

Sbarbati, A. and Osculati, F., 2006. Allelochemical communication in vertebrates: kairomones, allomones and synomones. *Cells Tissues Organs*, 183: 206–219.

Serra, M. and King, C.E., 1999. Optimal rates of bisexual reproduction in cyclical parthenogens with density-dependent growth. *Journal of Evolutionary Biology*, 12: 263-271.

Snell, T.W., 1998. Chemical ecology of rotifers. *Hydrobiologia*, 387/388: 267-276.

Snell, T.W., Fields, A.M. and Johnston, R.K., 2012. Antioxidants can extend lifespan of *Brachionus manjavacas* (Rotifera), but only in a few combinations. *Biogerontology,* 13: 261-275.

Spaak, P. and Boersma, M., 1997. Tail spine length in the *Daphnia galeata* complex: costs and benefits of induction by fish. *Aquatic Ecology,* 31: 89-98.

Tammilehto, A., Nielsen, T.G., Krock, B., Møller, E.F. and Lundholm, N., 2015. Induction of domoic acid production in the toxic diatom *Pseudo-nitzschia seriata* by calanoid copepods. *Aquatic Toxicology,* 159: 52-61.

Tollrian, R. and Harvell, C.D. (Eds.), 1999. *The Ecology and Evolution of Inducible Defenses.* Princeton University Press, New Jersey.

van Apeldoorn, M.E., van Egmond, H.P., Speijers, G.J.A. and Bakker, G.J.I., 2007. Toxins of cyanobacteria. *Molecular Nutrition and Food Research,* 51: 7-60.

van Donk, E. and van de Bund, W.J., 2002. Impact of submerged macrophytes including charophytes on phyto- and zooplankton communities: allelopathy versus other mechanisms. *Aquatic Botany,* 72: 261-274.

van Oosterhout, F. and Lürling, M., 2011. Effects of the novel 'Flock and Lock' lake restoration technique on *Daphnia* in Lake Rauwbraken (The Netherlands). *J. Plankton Res.,* 33: 255-263.

Wallace, R.L., Snell, T.W., Ricci, C. and Nogrady, T., 2006. Rotifera Part 1: Biology, ecology and systematics. Guides to the identification of the microinvertebrates of the continental waters of the world. Kenobi Productions Gent/Backhuys, The Netherlands.

Whitton, B.A., (Ed.), 2012. *Ecology of Cyanobacteria II: Their Diversity in Space and Time.* Springer, New York.

Wiedner, C., Visser, P.M., Fastner, J., Metcalf, J.S., Codd, G.A. and Mur, L.R., 2003. Effects of light on the microcystin content of *Microcystis* strain PCC 7806. *Applied and Environmental Microbiology,* 69: 1475-1481.

Wium-Andersen, S., Anthoni, U., Christophersen, C. and Houen, G., 1982. Allelopathic effects on phytoplankton by substances isolated from aquatic macrophytes (Charales). *Oikos,* 39: 187-190.

Wolfe, G.V., 2000. The chemical defense ecology of marine unicellular plankton: constraints, mechanisms, and impacts. *Biological Bulletin,* 198: 225-244.

Zamora Barrios, C.A., Nandini, S. and Sarma, S.S.S., 2015. Effect of crude extracts of *Dolichospermum planctonicum* on the demography of *Plationus patulus* (Rotifera) and *Ceriodaphnia cornuta* (Cladocera). *Ecotoxicology,* 24: 85-93.

Zaret, T.M., 1980. *Predation and Freshwater Communities.* Yale University Press, New Haven/London.

Znachor, P., Jurczak, T., Komárková, J., Jezberová, J., Mankiewicz, J., Kastovská, K. and Zapomelová, E., 2006. Summer changes in cyanobacterial bloom composition and microcystin concentration in eutrophic Czech reservoirs. *Environmental Toxicology and Chemistry,* 21: 236-243.

Section II
Wildlife

Chapter 12

Breeding Ecology of Yellow-Wattled Lapwing *Vanellus malabaricus* in the Kole Wetlands of Thrissur, Kerala

P. Greeshma and E.A. Jayson

ABSTRACT

Lapwings belong to the family Charadriidae of the avian order Charadriiformes. Yellow-wattled Lapwing is an uncommon resident bird inhabiting a variety of open lowland habitats like dry areas, bare lands, fallow fields and the fringes of wetlands. It has a shorter stature compared with that of Red-wattled Lapwing and is characterized by the presence of bright yellow fleshy lappets above and in front of eyes. The present study on Yellow-wattled Lapwing was carried out in the Kole wetlands of Thrissur, Kerala (Ramsar Site) (Latitude: 10°31'22" N and Longitude: 76°10'14" E); about 7 km to the West of Thrissur city. The study was conducted from 2015 January to June 2015 for a period of 6 months. The observations on nests, foraging and anthropogenic factors acting as the stress factors towards breeding and hatching success were made using spotting scope (10x- 45x) and binocular (7 X 50). Video clips were also been made for understanding of the behavioral aspects (Sony HDR PJ 410). Two types of nests; one with a collection of tiny pebbles in the completely open areas and second type in the middle of small grassy patches were recorded for the species. Adult birds never left its foraging cum breeding ground during any time of the day. Breeding period were from March to May. The eggs were laid in the nests which are highly camouflaged. Dumping of both organic and inorganic waste, man-made fires, cattle, stray dogs, usage of the area as a playground and bike racing are the major threats confronted by the species. Hatching success of Yellow-wattled Lapwing in this breeding season was 44 per cent.

***Keywords**: Yellow-wattled Lapwing, Nesting, Breeding success, Threats, Kerala, India.*

Introduction

Birds are an essential component of human inhabitations, mountains, oceans, Ice lands and virtually in each and every corner. However, few birds are characterized by their unusually veiled presence. Yellow-wattled Lapwing (*Vanellus malabaricus*) is one such bird (Gupta and Kaushik, 2012). The endemic Yellow-wattled Lapwing *Vanellus malabaricus* is found in most parts of the Indian Subcontinent (Sethi *et al.*, 2010). Yellow-wattled Lapwing is an uncommon resident bird inhabiting a variety of open lowland habitats like dry areas, bare lands, fallow fields and the fringes of wetlands. It has a shorter stature compared with that of Red-wattled Lapwing and is characterized by the presence of bright yellow fleshy lappets above and in front of eyes. Its habitats preferences include any sort of open ground, dry fields and the largest concentrations are found in and near wetlands fringes (Kumar, 2015). They make short distance movement in response to rain. It is obligate visual forager, meaning catch its prey at the substrate boundary layers, by picking small invertebrates from the surface or from low vegetation cover. Yellow-wattled Lapwing plays a prominent role in ecosystems. The Yellow-wattled Lapwing contributes to maintaining ecosystem food chain because they regulate and maintain the populations of many invertebrates (Pests) which are harmful for agricultural crops (Adesh and Amita, 2015). Only few workers have so far focused attention on Yellow-wattled lapwing (Jayakar and Spurway, 1965, 1968; Dhindsa, 1983; Santharam, 1995, Sethi *et al.*, 2010, Gupta and Kaushik, 2012). Even though Lapwing studies from Calicut University campus was reported (Vijayagopal, 1991) no one has studied Yellow-wattled Lapwing in the context of threats to its nesting in Kerala.

Study Area and Methods

The Kole Wetlands is one of the largest, highly productive and threatened wetlands in Kerala, declared as Ramsar Site in 2002. The Kole wetlands lies between 10° 20' and 10° 40' N latitudes and 75° 58' and between 76° 11' E longitudes. Avifaunal studies in various regions of Thrissur District concluded that the highest number of birds was reported from Kole wetlands (Jayson and Sivaperuman, 2005). The breeding cum foraging site of the Yellow-wattled Lapwing recorded during

Figure 12.1: Yellow Wattled Lapwing.

Figure 12.2: Nesting-cum-Foraging Ground.

the present study is located in the outskirts of Pullazhi Kole wetland (Latitude: 10°31'22" N and Longitude: 76°10'14" E); about 7 km to the West of Thrissur city.

The study was conducted from 2015 January to June 2015 for a period of 6 months. The observations on nests, foraging and anthropogenic factors acting as the stress factors towards breeding and hatching success were made using spotting scope (10x- 45x) and binocular (7 X 50). Video clips were also been made for understanding of the behavioral aspects (Sony HDR PJ 410). Field survey was carried to find the nest and the eggs. Nests were spotted and identified while the bird scraped the ground for nesting, and threatening predators at nest-sites. Searches for nests were done in all parts of the study area. In addition, local inhabitants such as children, farmers, and cattle-grazers were regularly queried regarding the occurrence of the nest(s) of Yellow-wattled Lapwing in their premises or nearby areas. During breeding season, daily visit was carried out and in each visit, for each nest the number of eggs, the species and any evidence of hatching success (chicks emerging) was recorded.

Results

Yellow-wattled Lapwings were found in groups of 8-10, never more. The Yellow-wattled Lapwings make their nest in the ground, in a peculiar fashion, which is of remarkable camouflage. The locations of nests in the study area were remarkably brilliant, that they preferred the most untouched area by human beings and other animals. Commonly 2 types of nesting pattern are seen, first with a collection of tiny pebbles (Sethi *et al.*, 2010) within which their well camouflaged eggs are laid and second type in the middle of grassy patch. Adult birds never left its foraging-cum-breeding ground during any time of the day.

They fed by pecking from the ground and also from the grassy patches in the barren land. Yellow-wattled Lapwings were observed running in a cursory on ground as retreating in response to human or other intrusions. Very often they were seen feeding in a normal way. Lapwing exhibits a very strong social and territorial behaviour. It was observed that one or two individuals of the entire group scan the surroundings while the rest of the group is feeding. Their call is a sharp, plaintive *ti-ee, ti-ee,* which lasts for few seconds and when they are alarmed they produce

Figure 12.3: Nest of Yellow-Wattled Lapwing.

Figure 12.4: Nest with Eggs Near the Glass Pieces.

Figure 12.5: Yong Lapwing Foraging in Ground.

Figure 12.6: Predated Yellow-wattled Lapwing Egg.

sharp, high-pitched *twit-twit-twit*. The main food items preferred by the bird were found to be ants, beetles, termites and other invertebrates. Breeding was observed from March to May. The eggs were laid in the nests which are highly camouflaged. The clutch size of this species ranges from 2 to 3. The colour of the eggs matches the soil and pebbles with at most perfection. Thus to re-spot the nest and eggs quickly was impossible.

A total of 11 eggs were spotted out from 5 nests seen in the study area. Of these 4 eggs hatches out leading to a hatching success of 44 per cent. Later 1 death was recorded and circumstantial evidences showed that 1 egg was predated and 6 egg loss (Table 12.1). When the nests are approached, the birds make distress calls and start to fly randomly above the nesting ground. Present study area is an open land surrounded with human habitations. Increased expansion of human dwellings together with the increased human intrusions leads to the shrinkage of the Lapwing habitat. Dumping of both organic and inorganic wastes, especially building materials permanently destroys the nesting area of the species. Further, the household waste attracts the stray dogs peeping into the area. The number of stray dogs was found to be increased in the study area from 12 to 19. The movement of the stray dogs becomes a threat to the lapwing for incubating their eggs. It was also observed that dogs chasing the incubating lapwing. Even though the fledglings of lapwings are so camouflaged to the small grassy patch, one death was reported.

Table 12.1: Productivity in Nests of the Yellow-Wattled Lapwing during 2015

Nests Observed	Eggs Laid	Eggs Hatched	Death	Egg Lost	Predation	Hatching Success (per cent)
05	11	04	01	06	01	44

Amongst other principal threats to eggs of yellow-wattled lapwing in the study area, natural dangers from birds of prey (*Corvus splendens, Centropus sinensis*) are looming large. In several occasions it was seen that the lapwings forming a group chasing the birds in the sky in a violent and agonistic flight. Rain and flooding is another threat, as the ground level is of varying heights with hikes and depressions

Figure 12.7: Stray Dogs in the Study Area. **Figure 12.8: Lapwing Habitat as Playground.**

the summer shower caused a great tragedy. Rain water carried the organic as well as inorganic wastes to the nesting areas of lapwing and re-dumped over there leading to the habitat destruction. After the rain, several eggs were misplaced and some were not seen. However all the impending dangers is the behavioural pattern of modern man. Prior to the usage of this area as playground, people set fire in several patches of the lapwing habitat; leading to the habitat fragmentation. It is well known that the ground-nesting birds are victims of high rate of their eggs and young ones. The exposed nests and its eggs together with the fledglings face a serious threat because of the unconscious activities of man. It's quite incidental that the summer vacation for children and the breeding season of lapwings comes together which pose a great threat to the activities of the lapwings. Also the man-made fires, sweep out the ground fauna and flora, which was the lapwing's foraging ground, leading to food depletion. Every nook and corner of the area was converted to cricket pitches and football courts, which disturbed the breeding lapwings in such an alarming rate. It was also seen that the presence of livestock deters the lapwings from settling on the ground for nesting and breeding. They never chose to move towards the grazing area. We observed several instances of bike racing and driving practices in the lapwing's foraging cum breeding ground. Bike racing alters the soil level and destroys the low green patches in the soil. Queries with local inhabitants revealed that people used to collect the eggs and consume it. Altogether Yellow-wattled lapwing is being pushed into a very difficult situation day by day.

Conclusions

During nest-site selection birds consider the proximity of feeding areas, shelter and protection or camouflage against predators. The gathering of Yellow-wattled Lapwings at above study site for feeding, nesting and breeding may be due to sufficient food availability and safe shelter from predators. Human threats to ground nesting birds are either direct or indirect. The small bushes provide safe habitat for hiding the fledglings should be conserved devoid of cattle grazing. Hart *et al.* (2002) negatively correlated that the breeding densities of lapwing negatively correlated with the presence of livestock, and they suggested the exclusion of livestock from some areas as a desirable option in order to increase the nesting success of lapwings.

Considering the difficult prevalent complex conditions, measures should be taken immediately to protect the lapwing habitat.

Acknowledgements

Thanks are due to the Director, KFRI for the facilities and infrastructure. We wish to record our sincere thanks to KSCSTE and Plan fund. Also the authors are highly thankful to Mr. Manoj K., Ms. Abhirami Suresh and the local people for their unconditional support during the field survey.

References

Adesh, K. and Amita, K., 2015. Unusual sighting of Yellow-wattled Lapwing (*Vanellus malabaricus*) in Lucknow District, Uttar Pradesh, India. *International Journal of Life Sciences*, 3(2): 181-184.

Ali, S., 1969. *Birds of Kerala*. Oxford University Press, pp. 132-133.

Dhindsa, M.S., 1983. Yellow-wattled Lapwing: a rare species in Haryana and Punjab. *Pavo*, 21(1-2): 103-104.

Grimmett, R., Inskipp, C. and Inskipp, T., 2011. *Birds of the Indian Subcontinent*. Oxford University Press, London, pp. 148-150.

Gupta, R.C. and Kaushik, T.M., 2012. Spectrum of threats to nests of Yellow-wattled Lapwing *Vanellus malabaricus* in Kurukshetra outskirts-a case study. *Journal of Applied and Natural Science*, 4(1): 75-78.

Hart, J.D., Milsom, T.P., Baxter, A., Kelly, P.F. and Parkin, W.K., 2002. The impact of livestock on Lapwing *Vanellus vanellus* breeding densities and performance on coastal grazing marsh. *Bird Study*, 49: 67-78.

Jayakar, S.D. and Spurway, H., 1965. The Yellow -wattled Lapwing Vanellus malabaricus (Boddaert), a tropical dryseason nester. II. Additional data on breeding biology. *Journal of Bombay Natural History Society*, 62: 1-14.

Jayakar, S.D. and Spurway, H., 1968. The Yellow-wattled Lapwing Vanellus malabaricus (Boddaert), a tropical dry season nester. III. Two further Seasons' breeding. *Journal of Bombay Natural History Society*, 65: 369-383.

Jayson, E.A. and Sivaperuman, C. 2005. Avifauna of Thrissur district, Kerala, India. *Zoo's Print Journal*, 20(2): 1774-1783.

Kumar, C., 2015. First record of a regularly occupied nesting ground of Yellow-wattled Lapwing, *Vanellus malabaricus* (Boddaert) in agricultural environs of Punjab with notes on its biology. *Journal of Entomology and Zoology Studies*, 3 (1): 129-134.

Santharam, V., 1980. Some observations on the nests of Yellow wattled Lapwing, stone Curlew, Blackbellied Finch-Lark and redwinged Bush-Lark. *Newsletter for Birdwatcher*, 20(6-7): 5-12.

Sethi, V.K., Bhatt, D. and Kumar, A., 2010. Hatching success in Yellow-wattled Lapwing Vanellus malabaricus. *Indian Birds*, 5(5): 139-141.

Vijayagopal, K., 1991. Comparative biology and ecology of the Red-wattled Lapwing and Yellow-wattled Lapwing and a preliminary comparative study of the vocalization of certain species of Indian birds (*Thesis*).

Chapter 13

A Taxonomic Review on the Genus *Pareumenes* de Saussure (Hymenoptera : Vespidae : Eumeninae) from the Indian Subcontinent

P. Girish Kumar and P.M. Sureshan

ABSTRACT

The potter wasp genus Pareumenes de Saussure, 1855, is reviewed from the Indian subcontinent. Three species with one additional subspecies were present, namely, P. bengalensis (Fabricius, 1804), P. brevirostratus (de Saussure, 1855), P. quadrispinosus acutus Liu, 1941 and P. quadrispinosus quadrispinosus (de Saussure, 1855). P. quadrispinosus acutus is recorded here for the first time from Andamans, Arunachal Pradesh and Tripura. A key to species from the Indian subcontinent and an updated checklist from Oriental region provided. The symbiotic associations of mites were reported in some specimens of P. quadrispinosus acutus.

Keywords: Review, Vespidae, Eumeninae, Pareumenes, New record, Key, Checklist, Ethology, Indian subcontinent.

Introduction

De Saussure (1855) described the potter wasp genus *Pareumenes* for four species; the type species *Eumenes quadrispinosus* de Saussure, 1855, was designated by Bequaert (1918). This potter wasp species (Vespidae: Eumeninae) is distributed at Ethiopian, Oriental and Palearctic Regions of the world. Six species with an additional four subspecies are recorded from the Oriental Region of which three species with an additional subspecies, namely, *P. bengalensis* (Fabricius, 1804), *P. brevirostratus* (de Saussure, 1855), *P. quadrispinosus acutus* Liu, 1941 and *P. quadrispinosus quadrispinosus* (de Saussure, 1855) are recorded from the Indian subcontinent. In this paper, we reviewed the genus from the Indian subcontinent. *P. quadrispinosus acutus* is recorded here for the first time from Andamans, Arunachal Pradesh and Tripura. A key to species from the Indian subcontinent and an updated checklist from Oriental region also provided. The symbiotic associations of mites were reported in some specimens of *P. quadrispinosus acutus*.

Material and Methods

The specimens were examined under LEICA M60 stereozoom microscope and images captured with the camera model LEICA DFC-450. The studied specimens were added to the 'National Zoological Collections' of the Western Ghat Regional Centre, Zoological Survey of India, Kozhikode (= Calicut), India (ZSIK).

Abbreviations used for the Museums: BMNH — British Museum (Natural History), London, England; OUM — Oxford University Museum, Oxford; UZMC — Universitetets Zoologiske Museum, Copenhagen, Denmark; ZSIK — Western Ghat Regional Centre, Zoological Survey of India, Kozhikode (= Calicut), India.

Abbreviations used for the terms: H = Head; M = Mesosoma; S = Metasomal sterna; T = Metasomal terga.

Results

Genus *Pareumenes* de Saussure, 1855

Pareumenes de Saussure, 1855: 133, division of genus *Eumenes* Latreille. Type species: *Eumenes quadrispinosus* de Saussure, 1855, by subsequent designation of Bequaert, 1918: 271.

Diagnosis: Forewing with prestigma longer than pterostigma; female with cephalic fovea; mesoscutum with deep prescutal grooves; mesepisternum without epicnemial carina; propodeum dorsally with elongate fovea from which carina runs to propodeal orifice; propodeum with dentiform projections above propodeal valvula; propodeal orifice rounded dorsally; axillary fossa narrower than long, slit-like; tegula with narrow posterior lobe which about equals parategula posteriorly; second submarginal cell acute basally; midtibia with one spur.

Distribution: Ethiopian, Oriental and Palearctic Regions.

Ethology: Symbiotic associations with mites were observed in *Pareumenes quadrispinosus acutus* Liu.

Key to Species and Subspecies of *Pareumenes* from the Indian Subcontinent

1. Head yellow except occiput transverse band on vertex enclosing ocelli brown .... *brevirostratus* (de Saussure, 1855)
— Head black with either yellow or yellow and reddish markings 2
2. Metasoma ferruginous red with yellow markings .... *bengalensis* (Fabricius, 1804)
— Metasoma black with yellow markings .... *quadrispinosus* (de Saussure, 1855).... 3
3. Base of petiole distinctly tinged with red; vertex and anterior half of mesoscutum less coarsely punctate .... *quadrispinosus quadrispinosus* (de Saussure, 1855)
— Base of petiole black, not tinged with red; vertex and anterior half of mesoscutum coarsely punctate .... *quadrispinosus acutus* Liu, 1941

1. *Pareumenes bengalensis* (Fabricius, 1804)

Polistes bengalensis Fabricius, 1804: 277 [male] - "Bengalia, Dom. Daldorff, Mus. Dom. Lund" (UZMC). - de Saussure, 1853: 41 (? *Icaria bengalensis*; doubtful species). - Dalla Torre, 1894: 123 (cat.).); 1904: 71 (doubtful species) [erroneously recorded from Senegal].

Pareumenes bengalensis; Schulz, 1912: 84 (type examined; probably identical to *P. brevirostratus* (de Saussure)). - van der Vecht, 1963: 17 (description of male type).

Diagnosis: Male: Clypeus strongly convex, somewhat flattened in the middle, slightly longer than wide, emarginate anteriorly with sharp tooth; clypeus with some scattered superficial punctures; vertex slightly raised above the level of eyes; apical antennal article hook-like, in curved position it reaches slightly beyond the apex of 10th segment; frons with a rather coarsely and densely punctate area between ocelli and upper level of eye emarginations; length of prescutal grooves about one fifth of the length of the mesoscutum; concavity of propodeum with narrowly triangular basal fovea, which is slightly longer than one third of the total length, behind the fovea with sharp median carina, flanked on each side by obliquely transverse ridges.

Colour description: Male: Head black, somewhat reddish on vertex, occiput and lower part of temple; mandible brown; antennae ferruginous except scape yellow beneath; the following parts yellow on the head: clypeus, frons except two dark lines running from ocelli to base of clypeus, a broad band on the temples along the outer orbits. Mesosoma reddish, suffused with black on pronotum, mesoscutum and scutellum; yellow markings approximately as in "*Pareumenes quadrispinosus intermedius*" but the pronotal band wider, the mesoscutal lines not narrowed anteriorly and the band on the metanotum reduced to two small irregular spots; mesepisternum with sub triangular spot below the tegulae. Metasoma ferruginous red with the following yellow markings: preapical band on T1 which is widened laterally, deeply incised, almost interrupted, in the middle; somewhat infuscated on the disk of T2, preapical wide band on T2; T3 with broad band, emarginate anteriorly; T4 with narrower band, twice angularly incised anteriorly; T5-T7 with rather large and transverse band; S1 with a line at lateral margins of posterior third; S2 and S3 with minute spot in posterior lateral angles. Legs ferruginous, apical two thirds of fore femora yellow on outer side, fore tibiae yellow with reddish line on inner side, mid and hind tibiae yellow on outer side. Wings yellowish hyaline.

Size (H+M+T1+T2): Male, 12-13 mm.

Distribution: India: West Bengal.

Remarks: No specimen of this species was available for our studies; hence the description of male was taken from van der Vecht (1963).

2. *Pareumenes brevirostratus* (de Saussure, 1855) (Figures 13.1–13.8)

Eumenes brevirostratus de Saussure, 1855: 136, pl. 7 fig. 1, female, male (in division *Pareumenes*) - "Les Indes Orientales" (? BMNH). - Smith, 1857: 23 (*brevirostratas*; cat.); 1871: 372 (*brevirostrata*; cat.). - Dalla Torre, 1894: 19 (*brevirostrata*; cat.). - Bingham, 1897: 334 (key), 337, female, male (*brevirostrata*; Sikhim; Calcutta; Madras). - Schulz, 1906: 214 (*brevirostrata*; in subgenus *Pareumenes* female from Malabar in Mus. Strassbourg). - Dover, 1921: 386 (Barkuda Island, Chilka Lake, India). - Dusmet, 1930: 102 (var. ? from Khandala, India, in Mus. Madrid).

Pareumenes brevirostratus; Dalla Torre, 1904: 19 (*brevirostrata*; cat.). - Dover, 1931: 252 (*brevirostrata*; Bombay; Travancore). - van der Vecht, 1963, Zool. Verh., Leiden 60: 19 (India: Allahabad, Odisha, Kerala, Nasik, Poona, Dharwar). – Girish Kumar and Sharma, 2015: 8122 (Chhattisgarh).

Pterochilus fulvipennis Cameron, 1898: 39, pl. 4 3a, 3b, female - [India] "Poona" (OUM); 1903: 165 (compared to *Montezumia flavobalteata* Cameron). - Dalla Torre, 1904: 58 (cat.)

Pareumenes fulvipennis; van der Vecht, 1937: 272 (holotype examined; additions to description) (in subgenus *Pareumenes*).

Eumenes fulvipennis; Cameron, 1907: 1008 (specimens from Deesa; in a footnote Cameron refers to the description and figures of *Pterochilus fulvipennis* Cameron, 1898, but his Deesa specimens belong indeed in the genus *Eumenes*).

Eumenes campaniformis var. *cameroni* Bequaert, 1928: 167 (selection of "holotype male" from Deesa (BMNH); also specimens from Bombay and Muscat) [replacement name for *Eumenes fulvipennis* Cameron, 1907 [!] non *Eumenes fulvipennis* Smith, 1857].

Diagnosis: *Female:* Clypeus convex, somewhat flattened in the middle, emarginate anteriorly with tooth; cephalic foveae present; metanotum sub angular posteriorly; apical teeth of propodeum sharper, forming an angle of about 60°. *Male:* Similar to female but smaller and slighter; apical antennal article hardly more than one third of the third segment, in curved position not quite reaching the apex of the tenth segment.

Figures 13.1–13.6: *Pareumenes brevirostratus* (de Saussure).

Figure 13.1–13.4 ♀: 13.1. Body profile; 13.2. Head frontal view;
13.3: Head and mesosoma dorsal view; 13.4: Metasoma profile view.
Figures 13.5 and 13.6 ♂: 13.5. Head frontal view; 13.6. Apical antennal articles.

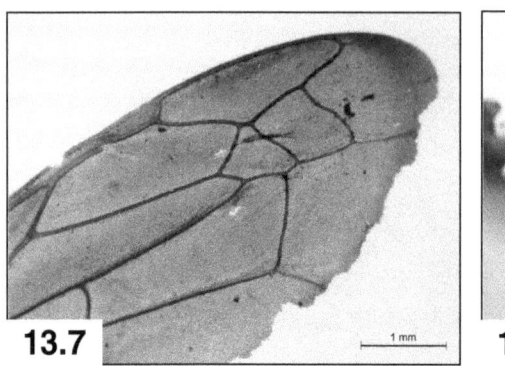 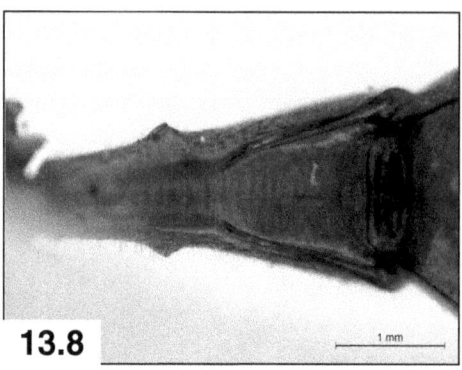

Figures 13.7 and 13.8: *Pareumenes brevirostratus* (de Saussure) ♂:
13.7. Apical half of forewing; 13.8. S1.

Colour description: *Female*: Head yellow except occiput and transverse band across vertex enclosing ocelli brown; mesosoma ferruginous except a broad transverse yellow band on pronotum; mandible, antenna and legs ferruginous; metasoma ferruginous with broad apical yellow bands on T2-T6. Wings flavohyaline. *Male*: Similar to female in colour pattern.

Size (H+M+T1+T2): Female, 17-20 mm; Male, 12 mm.

Material examined: INDIA: Kerala, Palakkad district, Walayar forest, 1nalin123, October 1963, V. K. Gupta and Party, Regd. No. ZSI/WGRS/I.R-INV.5022. Odisha, Barkuda Island, 1φ, 15-22.vii.1916, Coll. F. H. Gravely, Regd. No. ZSI/WGRS/I.R-INV.5023; Balasore district, Soro, 1φ, 9.x.1974, Coll. B. C. Saha and Party, Regd. No. ZSI/WGRS/I.R-INV.5024. West Bengal, Kolkata environs, 1φ, 10.i.1904, Coll. Brunetti, Regd. No. ZSI/WGRS/I.R-INV.5025.

Distribution: India: Chhattisgarh, Gujarat, Karnataka, Kerala, Maharashtra, Odisha, Sikkim, Uttar Pradesh, West Bengal.

3. *Pareumenes quadrispinosus* (de Saussure, 1855)

There are two subspecies of this species, namely, the nominotypical subspecies *P. quadrispinosus quadrispinosus* (de Saussure) and *P. quadrispinosus acutus* Liu are recorded from India.

Diagnosis: *Female*: Clypeus subcircular, length about equals to width, apical margin arcuately emarginate, lateral angles pointed; inner margin of mandible with four tooth; vertex with a shallow cephalic fovea, with two indistinct lateral pits and small tufts of short, dark brown hairs; mesoscutum wider than long; S1 enlarged gradually from base to apex with strong transverse carinae in the basal region and weak carinae in the apical region; vertex behind the ocellar triangle normal, not elevated; median concavity of propodeum shallow, with an anterior deep furrow and a posterior ridge ribbed on each side by a series of transverse carinae; lateroposterior angles of propodeum produced, spine-like; genal carina meeting on the occiput; subdiscoidal vein weakly sclerotized beyond the second recurrent nervure; lateral tubercles of T1 prominent, situated a little above the middle; T3 without apical extension.

3a. *Pareumenes quadrispinosus* acutus Liu, 1941 (Figures 13.9–13.18)

Pareumenes acutus Liu, 1941: 255 (key), 262, female, pl. 1 fig. 5, pl. 2 fig. 3, 13 (in subgenus *Pareumenes*) - "South China" (coll. Liu).

Pareumenes quadrispinosus acutus; van der Vecht, 1963: 22 (India: Sikkim; Assam: Khasia Hills). - Gusenleitner, 2006: 695 (India: Golaghat, Kalimpong, Melli Bazar).

Colour description: *Female*: Body black with the following yellow markings: clypeus except apical margin; frons between antennal insertions, extending up to anterior ocellus; ocular sinus; lower side of scape; long band on genae; large mark on pronotum; two longitudinal stripes on mesoscutum; tegula except brown spot at middle; parategula; two large spots on mesopleuron; two quadrate spots on scutellum; a band on posterior half of metanotum; two large spots on propodeum, posterior pair much larger and covering latero-posterior angles; apical band on T1

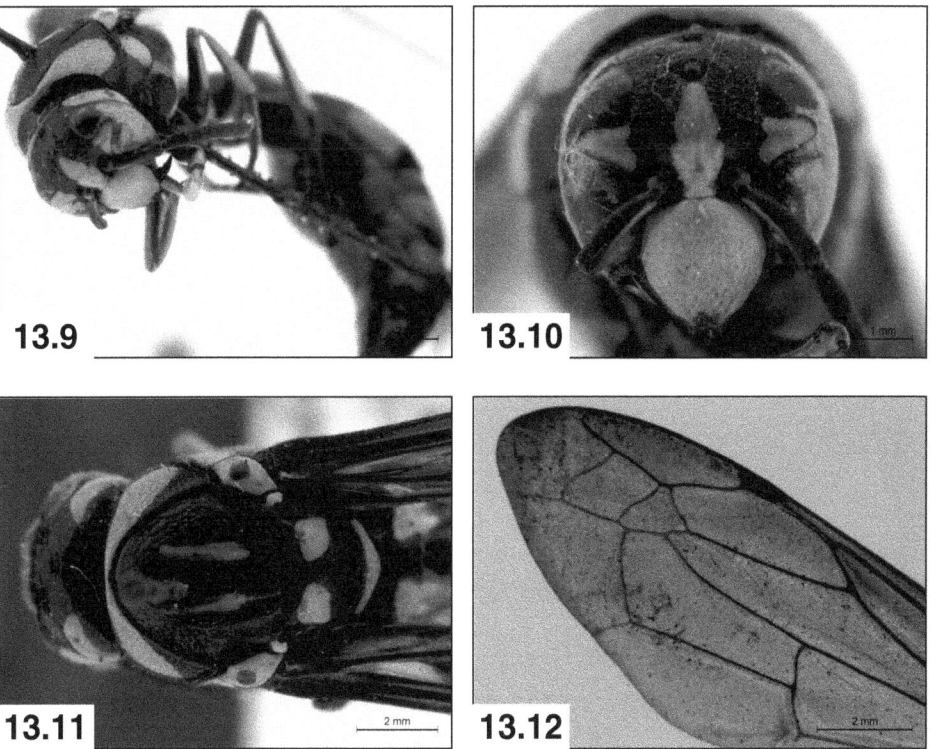

Figures 13.9–13.12: *Pareumenes quadrispinosus acutus* Liu ♀:
13.9: Body profile; 13.10: Head frontal view; 13.11: Head and mesosoma dorsal view; 13.12: Apical half of forewing.

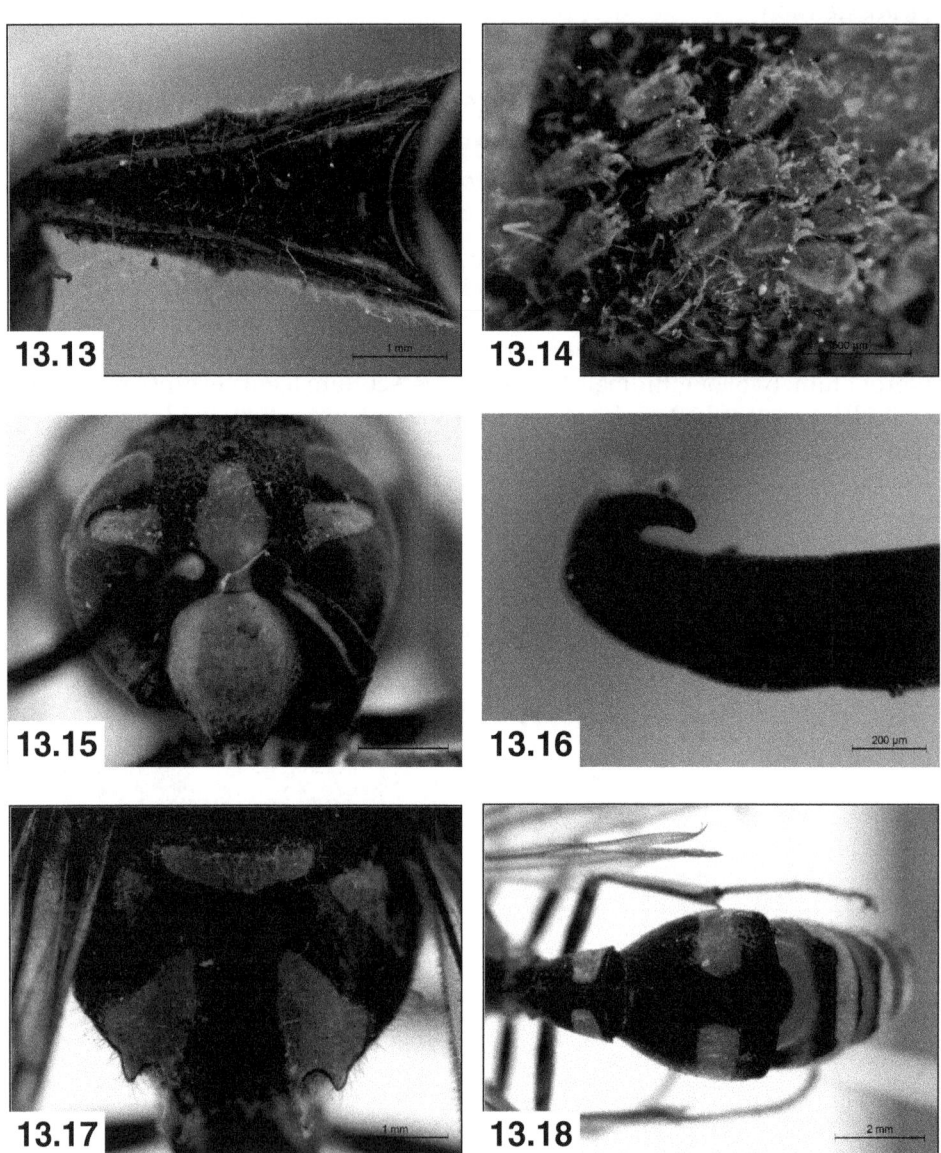

Figures 13.13–13.18: *Pareumenes quadrispinosus acutus* Liu.

Figures 13.13 and 13.14 ♀: 13.13. S1; 13.14: Mites inhabit on propodeum.

Figures 13.15–13.18 ♂: 13.15. Head frontal view; 13.16: Apical antennal articles; 13.17: Propodeum; 13.12: Metasoma dorsal view.

interrupted at middle; two lateral stripes at widened portion of S1; two large, widely separated lateral spots near posterior margin of T2; basal half and two small lateroposterior spots on S2; T3-T5 with apical band, bands on T3 and T4 incised anteriorly in the middle and at the sides, that on T5 abbreviated laterally; T6 with irregular median spot. Legs black with the following yellow mark: inner face of fore coxae; lower and apical portion of fore femora; fore tibiae and fore tarsi entirely; outer and apical third of inner face of mid femora; outer face of mid tibiae; two linear spots on the inner face of hind coxae; apex of hind femora; outer face of hind tibiae. Wings dark flavohyaline. *Male:* Similar to female in colour pattern.

Size (H+M+T1+T2): Female, 19-21 mm; Male, 16 mm.

Material examined: INDIA: Andamans, 1φ, exact collection locality and date of collection unknown, Coll. S. W. Kemp, Regd. No. ZSI/WGRS/I.R-INV.5026. Arunachal Pradesh, West Kameng district, Tipi, 1nalin123, 5.ix.1998, Coll. A. R. Lahiri and Party, Regd. No. ZSI/WGRS/I.R-INV.5027. Assam, Tinsukia district, Margherita, 1φ, date of collection and name of collector unknown, Regd. No. ZSI/WGRS/I.R-INV.5028. Sikkim, 2φ, exact collection locality, date of collection and name of collector unknown, Regd. Nos. ZSI/WGRS/I.R-INV.5029 and 5030; Sikkim, 1φ, exact collection locality and date of collection unknown, Coll. Niceville, Regd. No. ZSI/WGRS/I.R-INV.5031; Sikkim, Shamdang, 1φ, 7.ix.1909, name of collector unknown, Regd. No. ZSI/WGRS/I.R-INV.5032. Tripura, Dhalai district, Ambassa, 1φ, 25.v.1978, Coll. J. K. Jonathan and Party, Regd. No. ZSI/WGRS/I.R-INV.5033. West Bengal, Darjeeling district, Singla, 1φ, July,1912, Coll. Lord Carmichael, Regd. No. ZSI/WGRS/I.R-INV.5034.

Distribution: China; India: Andamans (**new record**), Arunachal Pradesh (**new record**), Assam, Meghalaya, Sikkim, Tripura (**new record**), West Bengal.

Ethology: Symbiotic associations with mites were observed in one female specimen from Andamans and another female specimen from Sikkim of *Pareumenes quadrispinosus acutus* Liu. In both specimens, mites inhabited at propodeum.

3b. *Pareumenes quadrispinosus quadrispinosus* (de Saussure, 1855)

Eumenes quadrispinosus de Saussure, 1855: 134, pl. VII fig. 2-2g, female, male (in division *Pareumenes*) - "Les Indes Orientales" (lectotype female from "India" in BMNH). - Smith, 1857: 23 (*quadrispinosa*; cat.); 1858: 108 (*quadrispinosa*; Malaya). - Wallace, 1871, in Smith: 296 (*quadrispinosa*; ethology). - Smith, 1871: 296 (ethology), 372 (*quadrispinosa*; cat.). - Dalla Torre, 1894: 31 (*quadrispinosa*; cat.). - Bingham, 1897, Fauna Br. India, Hym. 1: 334 (key), 336, fig. 94 (*quadrispinosa*; Mussooree, N.W. Himalayas; Sikhim; Calcutta; Central India; Madras; Tenasserim). - Rothney, 1903: 106 (*quadrispinosa*; Bengal). - Dover and Rao, 1922: 236 (Nilgiri Hills; Burma; Assam). - Bequaert, 1928: 172 (designation of lectotype). - ? Tosawa, 1934: 4 (key), 5, fig. 1 (*quadrispinosa*; Korea).

Pareumenes quadrispinosus; Dalla Torre, 1904: 19 (cat.). - Dover, 1925: 296 (Bhutan; Rungeet Valley; Tavoy) [*Eumenes eximius* Smith erroneously regarded as synonym]; 1931: 252 (Hong Kong; Hoabin [=Hòa Binh]; Sikkim; Tenasserim).

- Bequaert, 1928: 172 (type examined). - Giordani Soika, 1935: 137. - Liu, 1936: 102 (cat.). - van der Vecht, 1937: 269 (notes). - Sonan, 1938: 77 (*quarispinosus* [!]; Korea). - Iwata, 1942: 19, 38, 52 (ethology). - van der Vecht, 1963: 21 (misidentified by Piel, 1935 and Liu, 1941; distribution). - van der Vecht and Fischer, 1972: 123 (cat.). - Srinivasan and Kumar, 2010: 1319, image 9 (India: Arunachal Pradesh).

Pareumenes quadrispinosus quadrispinosus; Gusenleitner, 2011: 1363 (Laos).

This subspecies having head slightly more extensively black; the transverse black band on the vertex being connected by two irregularly curved lines with the antennal sockets; petiole darker except base distinctly tinged with red; vertex and anterior half of mesoscutum less coarsely punctate.

Distribution: Korea; China; India: Arunachal Pradesh, Assam, Karnataka, Meghalaya, Sikkim, Tamil Nadu, Uttarakhand, West Bengal; Bhutan; Myanmar; Laos; Vietnam; Malaysia.

Checklist of the Oriental species of the *Pareumenes* (*Pareumenes*) de Saussure

1. *P. bengalensis* (Fabricius, 1804) — India: West Bengal.
2. *P. brevirostratus* (de Saussure, 1855) — India: Gujarat, Karnataka, Kerala, Maharashtra, Odisha, Sikkim, Uttar Pradesh, West Bengal.
3. *P. nigerrimus* van der Vecht, 1963 — Indonesia: Sumba, Flores, Timor.
4. *P. obtusus* Liu, 1941 — China.
5. *P. pullatus* (Smith, 1864) — Indonesia: Moluccas.
6a. *P. quadrispinosus acutus* Liu, 1941 — China; India: Andamans (**new record**), Arunachal Pradesh (**new record**), Assam, Meghalaya, Sikkim, Tripura (**new record**), West Bengal, Vietnam.
6b. *P. quadrispinosus interjectus* van der Vecht, 1937 — Malaysia (including Sarawak); Indonesia: Sumatra.
6c. *P. quadrispinosus interruptus* Liu, 1941 — China.
6d. *P. quadrispinosus javanus* van der Vecht, 1937 — Indonesia: Java.
6e. *P. quadrispinosus quadrispinosus* (de Saussure, 1855) — Korea; China; India: Arunachal Pradesh, Assam, Karnataka, Sikkim, Tamil Nadu, Uttarakhand, West Bengal; Bhutan; Myanmar; Laos; Malaysia.

Acknowledgements

The authors are grateful to Dr. Kailash Chandra, Director-in-Charge, Zoological Survey of India, Kolkata, for providing facilities and encouragements.

References

Bequaert, J. C. 1918. A revision of the Vespidae of the Belgian Congo based on the collection of the American Museum Congo Expedition, with a list of Ethiopian diplopterous wasps. *Bull. Am. Mus. Nat. Hist.*, **39**: 1- 384.

Bequaert, J. 1928. A study of certain types of diplopterous wasps in the collection of the British Museum. *Ann. Mag. Nat. Hist.* (10) 2: 138-176.

Bingham, C. T. 1897. *The Fauna of British India, including Ceylon and Burma, Hymenoptera, I. Wasps and Bees*: Taylor and Francis, London, 579+ i- xxix.

Cameron, P. 1898. Hymenoptera orientalia. *Mem. Proc. Manchr. lit. Philos. Soc.*, 42 (11): 1-84. 1 pl.

Cameron, P. 1903. Descriptions of new genera and species of Hymenoptera taken by Mr. Robert Shelford at Sarawak, Borneo. *J. Straits Branch R. Asiat. Soc.* 39: 89-181.

Cameron, P. 1907. Description of a new genus and some new species of Hymenoptera captured by Lieut. Col. C.G. Nurse at Deesa, Matheran and Ferozepore. *J. Bombay Nat. Hist. Soc.*, **17**: 1001-1012.

Dalla Torre, K. W. Von. 1894. *Catalogus Hymenopterorum* **9**, Vespidae (Diploptera), Leipzig, 181 p.

Dalla Torre, K. w. Von. 1904. *Vespidae, Genera Insectorum*, **19**: 1-108.

Dover, C. 1921. The wasps and bees of Barkuda Island. *Rec. Indian Mus.*, **22**: 381-391.

Dover, C. 1925. Further notes on the Indian Diplopterous wasps. *J. As. Soc. Bengal*, new ser. Vol. 22, p. 289-305.

Dover, C. 1931. The Vespidae in the F.M.S. Museums. *J. Fed. Malay St. Mus.*, **16**: 251-260.

Dover, C. and Rao, H.S. 1922. A note on the Diplopterous Wasps in the Collection of the Indian Museum. *J. As. Soc. Bengal, New Series*, 18: 235-249.

Dusmet, J. M. 1930. Himenópteros de la India inglesa cazados por el P. Ignacio Sala de Castellarnau, S. J. I. Serie. Véspidos e Euménidos. *Bol. Soc. Entomol. Esp.*, 13(6-8): 99-107.

Fabricius, J. C. F. 1804. *Systema Piezatorum Secundum, Ordines, Genera, Species, Adiectis Synonymis, Locis, Observationibus, Descriptionibus*. Brunschweig, XIV+[15]-[440]+[1]-30pp.

Giordani Soika, A. 1935. Richerche sistematiche sugli *Eumenes y Pareumenes* dell'-Archipelago Malese e della Nova Guinea. *Ann. Mus. Stor. Nat. Genova*, **57**: 114-151.

Girish Kumar, P. and Gaurav Sharma (2015). Taxonomic studies on vespid wasps (Hymenoptera: Vespoidea: Vespidae) of Chhattisgarh. *Journal of Threatened Taxa*, 7(14): 8096-8127; http://dx.doi.org/10.11609/jott.2426.7.14.8096-8127

Gusenleitner, J. 2006. Uber Aufsammlungen von Faltenwespen in Indien (Hymenoptera, Vespidae). *Linzer boil. Beitr.*, **38** (1): 677-695.

Gusenleitner, J. 2011. Eine Aufsammlung von Faltenwespen aus Laos im Biologiezentrum Linz (Hymenoptera: Vespidae: Vespinae, Stenogastrinae, Polistinae, Eumeninae). *Linzer boil. Beitr.* **43** (2): 1351-1368.

Iwata, K. 1942. Comparative studies on the habits of solitary wasps. *Tenthredo*, 4: 1-146.

Liu, C. L. 1936. A bibliographic and synonymic catalogue of the Vespoidea of China, with a cross-referring index for the genera and species (1). *Peking Nat. Hist. Bull.*, 11: 91-114.

Liu, C. L. 1941. Revisional studies of the Vespidae of China. I. The genus *Pareumenes* Saussure, with description of six new species. Notes dEnt. Chin., 8: 245-289, 2 pls.

Piel, O. 1935. Biologie de *Pareumenes quadrispinosus* Saussure (Hymènoptéres Vespides) et de ses parasites, en particulier : *Calosota chinensis* Ferriere. *Notes Entomol. Chin.*, 2 : 105-139.

Rothney, G. A. J. 1903. The aculeate Hymenoptera of Barrackpore, Bengal. *Trans. Royal Ent. Soc. London*, 93-116.

Saussure, H. De. 1852-53. *Monographie des Guepes Solitaires ou de la Tribu des Eumeniens*. Etudes sur la famille des Vespides. I- Paris, 6-50-286 pp.+ 21 pls.

Saussure, H. de. 1854-1856. *Études sur la famille des vespides. Toisième partie comprenant la Monographie des Masariens et un supplement a la Monographie des Euméniens*. V. Masson, Paris and J. Kessmann and J. Cherbuliez, Genève, 352 pp. + 15 pls. (1854) 1-48 + pl. 1-5; (1855) 49-288 + pl. 6-14; (1856) 289-352 + pl. 15, 16. [Dates of publication after Griffin 1939].

Schulz, W. A. 1906. Spolia Hymenopterologica. Insel Creta. 355 pp. Paderborn.

Schulz, W. A. 1912. Aelteste und alte Hymenopteren skandinavischer Autoren. *Berl. Entomol. Zeitschr.*, **57**: 52-102.

Smith, F. 1857. Catalogue of Hymenopterous insects in the collection of the British Museum.**5**: 1-147.

Smith, F. 1858. Catalogue of the Hymenopterous insects collected at Sarawak, Borneo; Mount Ophir, Malacca; and at Singapore, by A. R. Wallace. *J. Proc. Linn. Soc. (Zool.)*, **2**: 42-130.

Smith, F. 1871. A catalogue of the Aculeate Hymenoptera and Ichneumonidae of India and eastern Archipelago, with introductary remarks by A.R. Wallace. *J. Proc. Linn. Soc. London Zool.*, **11**: 285-415.

Sonan, J. 1938. Notes on the Vespoidea in Japan (Hymenoptera). *Trans. Nat. Hist. Soc. Formosa*, 28: 77-81.

Srinivasan, G. and Girish Kumar, P. 2010. New records of potter wasps (Hymenoptera: Vespidae: Eumeninae) from Arunachal Pradesh, India: five genera and ten species. *J. Threatened Taxa*, **2** (12): 1313-1322. http://dx.doi.org/10.11609/JoTT.o2468.1313-22

Tosawa, N. 1934. On *Eumenes* of Japan Empire. *Trans. Kansai Entomol. Soc.*, **5**: 3-16, pl. 1.

Vecht, Van Der, J. 1937. Descriptions and records of Oriental and Papuan solitary Vespidae. *Treubia*, 16: 261-293.

Vecht, J., van der. 1963. Studies on Indo-Australian and East Asiatic Eumenidae (Hymenoptera: Vespoidea). *Zool. Verh., Leiden,* **60**: 1-116.

Vecht, J. van der and Fischer, F. C. J. 1972. Palearctic Eumenidae. *Hym. Cat. (n. ed.)*, **8**: i-v+199.

Chapter 14

Responses of Serum Luteinizing Hormone (LH) and Testosterone (T) Levels as Correlated with Testicular Morphology of Albino *Mus norvegicus* Induced by Sublethal Heroin Administration

Kaminidevi K. Bhoir, S.A. Suryawanshi
and A.K. Pandey

ABSTRACT

Effect of sublethal (0.50 LD_{50}; 13.5 mg/kg/day) heroin administration on serum luteinizing hormone (LH), testosterone and testicular morphology of Mus niorvegicus (Wistar strain) was investigated. Serum LH of control rat varied between 26.86±5.06 and 28.26±7.17 ng/ml while testosterone from 2.16±0.58 to 2.30±0.32 µg/100 ml during the experimental period. Sublethal heroin administration induced significant (P<0.01) decline in serum LH level at 24 (9.66±0.81 ng/ml) and 96 hours (9.54±0.45 ng/ml) with minimal value on day 30 (7.84±0.42 ng/ml). Serum testosterone level of the heroin treated rats also depicted a decline (P<0.05) at 24 (1.48±0.08 µg/100 ml) and 96 hours (1.40±0.05 µg/100 ml) while minimum value (P <0.001) was recorded on day 30 (0.81±0.13µg/100 ml). Testis of control rat showed convoluted seminiferous tubules consisting of basement membrane and lining of stratified epithelium which comprised Sertoli cells and spermatogenic cells. Sertoli cells were tall, irregularly columnar and extended from basal lamina to the lumen. The spermatogenic cells exhibited uniform cellular arrangement with all the five maturation stages. Seminiferous tubules of heroin treated rats exhibited massive degenerative changes in spermatogonial cells, primary *spermatocytes, secondary spermatocytes and spermatids. Intercellular space between the cells increased and number of Sertoli cells was reduced. The lumina showed debris of spermatogenic cells with scanty spermatozoa. Leydig cells of the heroin treated rats showed atrophy and vacuolization.*

Keywords: Heroin, Luteinizing hormone, Testosterone, Testis, Mus norvegicus.

Introduction

Chronic abuse of heroin has diverse effects on various body systems due to widespread distribution of specific receptors in many tissues and organs (Martin, 1984; Sawynok, 1986; Cami and Farre, 2003). The drug (diacetylmorphine) is initially metabolized to 6-acetylmorphine and subsequently to morphine in human body (Sawynok, 1986; Sporer, 1999). Despite of long history of clinical therapeutic use and protracted abuse by addicted subjects (Sawynok, 1986; Cami and Farre, 2003), little is known regarding possible influences of this drug on the endocrine system of mammals (George et al., 2005; Brown et al., 2006; Al-Gommer et al., 2007; Bhoir et al., 2007, 2009; Barai et al., 2009a, b). There are clinical evidences suggesting inhibition of some parameters of sexual function in human addicted to heroin, most notably impaired libido, impotency and delayed ejaculation (Mumford and Kumar, 1979; McKendry et al., 1983; Wieland and Yunger, 1985). Plasma level of luteinizing hormones (LH) was normal in both methadone maintained as well as active heroin addicts (Cushman, 1973). Chronic narcotic administration produced a marked atrophy of the secondary sex organs and suppression of plasma testosterone level in male rat (Cicero et al., 1976). About 87 per cent reduction in serum testosterone level was associated with the atrophy of seminal vesicle, prostrate and epididymis. Further, sexual difficulties were also frequently reported by the persons addicted to narcotics (Cushman, 1973; Mumford and Kumar, 1979; Daniell, 2002). It was suggested that the long-term methadone use impairs functions of secondary sexual organs in human beings (Cicero et al., 1975). There exist reports demonstrating lower testosterone level in heroin and methadone users (Azizi et al., 1973; Mendelson et al., 1975a, b, 1980; Rasheed and Tareen, 1995; Daniell, 2002; Blesener et al., 2005). Since these observations are of shorter duration extended not beyond 10 days, an attempt was made to record the concurrent changes occurring in serum luteinizing hormone (LH) and testosterone (T) levels as well as testicular morphology of albino *Mus norvegicus* induced by prolonged sublethal heroin administration.

Materials and Methods

Healthy male *Mus norvegicus* (Wistar strain) weighing 150-200 gm were procured from Hoffkin Institute, Parel, Mumbai and housed in specially made plastic cages. They were acclimatized under the ambient laboratory conditions (temperature $28\pm2°C$; photoperiod 14L:10 D) for 10 days, fed *ad libitum* on rat feed (Lipton, Hindustan Lever Ltd., Bangalore) and clean water was provided for drinking. Care was taken to ensure that the rats were treated in the most humane and ethically accepted manner. 60 male rats were randomly selected and divided into two equal groups - experimental and control. Heroin (85 per cent pure) was dissolved initially in small quantity of alcohol and diluted with physiological saline to prepare the test dose of 0.50 LD$_{50}$ (13.5 mg/kg/day). The drug was administered through subcutaneous (s.c.) route to the experimental rats while the control rats received equal volume (0.2 ml/kg body weight) of the physiological saline. Estimation of LH and testosterone were carried out by radioimmunoassay (RIA) techniques at Tata Cancer Research Centre of Bhabha Atomic Research Centre (BARC), Mumbai (India).

The animals were killed on day 30 and their testes were surgically removed, washed in normal saline and fixed immediately in Bouin's fluid for light microscopy. After 24 hours, the tissues were dehydrated in ascending series of alcohol, cleared in xylene and embedded in paraffin wax at 60°C. The sections were cut at 5 µm and stained in hematoxylin-eosin (H and E) and buffered toluidine blue for light microscopic studies. For electron microscopy, the tissues were removed immediately after the sacrifice and sliced into 1 mm pieces to allow better penetration of fixative chemical (3 per cent ice-cold glutaraldehyde) for 12 hours followed by 4 hours in 0.1.M cacodylate buffer. They were rinsed in buffer and post-osmicated in 1 per cent osmium tetraoxide (OsO$_4$) for 1-2 hours. The tissues were then dehydrated in ascending alcohol grades followed by propylene oxide and embedded in resin polymerized at 60°C.

The blocks were prepared in araldite. 1 µm thin sections were cut with glass knife on LKB-2000 ultramicrotome. Sections were mounted on glass slide and stained with buffered toluidine blue for light microscopic studies. Ultrathin sections of the selected blocks were cut with glass knife, picked up on copper grids and stained with uranyl acetate and lead citrate for final observation under ZEIM-EM-109 electron microscope.

Results and Discussion

Effects of sublethal heroin administration on serum LH and testosterone (T) levels of albino *Mus norvegicus* have been summarized in Table 14.1. Serum LH of control rat varied between 26.86±5.06 and 28.26±7.17 ng/ml while testosterone from 2.16±0.58 to 2.30±0.32 µg/100 ml during the experimental period. Sublethal heroin administration induced a significant (P<0.001) decline in serum LH level at 24 (9.66±0.81 ng/ml) and 96 hours (9.54±0.45 ng/ml) with minimal value on day 30 (7.84±0.42 ng/ml). Serum testosterone level of the heroin treated rats also depicted a decline (P<0.05) at 24 (1.48±0.08 µg/100 ml) and 96 hours (1.40±0.05 µg/100 ml) while minimum value (P <0.001) was recorded on day 30 (0.81±0.13µg/100 ml).

Table 14.1: Effect of Sublethal Heroin Administration on Serum LH (ng/ml) and Testosterone (T) (µg/100 ml) Levels of Albino *Mus norvegicus*

Duration	Serum LH		Serum Testosterone	
	Control	Experimental	Control	Experimental
24 hours	28.26±7.17	9.66±0.81[b]	2.16±0.58	1.48±0.08[a]
96 hours	26.86±5.06	9.54±0.45[b]	2.25±0.35	1.40±0.05[b]
30 days	27.78±4.97	7.84±0.42[b]	2.30±0.32	0.80±0.13[b]

Values are mean±S D of 5 animals. Significant responses: [a] P<0.05, b P<0.001.

Testis of control *Mus norvegicus* exhibited convoluted seminiferous tubules, the wall of which was consisted of basement membrane and a lining of stratified epithelium. The epithelium was consisted of Sertoli cells (or supporting cells) and the spermatogenic cells. Sertoli cells were tall, irregularly columnar and extended from basal lamina to the lumen. The spermatogenic cells exhibited uniform cellular arrangement with five maturation stages - spermatogonia, primary spermatocytes, secondary spermatocytes, spermatids and spermatozoa (Figures 14.1 and 14.2). The early spermatids were seen with acrosome formation. The mitochondria were dispersed throughout the cytoplasm and Sertoli cells were also seen with normal architecture. Leydig cells were seen in the interlobular area of the seminiferous tubules. The spermatogenetic cell depicted normal cytoarchitecture (Figure 14.5).

Seminiferous tubules of rats treated with heroin showed massive degenerative changes in the spermatogonial cells (Figure 14.3). Though spermatogonia, resting on basement membrane, were in active phase of division but number of these cells was reduced (Figure 14.3). Primary spermatocytes, secondary spermatocytes and spermatids also exhibited degenerating changes and the intercellular space between these cells was increased. Number of Sertoli cells was also reduced but spermatocytes as well as spermatids were seen attached to it. The lumina showed debris of spermatogenic cells with scanty spermatozoa (Figure 14.4). Leydig cells located in the intertubular space showed atrophy and vacuolization (Figure 14.4). The ultrastructural observations of the testis of heroin treated rat revealed Leydig cells in the angular interstices between the seminiferous tubules with indented nucleus. The cytoplasm showed hypertrophied mitochondria, Golgi with dilated cisternae, multivesicular bodies and membrane limited lysosomes (Figure 14.6).

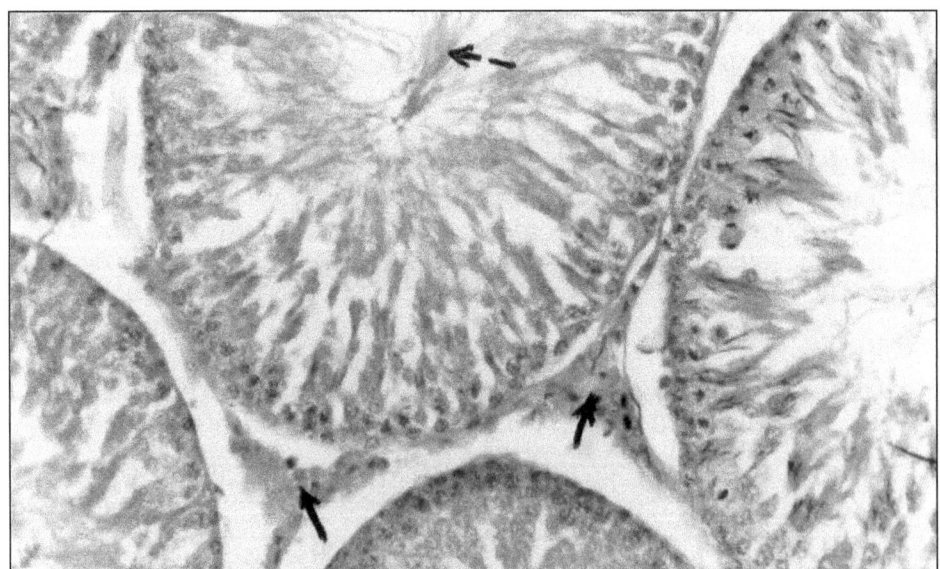

Figure 14.1: Seminiferous Tubules of Control *Mus norvegicus* Exhibiting different Stages of Spermatogenesis with Spermatozoa in the Lumen (Arrow) and Leydig Cells in the Intertubular Spaces (Broken arrow). H and E. x 250.

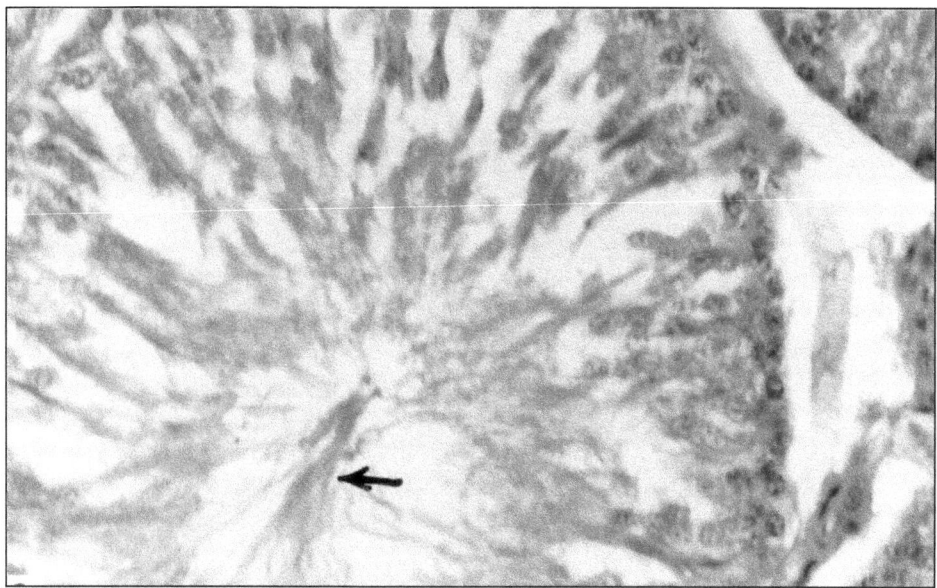

Figure 14.2: Magnified View of Seminiferous Tubules of Control Rat Depicting Germinal Epithelium and different Stages of Spermatogenesis. Mark spermatozoa in the lumen (arrow). H and E. x 400.

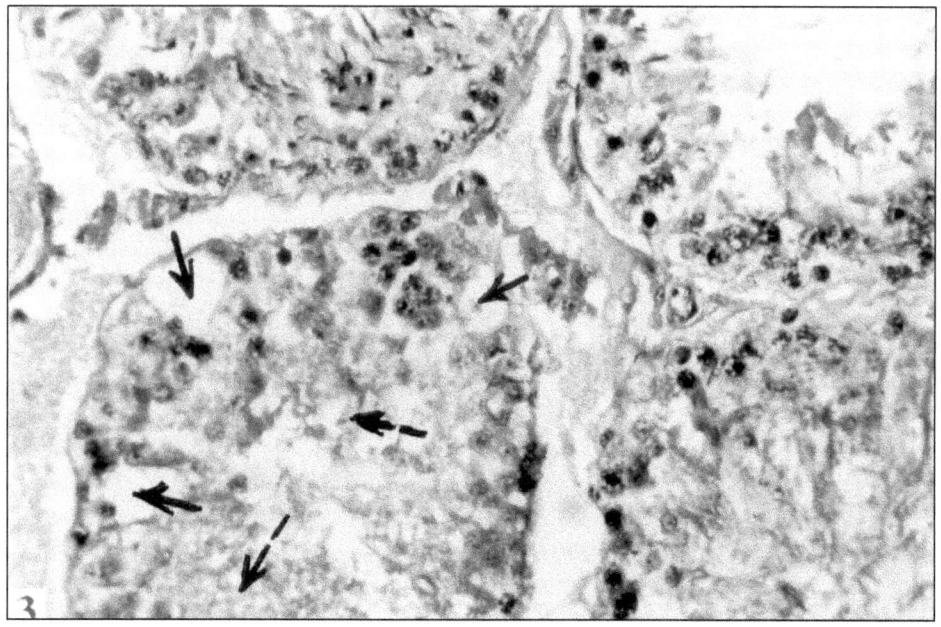

Figure 14.3: Testis of Heroin Treated Rat on Day 30 Showing Degenerative Changes in Seminiferous Tubules. Mark the lumen with cellular debris and scanty spermatozoa. H and E. x 250.

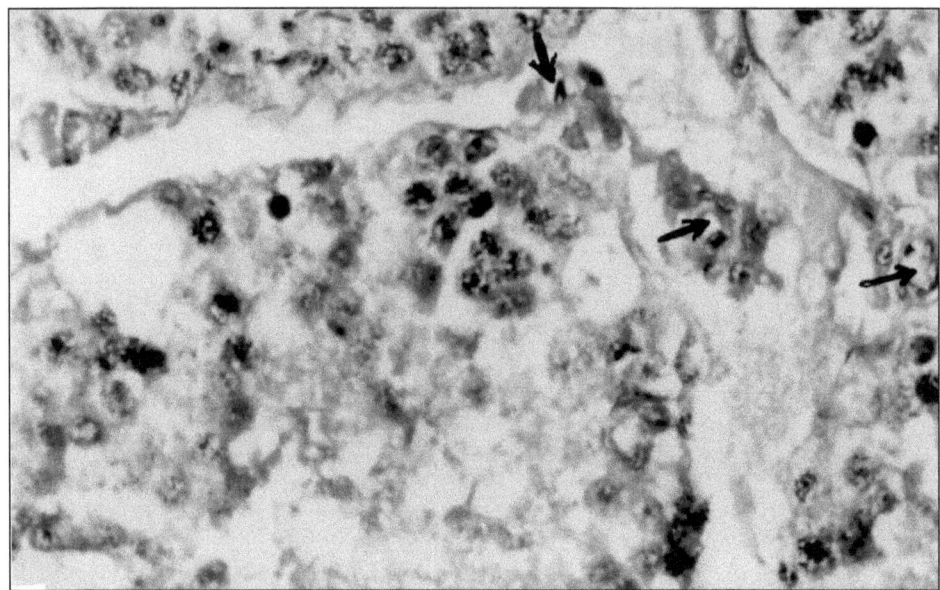

Figure 14.4: Seminiferous Tubules of Heroin Treated Rat on Day 30 Showing Massive Degeneration of Cpermatogenic Cells. Mark the atrophied Leydig cells with vacuolated cytoplasm (arrow). H and E. x 400.

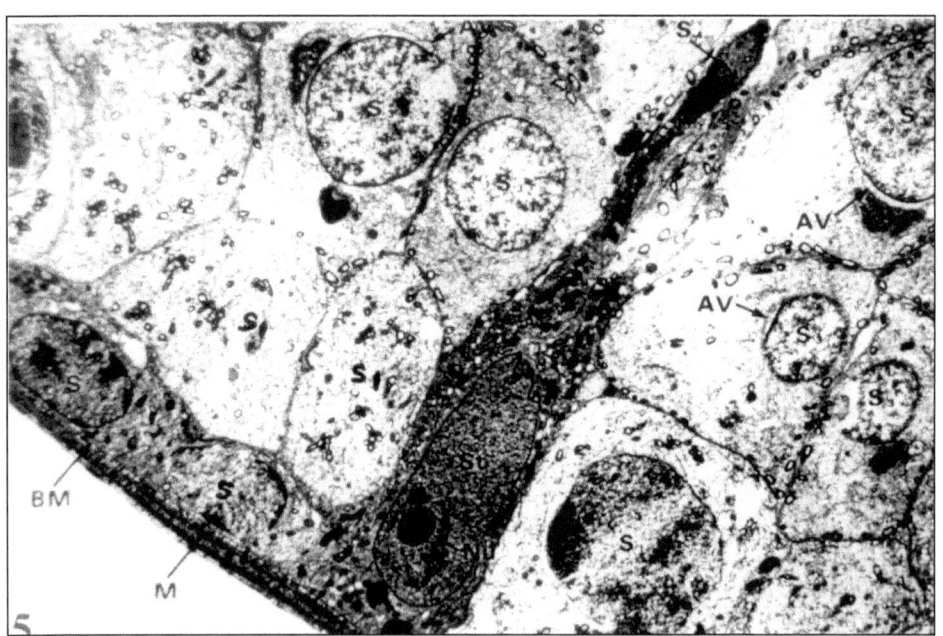

Figure 14.5: Electron Micrograph of Control Testis Showing Basement Membrane of Seminiferous Tubules (BM), Spermatogonia (S), Primary Spermatocyte (S1), Spermatid (S3), Acrosomal Vesicle (Av), Spermatozoon (S4) and Sertoli Cell (St). x 3,400.

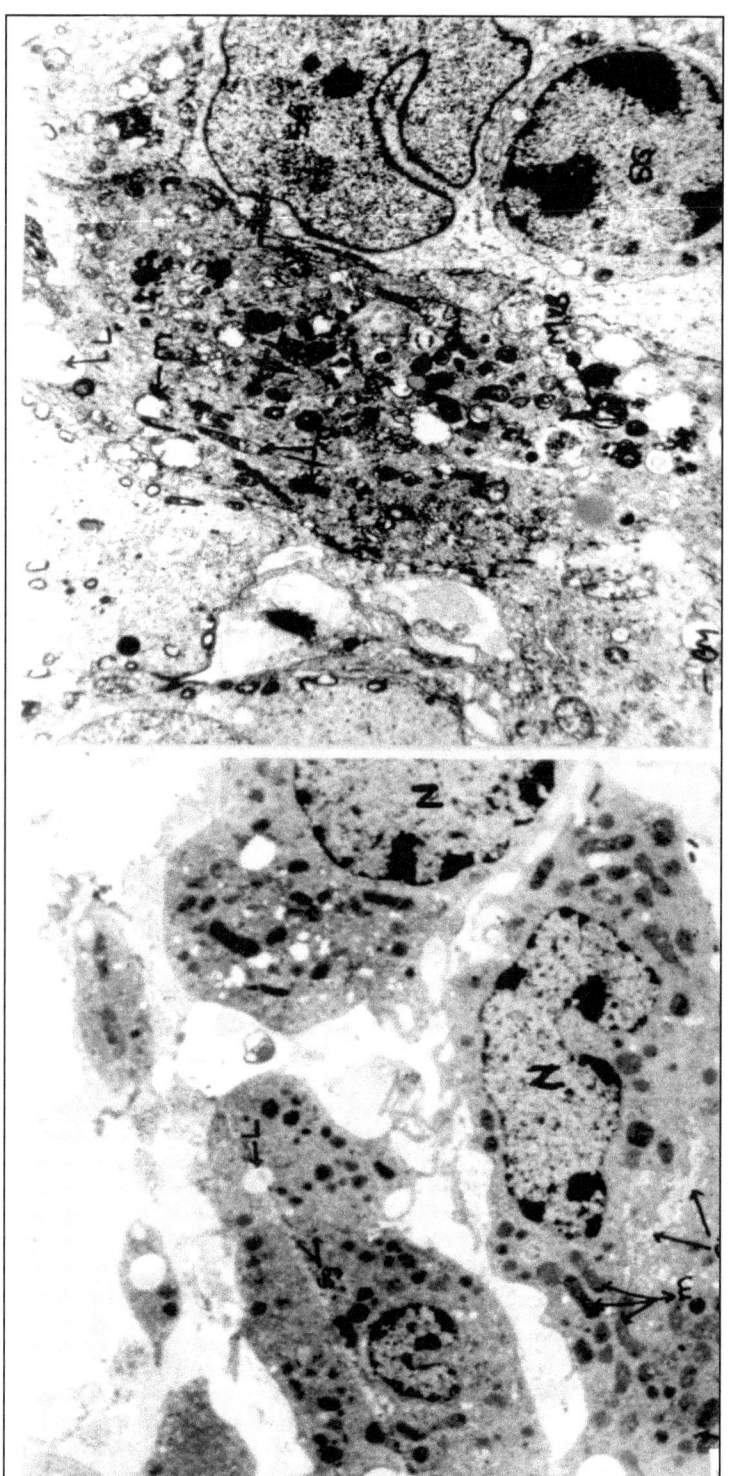

Figure 14.6: Leydig Cell of Heroin Treated Rat for 30 Days Showing Nucleus (N), Mitochondria (m), Multivesicular Bodies (Mvb), Secretory Granules (sg), Golgi Bodies (Gb) and Lysosome (L). x 6,500.

Figure 14.7: Sertoli Cell of Heroin Treated Rat on Day 30 Showing Multivesicular Bodies (Mvb), Mitochondria (m), Rough Endoplasmic Reticulum (RER), Lysosome (L), Spermatogonia (SG) and Basement Membrane (Bm). x 4,500.

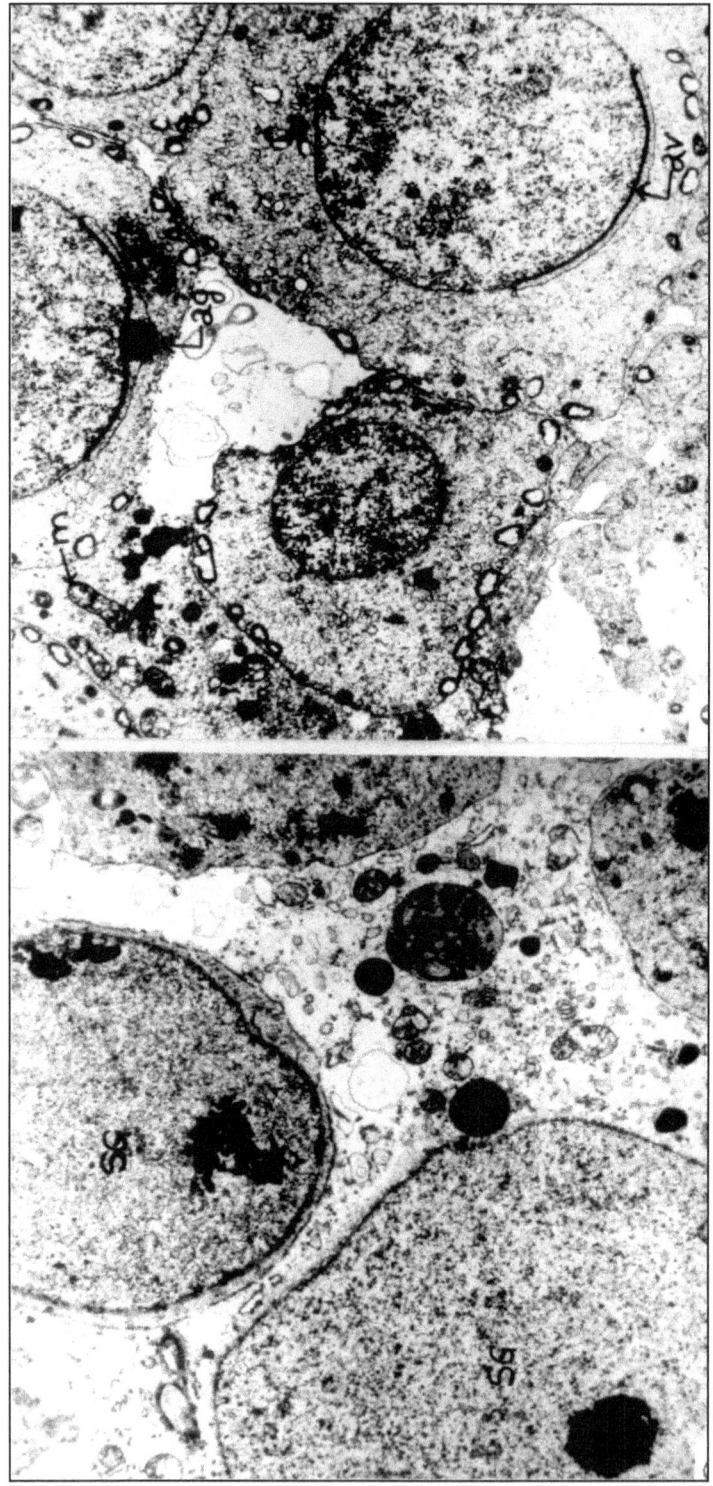

Figure 14.8: Seminiferous Tubule of Heroin Treated Rat on Day 30 Showing Spermatogonia (SG). x 4,500.

Figure 14.9: Seminiferous Tubule of Heroin Treated Rat with Spermatogonia Exhibiting Acrosomal Vesicle (av), Acrosomal Granules (ag) and Mitochondria (m). x 6,400.

The myoid layer of lamina propria was clearly seen. Sertoli cells were with infolded nuclear membrane. The cytoplasm of Sertoli cells showed numerous elongated mitochondria, lipid droplets of varying sizes and density, endoplasmic network, scattered multi-vesicular bodies. Onset of vacuolization was prominent at some places (Figure 14.7). Spermatogonia were firmly resting on the lamina propria, separated by continuous tight junctional complexes (Figure 14.8).

In the lumen, formation of spermatids and acrosomal granule or the acrosomal cap with well developed acrosomal membrane were visible (Figures 14.9 and 10). The lumen also showed spermatocytes with different regions and axonomal filaments (Figure 14.10). Extreme cytoplasmic degeneration in lumen of seminiferous tubule showed lysosomal activity. Prelysosomal vesicular structures and many hypertrophied mitochondria were observed (Figure 14.10).

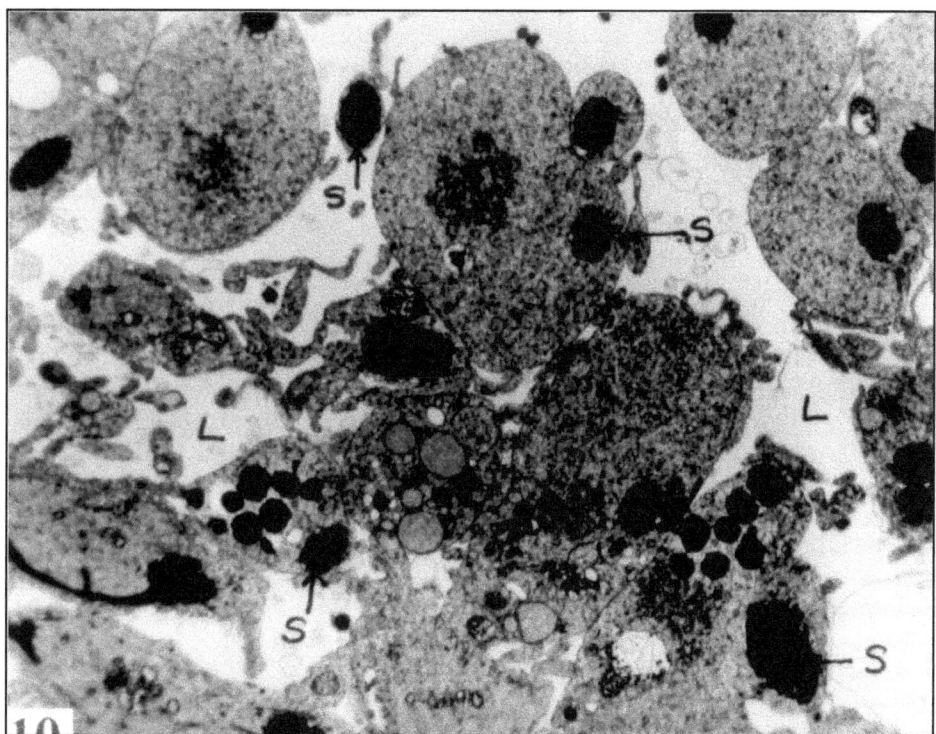

Figure 14.10: Testis of Heroin Treated Rat on Day 30 Showing Lumen (L) of Seminiferous Tubules with Sperm Formation (S). x 5,000.

Cytoarchitecture of testis of albino *Mus norvegicus* resembled to those described for other mammalian species (Guraya, 1980; Nagahama, 1986). The effects of narcotics on hormonal and sexual physiology were not well understood (Brambilla *et al.*, 1977; Wang *et al.*, 1978; Malik *et al.*, 1992; Daniell, 2002; George *et al.*, 2005; Hezazian *et al.*, 2007). It was found that long-term methadone administration in human males markedly impaired the function of secondary sex organs and depressed testosterone level (Cicero *et al.*, 1975). Methadone induced reduction in serum testosterone level had been recorded in human (Azizi *et al.*, 1973; Mendelson *et al.,* 1975a, b). Martin *et al.* (1973) observed that methadone decreases gonadotropin level while Cushman (1973) found reduction in luteininzing hormone (LH) and testosterone level in male heroin or methadone users. Chronic cocaine abuse was associated with significant decrease in libido and reproductive function (Washton *et al.,* 1985). Impotence and gynecomastia have been observed in male cocaine users while major derangement in menstrual cycle function has been recorded in case of women leading to amenorrhea and infertility (Siegel *et al.,* 1982; Cocores *et al.,* 1986). Though cocaine administration did not induce significant change in LH and testosterone levels in man and rhesus monkey ((Mendelson *et al.,* 1989; Mello *et al.,* 1993), there exists the possibility that opioids may effects on the gonadal portion of the hypothalamo-pituitary-gonadal (HPG) axis (Brambilla *et al.,* 1979; Adams *et al.,* 1993). Brambilla *et al.* (1977, 1979), Wang *et al.* (1978), Bolelli *et al.* (1979), Mendelson *et al.* (1980) and Malik *et al.* (1992) observed suppression in the levels of LH, FSH (follicle stimulating hormone) and testosterone levels in the human subjects addicted to heroin. Though there are indications of the involvement of hypothalamus and higher centres of brain in heroin-induced alterations of reproductive physiology (Brambilla *et al.,* 1979), the observed decline in LH and testosterone (T) levels concomitant with atrophy and vacuolization in Leydig cells as well as degenerative changes in seminiferous tubules of *Mus norvegicus* suggest that heroin induced changes is mediated through hypophysial-gonadal axis in the rat.

Acknowledgements

The senior author is thankful to the N.M. Wadia Trust, Mumbai for partial financial assistance to carry out this work. We are grateful to Hon'ble Justice Mrs. K.K. Baam, the then High Court Judge, Bombay for permitting us to work on heroin and to Mr. Rahul Rai Sur, the then Deputy Commissioner of Police, Narcotics Cell, Greater Mumbai for the procurement of the drug. Help extended by Dr. K.N.S. Panikar, Mr. Narayan, Mr. Harish Matal, Ms. V. Kailaje and Mrs. Vinita of Tata Cancer Research Centre, Mumbai in hormone assays and electron microscopy is acknowledged.

References

Adams, M.L., Sewing, B., Forman, J.B., Meyer, E.R. and Cicero, T.J., 1993. Opioid induced suppression of rat testicular function. *J. Pharmacol. Exp. Ther.,* 266: 323-328.

Al-Gommer, O., George, S. and Haque, S., 2007. Sexual dysfunctions in male opiate users: a comparative study of heroin, methadone and buprenorphine. *Addict. Disord. Their Treat.,* 6: 137-143.

Azizi, F., Vagenakis, A.G., Longcope, C., Ingbar, S.H. and Braverman, L.E., 1973. Decreased serum testosterone concentration in male heroin addicts. *Steroids,* **23**: 467-472.

Barai, S.R., Suryawanshi, S.A. and Pandey, A.K., 2009a. Levels of plasma sodium and potassium levels as well as alterations in adrenal cortex of *Rattus norvegicus* to sublethal heroin administration. *J. Environ. Biol.,* 30: 253-258.

Barai, S.R., Suryawanshi, S.A. and Pandey, A.K., 2009b. Responses of plasma calcium and inorganic phosphate levels, parathyroid gland and calcitonin-producing C cells of *Rattus norvegicus* to sublethal heroin administration. *J. Environ. Biol.*, 30: 917-922.

Bhoir, K.K., Suryawanshi, S.A. and Pandey, A.K., 2007. Responses of serum adrenocorticotropic hormone (ACTH) and cortisol levels as well as adrenal cortex of *Rattus norvegicus* to sublethal heroin administration. *J. Ecophysiol. Occup. Hlth.*, 7: 185-191.

Bhoir, K.K., Suryawanshi, S.A. and Pandey, A.K., 2009. Effects of sublethal heroin administration on serum thyroid stimulating hormone (TSH), thyroid hormones (T_3, T_4) and thyroid gland of *Rattus norvegicus*. *J. Environ. Biol.*, 30: 989-994.

Blesener, N., Albrecht, S., Schwager, A., Wecbecker, K., Litchermann, D. and Lingmuller, D., 2005. Plasma testosterone and sexual function in men receiving buprenorphine maintenance for opioid dependence. *J. Clin. Endocrinol. Metab.*, 90: 203-206.

Bolelli, G., Lafisca, S., Flamigni, C., Lodi, S., Franceschetti, F., Filicori, M. and Mosca, R., 1979. Heroin addiction: relationship between the plasma levels of testosterone, dihydrotestoterone, androstenedione, LH, FSH and the plasma concentration of heroin. *Toxicology*, 15: 19-29.

Brambilla, F., Sacchetti, E. and Brunetta, M., 1977. Pituitary-gonad function in heroin addicts. *Neuropsychobiology*, 3: 160-166.

Brambilla, F., Resele, L., de Maio, D. and Nobile, P., 1979. Gonadotropin response to synthetic gonadotropin hormone-releasing hormone (GnRH) in heroin addicts. *Am. J. Psychiat.*, 136: 314-317.

Brown, T.T., Wisniewski, A.B. and Dobs, A.S., 2006. Gonadal and adrenal abnormalities in drug users: cause of consequence of drug use behaviour and poor health. *Am. J. Infect. Dis.*, 2: 130-135.

Cami, J. and Farre, M., 2003. Drug addiction. *N. Engl. J. Med.*, 349: 975-986.

Cicero, T.J., Bell, R.D., Wiest, W.G., Allison, J.H., Polakoski, K. and Robins, E., 1975. Function of the male sex organs in heroin and methadone users. *N. Engl. J. Med.*, 292: 882-887.

Cicero, T.J., Mayer, E.R., Bell, R.D. and Koch, G.A., 1976. Effect of morphine and methadone on serum testosterone and luteinizing hormone levels and on the secondary sex organs of the male rat. *Endocrinology*, 98: 367-372.

Cocores, J.A., Dackis, C.A. and Gold, M.S., 1986. Sexual dysfunction: secondary to cocaine abuse in two patients. *J. Clin. Psychiat.*, 47: 384-385.

Cushman, P. Jr., 1973. Plasma testosterone in narcotics addiction. *Am. J. Med.*, 55: 452-458.

Daniell, H., 2002 Hypogonadism in men consuming sustained-action oral opioids. *J. Pain*, 3: 377-384.

George, S., Murali, V. and Pullickal, R., 2005. Review of neuroendocrine correlates of chronic opiate misuse: dysfunctions and pathophysiological mechanisms. *Addict. Disord. Their Treat.*, 4: 99-109.

Guraya, S.S., 1980. Recent progress in the morphology, histochemistry, biochemistry, and physiology of developing and maturing mammalian testis. *Int. Rev. Cytol.*, 62: 187-309.

Hejazian, S.H., Dasthi, M.H. and Rafati, A., 2007. The effect of opium on serum LH, FSH and testosterone concentration in addicted men. *Iranian J. Reprod. Med.*, 5: 35-38.

Malik, S.A., Khan, C., Jabbar, A. and Iqbal, A., 1992. Heroin addiction and sex hormones in males. *J. Pak. Med. Assoc.*, 42: 210-212.

Martin, W.R., 1984. Pharmacology of opioids. *Pharmacol. Rev.*, 35: 282-323.

Martin, W.R., Jesinski, D.R., Haertzen, C.A., Kay, D.C., Jones, B.E., Mansky, P.A. and Carpenter, R.W., 1973. Methadone: a re-evaluation. *Arch. Gen. Psychiat.*, 28: 286-295.

McKendry, J.B., Collins, W.E., Silverman, M., Krul, L.E., Collins, J.P. and Irvine, A.H., 1983. Erectile impotence: a clinical challenge. *Can. Med. Assoc. J.*, 128: 653-663.

Mello, N.K., Lukas, S.E., Mendelson, J.H. and Drieze, J., 1993. Naltrexone-buprenorphine interactions: effects on cocaine self-administration. *Neuropsychopharmacology*, 9: 21-224.

Mendelson, J.H., Mendelson, J.E. and Patch, V.D., 1975a. Plasma testosterone levels in heroin addiction and during methadone maintenance. *J. Pharmacol. Exp. Ther.*, 192: 211-217.

Mendelson, J.H., Meyer, R.E., Ellingboe, J., Mirin, S.M. and DcDaugle, M., 1975b. Effect of heroin and methadone on plasma cortisol and testosterone. *J. Pharmacol. Exp. Ther.*, 195: 296-302.

Mendelson, J.H., Ellingboe, J., Kuehnle, J.C. and Mello, N.K., 1980. Heroin and naltrexone effects on pituitary-gonadal hormones in man: interaction of steroid feedback effects, tolerance and supersensitivity. *J. Pharmacol. Exp. Ther.*, 214: 503-506.

Mendelson, J.H., Mello, N.K., Toeh, S.K., Ellingboe, J. and Cochin, J., 1989. Cocaine effects on pulsatile secretion of anterior pituitary, gonadal and adrenal hormones. *J. Clin. Endocrinol. Metab.*, 69: 1256-1260.

Mumford, L. and Kumar, R., 1979. Sexual behaviour of morphine-dependent and abstinent male rats. *Psychopharmacology*, 65: 179-185.

Nagahama, Y., 1986. Testis. In: *Vertebrate Endocrinology: Fundamentals and Biomedical Implications. Vol. 1. Morphological Considerations* (Eds.) Pang, P.K.T. and Schreibman, M. P. Academic Press, San Diego and New York, pp. 399-437.

Rasheed, A. and Tareen, I.A.K., 1995. Effects of heroin on thyroid function, cortisol and testosterone levels in addicts. *Polish J. Pharmacol.*, 47: 441-447.

Sawynok, J., 1986. The therapeutic use of heroin: a review of the pharmacological literature. *Can. J. Physiol. Pharmacol.*, 64: 1-6.

Siegel, P., Hinson, R.E., Krank, H.D. and McCully, J., 1982. Heroin "overdose" deaths: contribution of drug-associated environmental cues. *Science*, 216: 436-437.

Sporer, K.A., 1999. Acute heroin overdose. *Ann. Intern. Med.*, 130: 584-590.

Wang, C., Chan, V. and Yeung, R.T., 1978. The effect of heroin addiction on pituitary-testicular function. *Clin. Endocrinol.*, 9: 455-461.

Washton, A.M., Gold, M.S. and Pottash, A.C., 1985. The 800 Cocaine Helpline: Survey of 500 callers. In: *Problems of Drug Dependence* (Ed.) Harris, L.S. NIDA Research Monograph Series 55. Government Printing Office, Washington, D.C., pp. 224-230.

Wieland, W.F. and Yunger, M., 1985. Sexual effects and side effects of heroin and methadone. In: *Proceedings of the Third International Conference on Methadone Treatment*. PHS Pub. No. 2172. National Institute of Mental Health, Bethesda. 50 p.

Chapter 15

Wildlife Diversity of Odisha, India and their Conservation

Sudhakar Kar

The wildlife diversity of Odisha, one of the coastal states of Indian peninsula is considered to be very rich and diverse. Different types of forests and terrain conditions provide ideal habitats for a large variety of wildlife, both aquatic and terrestrial including the arboreal as well as the transitional amphibious forms.

In our state the Protected Area network constitute nineteen Sanctuaries, one National Park and a proposed National Park that covers 8352.19 Sq.km area, which is 5.36 per cent of the state's geographical area.

Bhitarkanika was constituted as the first sanctuary in the state under the provisions of Wildlife (Protection) Act, 1972, and was notified in April 1975. Other Sanctuaries to follow were Satkoshia Gorge (1976), Hadgarh (1978), Similipal and Nandankanan (1979), Baisipalli and Kotagarh (1981), Chandaka, Karlapat and Khalasuni (1982), Kuldiha and Balukhand (1984), Lakhari and Debrigarh (1985), Badrama and Chilika-Nalaban (1987), Sunabeda (1988), Gahirmatha Marine Sanctuary (1997) and Kapilash (2011).

The Bhitarkanika National Park was finally notified in the year 1998, and the notification of Similipal proposed National Park was issued in two phases in 1980 and in 1986. The Wildlife (Protection) Act 1972, the Forest Conservation Act, 1980 and the Environmental (Protection) Act 1986 as amended from time to time have provided necessary legal support for conservation of the forests as well as Wildlife.

Different ecological niches display interesting groups of wildlife in Odisha. There are 25 species of amphibians, 110 species of reptiles, 473 species of birds and 86 species of mammals so far identified and listed in our state in addition to large varieties of fish (about 400 species- fresh water, estuarine and marine) and

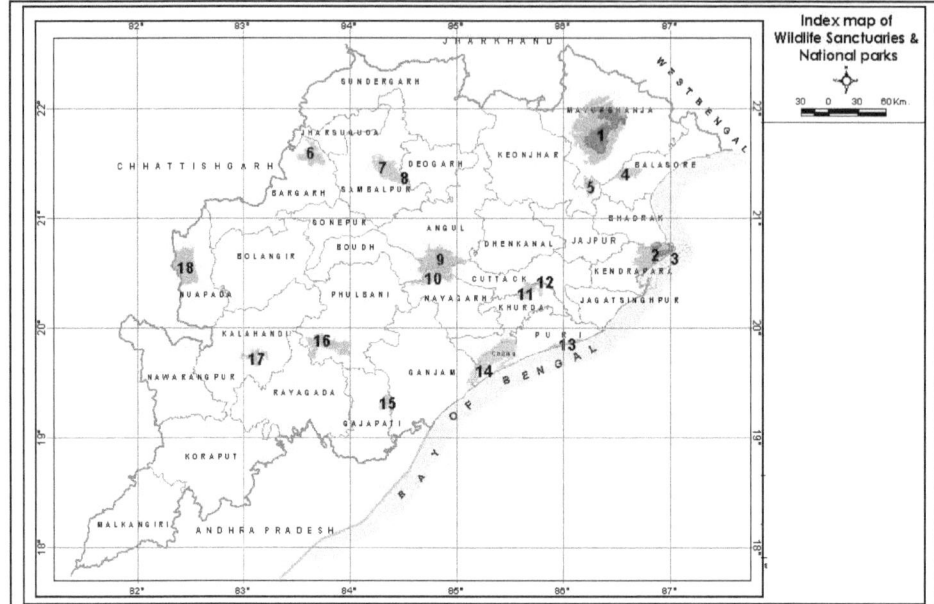

Figure 15.1

invertebrate fauna. According to IUCN RED DATA BOOK-1997 the threatened species of Reptiles, Birds and Mammals in Odisha include 17 species, 15 species and 22 species, respectively.

Figure 15.2: Meadow at Chahala in Similipal Tiger Reserve.

Some of the major wildlife species of Odisha are Tiger, Leopard, Jungle cat, Fishing cat, Leopard cat, Hyena, Wolf, Wild dog, Fox, Jackal, Otter, Pangolin, Porcupine, Wild boar, Bear, Ratel, Elephant, Bison, Sambar, Spotted deer, Barking deer, Mouse deer, Blackbuck, Chowsingha, Blue bull, Hanuman langur, Rhesus macaque, Giant squirrel, three species of Indian crocodilians (Gharial, Estuarine crocodile and Mugger crocodile), King cobra, Python, Star tortoise, Land tortoise, Olive Ridley sea turtle, Peafowl, Woodpecker, Hornbill, Kingfishers, Orioles, White bellied sea eagle, Golden eagle, Brahminy kite, Pariah kite, Red jungle fowl, Vultures and a variety of long and local migratory birds.

The state has rich floral diversity with around 3000 species of plants. This includes 132 species of orchids, sixtyfive species of mangroves and a host of associates occurring in the coastal swamps, particularly in Bhitarkanika National Park.

Figure 15.3: Bhitarkanika River with Luxuriant Overhanging Mangrove Vegetation.

For *ex-situ* conservation of different species of wildlife, presently there are one large Zoo and 11 Mini Zoos in the state. The large Zoo, Nandankanan and seven Mini Zoos such as 1. Kuanria in Nayagarh district, 2.Taptapani in Ganjam district, 3.Kapilash in Dhenkanal district, 4.Motijharan in Sambalpur district, 5.Papadahandi in Nawarangpur district, 6.Harishankar in Bolangir district and, 7. Gharial Research and Conservation Unit at Tikarpada in Angul district are managed by the Forest Department.

The Mini Zoo located in the University campus at Bhanjavihar, Berhampur in Ganjam district is managed by the Berhampur University. The rest three Mini Zoos in the State are owned by the private agencies/organizations. These include

Figure 15.4: One of the Largest Male Estuarine Crocodiles Basking in the Winter Sun.

Figure 15.5: Community Nesting of White Ibis.

Figure 15.6: A Flock of Greater Flamingo in Nalaban Wildlife Sanctuary/ Chilika Lagoon.

(1) Indira Gandhi Park Mini Zoo, Rourkela, (2) Hindustan Aeronautics Limited (HAL) Mini Zoo, Sunabeda, and (3) Municipal Corporation Mini Zoo, Cuttack.

The Nandanakanan Zoological Park along with the adjoining forests was notified as Nandankanan Wildlife Sanctuary on 3rd August 1979. It covers an area 4.37 sq km which include Kanjia Lake (0.66 sq km), Zoological Park (0.75 sq km) and Botanical Garden (2.96 sq km). The Zoo has the distinction of breeding endangered Gharials successfully in captivity in 1980. This was the world record for breeding of Gharials for the first time in captivity. The White Tiger Safari, Lion Safari, Bear Safari, Aquarium, Aviary, Orangutan, Chimpanzee, Gharial breeding facility, Toy Train, Boating and Aerial Ropeway are some of the prime attractions of this Zoo.

After the birth of white tiger cubs in Nandankanan in January, 1980 to parents' appearing normal in colour, a systematic breeding programme was implemented. The Zoo holds the largest collection of white tigers in any zoo in the country. These white tigers have been used to procure rare animals from other zoos in the country and abroad in exchange. At present Govt. of India have also accorded conservation breeding programme of Pangolins and Vultures to Nanadankanan.

Similipal was chosen as one of the nine prime locations for tiger conservation under the 'Project Tiger' launched in the year 1973 in the country. New area/habitat with potentiality for tiger conservation in the state such as Satkoshia Gorge Wildlife

Figure 15.7: Gharial Captive Breeding Facility at Nandankanan Zoological Park.

Sanctuary and Baisipalli Sanctuary have been jointly notified as second Tiger Reserve during the year 2008, and Sunabeda Wildlife Sanctuary has been identified to be designated as the third Tiger Reserve under the Project Tiger.

The Crocodile Conservation and Research programme was launched in the state as well as in the entire country with the support of FAO/UNDP and Government of India in 1974-75. The technical expertise was provided by a FAO expert, Dr.H.R.Bustard, who is a renowned Crocodile and Seaturtle expert of the world. Research and Conservation Centers have been established at various strategic locations such as (i) Tikarpada in Satkoshia Gorge Sanctuary for Gharials (ii) Dangmal in Bhitarkanika National Park for Estuarine Crocodiles (iii) Ramatirtha for Mugger or Marsh Crocodile Research and Rehabilitation Centre in Similipal Sanctuary, and (iv) Nandankanan Zoo for Captive Breeding of three crocodilian species.

The Estuarine crocodile, designated as 'endangered in 1975' has now a viable population of above 1600 in the river systems of Bhitarkanika Sanctuary and associated areas/habitats, the most promising place for these crocodiles in the entire country. The sighting of crocodile nests in the sanctuary has gone up from only 6-7 nests in the eighties to over 70. At present, Bhitarkanika river systems hold the largest population of Estuarine crocodiles in the wild in comparision to all the distributional ranges in India. It is to be noted that above 80 per cent of the total Indian population of *Crocodylus porosus* are available in Bhitarkanika mangrove ecosystem of Odisha, India. The four decade long "Rear and Rehabilitation" programme of Estuarine crocodiles in Bhitarkanika river systems is a resounding success.

Elephant conservation programme under Project Elephant was launched in the year 1991. Three Elephant Reserves (ERs), namely Mayurbhanj ER, Mahanadi ER and Sambalpur ER were notified in the years 2001 and 2002 to offer more focused protection to about 50 per cent of the state's elephant population using 8509 sq.km. The ER-Network is being expanded to 14884 sq.km of forest habitat to offer protection to over 90 per cent of the 1954 elephants (as per 2015 census) now inhabiting the state.

Figure 15.8: A Herd of Elephants in Similipal Tiger Reserve.

The Olive Ridley Sea turtle conservation was started in the year 1976. About 50 per cent of the world population of Olive Ridleys which is about 90 per cent of the Indian population of sea turtles use the Odisha coasts for nesting. The nesting intensities of sea turtles are effectively monitored at Gahirmatha Marine Sanctuary(known as world's largest rookery and Rushikulya rookery. During 2014-15 nesting season, 4.13 lakh and 3.09 lakh nesting female turtles laid eggs at Gahirmatha and Rushikulya rookery, respectively. The State Wildlife organization, State Fisheries Department, Home Department and the Indian Coast guards have been involved in a collaborate offshore and on-shore Sea turtle protection activities. Due to effective protection activities Sea Turtle mortalities have been reduced considerably.

Figure 15.9: Mass Nesting (Arribada) of Olive Ridleys at Gahirmatha Coast.

The constitution of Similipal Biosphere Reserve over an area of 5569 sq.km in 1994 is an additional support to the management inputs started in the year 1973 under Project Tiger and the sustenance of the same under the State Wildlife Organisation.

The picturesque Chilika wetland is one of the largest coastal lagoons in Asia and a Ramsar site. Chilika is very rich in biodiversity and its prime attraction is the annual congregation of about a million of long distance migratory water birds (above 100 species) which include variety of ducks, geese and shore birds. *Barkudia insularis,* a limbless lizard/skink, is endemic to Chilika and named after the "Barakuda" island of the lagoon. Irrawaddy dolphin (*Orcaella* brevirostris), locally known as Bhuasuni Magar or Kherra is a popular aquatic mammal in the Lagoon. As per the winter 2015 census, there were 144 Irrawaddy dolphins in the Lagoon.

The Blackbuck of Balippadar-Bhetnoi areas within Buguda and Aska police stations of Ganjam district are protected socio-religiously by the local people. This is one of the best examples of people's participation in wildlife conservation of the state as well as the country. The belief that the presence of Blackbuck in the paddy fields brings prosperity to the local villagers has contributed a lot for conservation of his endangered wildlife. The estimated population of Blackbuck in Balipadar-Bhetnoi and associated areas in Ganjam district was above 3000 (as 2015 estimation). The Forest Department, Govt. of Odisha is taking all possible steps to declare this

Blackbuck habitat as Community Reserve as per the provision of the Wildlife (Protection) Act, 1972 (as amended).

Due to effective protection measures over last few years, the wildlife population has increased in most of the Protected Areas and also in other wildlife areas of the state. It is not becoming possible to conduct census of all the species of wildlife in the state but the census result of a few endangered/flagship species indicate that the population is on the increasing trend.

The wildlife resource, the modest infrastructure facilities in the field and a small group of scientists, constitute a unique combination in India. The crocodile project, sea turtle project, the project tiger, the project elephant and the captive breeding programme of wildlife at Nandananan are the main producers of research data. These days many graduate and Post-graduate Departments from state and national institutions and universities have also produced dissertations relating to wildlife and nature conservation.

In-house capability for research on wildlife was acquired and developed along with the launching of crocodile conservation programme and it has expanded to all other faculties of wildlife management and conservation projects. Three Senior Scientists, with international repute were involved in study of ecology and biology of flagship wildlife species such as Crocodilians, Sea turtles, Elephants, Tigers, etc over three and half decades and provided scientific inputs for conservation and management of wildlife resource of the state.

Besides a very good documentation of scientific data/information on biology and behaviour, disease and treatment aspects, etc of captive animals have been done at the Nandankanan Zoological Park. To strengthen the research base, eleven Research Fellows have been inducted in the Wildlife Organisation, out of which eight Research Fellows have been engaged at eight Forest Circles of the state to study on elephants' ecology and biology, and also man-elephant interface.

Protection measures have been effectively worked out through various ways and means including VHF network system, deployment of mobile squad, Sabuja Bahini and intelligence collection, *etc.* within the limited manpower and resources. Construction of watch towers at strategic locations inside the wildlife habitats, establishment of anti-poaching camps and provision of saltlicks have helped to improve wildlife status.

Wildlife crime cell operating in the State Wildlife Headquarters has been strengthened to tackle death of wild animals due to poaching, poisoning and deliberate electrocution. The Cell also monitors such offence cases relating to arrest of offenders, seizure of stolen tusks, and various body parts of wild animals.

Steps are being taken to involve the public in the management of wildlife for tackling the man-animal conflict and to reduce pressure on wildlife habitats/forests through eco-development programme and beneficiary oriented schemes. Wildlife education and interpretation programme have been imparted to the public through suitable interpretive methods including audio-visuals.

Figure 15.10: Ecotourism Facility in Bhitarkanika National Park.

Eco-friendly tourism/Eco-tourism is also being promoted in the protected areas such as Bhitarkanika, Similipal, Satkoshia, Baisipalli, Kuldiha, Debrigarh, Chandaka, Kuldiha, Hadgarh, *etc.* for generating support for conservation of wildlife resource of the state.

Acknowledgements

I am grateful to the Wildlife Organisation of the Odisha Forest Department for their kind support and encouragement. Photographs used in the text belong to the author and friends. I am thankful to my friends who are kind enough to provide photographs.

References

Acharjyo, L.N., Kar, S.K. and Patnaik, S.K., 1996a. Role of Nandankanan Biological Park, Odisha in conservation of the Gharial (*Gavialis gangeticus*). *Tiger Paper*, 23(3): 5-8.

Acharyo, L.N., Kar, S.K. and Patnaik, S.K., 1996b. Studies on captive breeding of the Gharial, *Gavialis gangeticus* (Gmelin) in Odisha. *J. Bombay Nat. Hist. Soc.*, 93(2): 210-213.

Behura, B.K., 1999. *Bhitarkanika: The Wonderland of Odisha*. Nature and Wildlife Conservation Society of Odisha, Bhubaneswar, pp. 1-63.

Bustard, H.R., 1976. World's largest Seaturtle rookery. *Tigerpaper*, pp. 3.

Kanungo, B.C., 1976. An integrated Scheme for Conservation of Crocodiles in Odisha with Management Plan for Satkosia Gorge and Bhitarkanika Sanctuaries. Forest Department, Govt. of Odisha, Cuttack, pp. 1-128.

Kar, S.K., 1981. Studies on the saltwater crocodile, *Crocodylus porosus* Schneider. Ph.D. Thesis, Utkal University, Bhubaneswar, Odisha, India.

Kar, S.K., 1984. Conservation future of the Saltwater Crocodile, *crocodylus porosus* Schneider in India. IUCN publications New Series. Proceedings of the 6th Working Meeting of the Crocodile Specialist group, pp. 29-32.

Kar, S.K., 2001. Balipadar's Blackbuck (an insight into myth and realty of human-Blackbuck relationship). Wildlife Wing, Forest Department, Govt of Odisha. pp. 1-41.

Kar, S.K., 2016. Chilika: The unique winter home of migratory waterbirds. In Souvenired. Pubulished by the Chilika Development Authority on the occasion Chilika Mahotsav (held at Satpada from 11-13th Jan. 2016). pp.64-66.

Kar, S.K., Acharjyo, L.N. and Patnaik, S.K., 1998. Rehabilitation of crocodilians in Odisha. *Cobra*, 33: 23-26.

Mishra, Ch.G., Patnaik, S.K., Sinha, S.K., Kar, S.K., Kar, C.S. and Singh, L.A.K., 1996. *Wildlife Wealth of Odisha*. Wildlife Wing, Odisha forest Department, Government of Odisha, Bhbaneswar, pp. 1-185.

Mohanty, S.C., Kar, C.S., Kar, S.K. and Singh, L.A.K., 2004. *Wild Odisha-2004*. Wildlife Organisation, Forest Department, Govt. of Odisha, Bhubaneswar. pp. 1-81.

Patnaik, S.K., Kar, C.S. and Kar, S.K., 2001. *A Quarter Century of Sea Turtle Conservation in Odisha*. Wildlife Wing, Forest Department, Government of Odisha, Bhubaneswar, pp. 1-34.

Patnaik, S.K., Kar, C.S. and Kar, S.K., 2012. Mangrove ecosystem of Odisha with special reference to Bhitarkanika. In: *Coastal Tract of Odisha: Geology, Resources and Environment*, (Ed.) N.K. Mahalik, pp. 217-240.

Mohanty, S.C., Singh, L.A.K., Kar, S.K., Kar, C.S., and Nair, M.V., 2006. *Nesting Animals of Odisha-2006*. Wildlife Organisation, Forest Department, Govt. of Odisha, Bhubaneswar, pp. 1-60.

Singh, L.A.K., Kar, S.K., Kar, C.S., Nair, M.V., Mishra, S.K. and Mohanty, S.C., 2007. *Wildlands of Odisha-2007*. Wildlife Organisation, Forest Department, Govt. of Odisha, Bhubaneswar, pp. 1-68.

Wildlife Odisha, 2015. Wildlife Organisation, Forest and Environment, Forest and Environment Department, Bhubaneswar, pp. 1-72.

Chapter 16

Ecology and Conservation of Mammals of Oak Forest of Central Himalaya, India

Aisha Sultana and Mohammad Shah Hussain

ABSTRACT

This study was undertaken to assess the mammal species diversity, their distribution in relation to habitat in the oak forest of Central Himalaya of India to assist in habitat management and species conservation. The study was conducted in 23 oak forest patches of six districts of Central Himalaya. Direct and indirect methods of mammal survey using random search method and systematic quadrate sampling were applied. Total 18 mammal species were found in the study area. The key mammal species are Himalayan Tahr (Hemitragus jemlahicus), Serow (Capricornis sumatraensis), Musk deer (Moschus leucogaster), Sambar (Rusa unicolor), Common leopard (Panthera pardus) and Red giant flying squirrel (Petaurista petaurista). Snow leopard was recorded by indirect evidence. Maximum group size was calculated for Himalayan Tahr (6.83 ± 5.12) and minimum for Serow (1.05 ± 0.23). A significant difference was found in the occurrence of ungulate species at different altitudes. Different habitats were also used by species in terms of tree cover, shrub cover, slope. Their specific habitat protection and conservation were discussed in the light of results.

Keywords: Mammals, Central Himalaya, India, Habitats, Conservation, Protection

Introduction

Inventorying biodiversity and monitoring efficacy of measures for its conservation have emerged as important scientific challenges in recent years. These are also commitments for India and over 160 other countries worldwide that are parties to the Convention on Biological Diversity (UNEP, 1992). The great mountain chain of the Himalayas with its associated hill tracts in the Kumaon region is the biological treasure of the country. The tremendous altitude ranges of these hills, large variation in rainfall and being dry Pine forests to the wet broad-leaved Oak forests to sub-alpine and alpine zone have created an unparalleled diversity of plants and animals in a variety of habitats. This region sits at the junction of Sino-Himalayan and Sino-Indian biogeographic realms (Legris and Meher Homji, 1968). Due to highly dissected mountain topography and intense geographical isolation there is also a high degree of endemism in this region (Pandit *et al.*, 2007).

Information from other parts of Himalayas is existing (Basu, 2004; Bagchi *et al.*, 2006; Mishra *et al.*, 2006). About 300 mammal species have been recorded in the Himalayas, including a dozen that are endemic to the hotspot. Among the endemic species are the golden langur (*Trachypithecus gee*i, EN), restricted to a small area in the Eastern Himalaya, the Himalayan tahr (*Hemitragus jemlahicus*, VU) and the pygmy hog (*Sus salvanius*, CR), which has its stronghold in the Manas National Park. The only endemic genus in the hotspot is the Namadapha flying squirrel (*Biswamoyopterus biswasi*, CR), described only from a single specimen from Namdapha National Park.

The mammalian fauna in the lowlands is typically Indo-Malayan, consisting of langurs (*Semnopithecus* spp.), Asiatic wild dogs (*Cuon alpinus*, VU), sloth bears (*Melursus ursinus*, VU), gaurs (*Bos gaurus*, VU), and several species of deer, such as muntjac (*Muntiacus muntjak*) and sambar (*Cervus unicolor*). In the mountains, the fauna transitions into Palearctic species, consisting of snow leopard (*Uncia uncia*, EN), black bear (*Ursus thibetanus*, VU), and a diverse ungulate assemblage that includes blue sheep (*Pseudois nayaur*), takin (*Budorcas taxicolor*, VU), and argali (*Ovis ammon*, VU).

The alluvial grasslands support some of the highest densities of tigers (*Panthera tigris*, EN) in the world, while the Brahmaputra and Ganges rivers that flow along the foothills also support globally important populations of the freshwater Gangetic dolphin (*Platanista gangetica*). Some of the world's last remaining populations of wild water buffalo (*Bubalus bubalis*, EN) and swamp deer (*Cervus duvaucelii*, VU) are restricted to protected areas in southern Nepal and northeastern India.

Topographic features (*e.g.* altitude, slope, landscape ruggedness) in combination with habitat (vegetation types) are often used to predict the distribution of species (Guisan and Zimmerman, 2000; Hirzel, Helfer and Metral, 2001; Seoane, Bustamante and Díaz-Delgado, 2004). The aim of this study was to document mammal species and their habitat requirements for their long term conservation in Central Himalaya. No detailed information about mammals from these fragmented oak forest existed earlier.

Materials and Methods

The study was conducted in the six districts; Almora, Bageshwer, Champawat, Naini Tal, Pithoragarh and UdhamSingh Nagar of Kumaon region in Central Himalaya (28° 43' and 30° 30' N latitude and 78° 44' and 80° 45' E longitude) in the Uttarakhand state of India. The data were collected in 23 oak forest stands, varying in

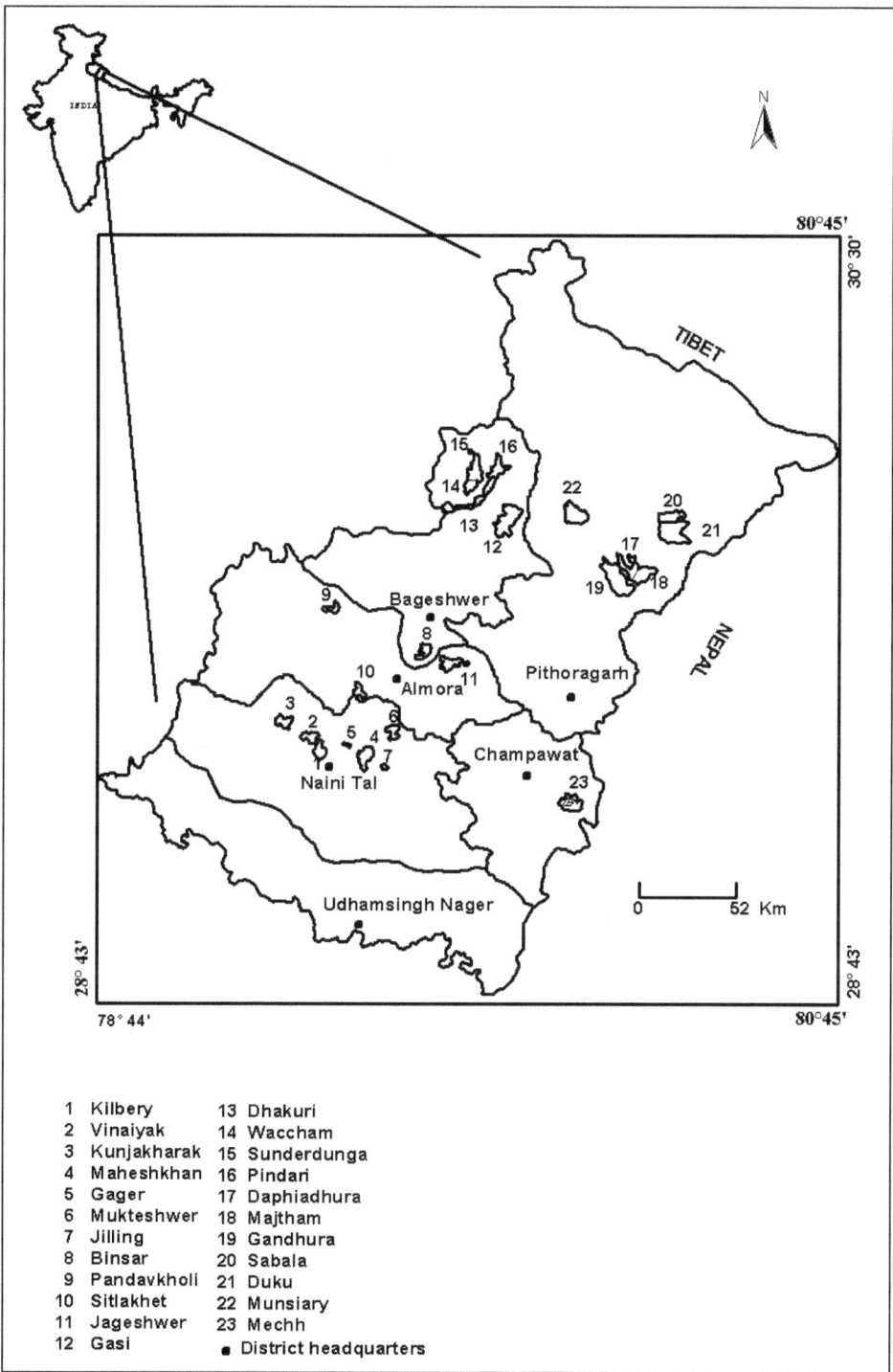

Figure 16.1: Location of the Surveyed Oak Patches of Kumaun Himalaya, Uttarakhand, India.

sizes and altitude ranges during the summer (March- Mid July) and winter months (September-December) (Figure 16.1). About 90 per cent of the area of Kumaon is mountainous. The Kali, Dhauli Ganga, Gori Ganga, Saryu and Pindar form the river system of Kumaon. The monsoon starts at the end of June and ceases by middle of September (Singh and Singh, 1987). The area represents from temperate (1500-3500m) to alpine (>3500m) climatic condition. The boundary of Kumaon and oak forest patches were digitized from the topo sheets of Survey of India.

The extant oak forests in all six districts were extensively surveyed using existing forest trail. Monitoring of trails and random searches were done in morning and evening hours to document as many number of species as possible. From trails and ridge lines, slopes were scanned with binoculars and spotting scope for large animals. Direct and indirect evidences were systematically recorded. Encounters included records of animals that were not seen but were identified from their alarm calls or other characteristic sounds and also from pellet group. Pellet group is a reliable indicator of animal presence and has been the basis of evaluation of habitat use of ungulates by various workers (Eisenberg and Lockhart, 1972; Dinerstein, 1980) or latrine site. The data were collected on following parameters whenever an animal or group of animals was encountered: location, altitude, slope, aspect, tree cover, shrub cover, and ground cover, Number of individuals in a group, sex (if identifiable) and activity.

Tree cover within an area of approximately 10 m radius from an animal's location, were recorded according to the following scale: 0-25 per cent, 25-50 per cent and >50 per cent while the shrub cover was assessed in nominal categories: 0-25 per cent, 25-50 per cent, 50-75 per cent and >75 per cent.

Table 16.1: Different Classes of Habitat Variables Used in Assessment of Mammal Species of Kumaun Himalaya

Altitude (Meters)	Tree Cover (per cent)	Shrub Cover (per cent)	Ground Cover (per cent)	Slope (º)	Aspects of Slope
1800-2000	0-25	0-25	0-25	0-15	North
2001-2200	26-50	26-50	26-50	16-30	East
2201-2400	51-75	51-75	51-75	31-45	South
2401-2600	>75	>75	>75	46-60	West
2601-2800	-	-	-	61-75	North-East
2801-3000	-	-	-	76-90	South-East
3001-3200	-	-	-	-	South-West
>3200	-	-	-	-	North-West

Mean group size was calculated for five ungulate species *i.e.* Indian muntjac (*Muntiacus muntjak*), Goral (*Naemorhedus goral*), Serow (*Naemorhedus sumatraensis*), Himalayan tahr (*Hemitragus jemlahicus*), Sambar (*Cervus unicolor*) and Wild pig (*Sus scrofa*) as the sample size was adequate for these mammals only. Kruskal-Wallis One way ANOVA (Zar, 2010) was performed to test for significant difference in the association of different mammal species (ungulates) with altitude, slope, tree cover, shrub cover, ground cover and aspects of slope. Each variable was divided into several nominal classes with different range between minimum and maximum values (Table 16.1). The order of preference with respect to the different classes was compared between pairs of species choosing Spearman's rank correlation coefficient (Zar, 2010). A high value indicates that the ranked order of preferences in one species is strongly correlated with that in other species. A strongly negative value shows that there is little potential for competition occurring between a pair of species.

Chi-square contingency test was performed to observe differences in habitat use of different ungulate species and in order to decide which ungulate species was using which particular habitat (Fowler and Cohen, 1986). To see the similarity in sites of Kumaon Himalaya on the basis of mammal species, Hierarchical Cluster analysis by nearest neighbour method was performed on binary measure (presence/absence) of species for all sites. This clustering tends to produce straggly clusters, which quickly agglomerate very dissimilar samples (Gauch, 1989). All statistical tests were performed by using IBM SPSS software programme version 23.0.

Results

Group Size of different Ungulate Species

A total of 18 mammal species were encountered during the surveys of Kumaon Himalaya (Table 16.2). Out of this, mean group size was calculated only for different ungulate species. Maximum group size was calculated for Himalayan tahr (*Hemitragus jemlahicus*) (6.83 ± 5.12) and minimum for Serow (*Capricornis*

Table 16.2: List of Mammal Species Sighted during Surveys in Kumaon Himalaya

Species	Altitude Range (meters)
Hanuman langoor *Presbytis entellus*	1730-2700
Rhesus macaque *Macaca mulatta*	1520-2860
Indian muntjac *Muntiacus muntjak*	1880-2920
Musk deer *Moschus moschiferus*	3100-3200
Wild pig *Sus scrofa*	2210-2780
Goral *Nemorhaedus goral*	1800-2800
Serow *Capricornis sumatraensis*	2280-2850
Himalayan tahr *Hemitragus jemlahicus*	2400-3000
Sambar *Cervus unicolor*	1820-2570
Blue sheep *Pseudois nayaur*	3600-3900
Leopard *Panthera pardus*	2100-2800
Himalayan black bear *Selenarctos thibetanus*	2040
Jackal *Canis aureus*	1900
Yellow-throated martin *Martes flavigula*	1820-2930
Himalayan mouse hare *Ochotona roylei*	2670-2940
Indian porcupine *Hystrix indica*	1770-2240
Red giant flying squirrel *Petaurista petaurista*	2150

sumatraensis) (1.05 ± 0.23) with Goral (*Nemorhaedus goral*) (1.74 ± 1.32) and Indian muntjac (*Muntiacus muntjak*) (1.12 ± 0.40) having intermediate values for group size (Table 16.3).

Table 16.3: Mean Group Size of Ungulate Species Encountered during the Surveys of Kumaon Himalaya

Species	Mean	S.D.
Indian muntjac	1.12	0.4
Goral	1.74	1.32
Serow	1.05	0.23
Himalayan tahr	6.83	5.12
Sambar	1.5	1.73

Altitudinal Distribution of Mammals

There was significant difference in distribution of ungulate species with respect to altitude ($F = 33.82$, $p < 0.001$) and Spearman's rank correlation showed significant negative correlation in Indian muntjac and Goral occurrence ($r_s = -0.392$, $p < 0.04$). Sambar was never encountered above 2570 m of altitude and below 1820 m while Serow was never found below 2280 m and the occurrence was significantly correlated ($r_s = 0.581$, $p < 0.04$). Himalayan tahr and Goral were also significantly correlated ($r_s = 0.806$, $p < 0.05$) and there existed high competition at similar altitude ranges while Himalayan tahr and Serow were not significantly correlated in altitude ranges (Figure 16.2).

Slope

No significant difference was observed in the association of different mammal species with slope. Himalayan tahr was always encountered at the cliff and steeper slopes than Indian muntjac and Serow and they are not significantly correlated too. Serow and Sambar were found on medium slopes but there was no significant correlation between them (Figure 16.3).

Habitat Types

Chi-square contingency test showed a significant difference in the use of different habitats by various ungulate species ($\Sigma\chi^2 = 111.20$, $p < 0.01$). The pin pointing associations showed that Himalayan tahr used grassy slope more than expected while Goral were found more in Oak-Scrub and Oak-Pine habitats than expected.

Tree Cover

Tree cover utilisation also differed significantly ($F = 24.81$, $p < 0.00$) among the mammals. Himalayan Tahr was never found in areas having >50 per cent tree cover. No significant correlation was found between Indian muntjac, Goral and Serow. Sambar was never encountered in places having less than 60 per cent tree cover while Goral and Himalayan Tahr often encountered in places with no tree cover.

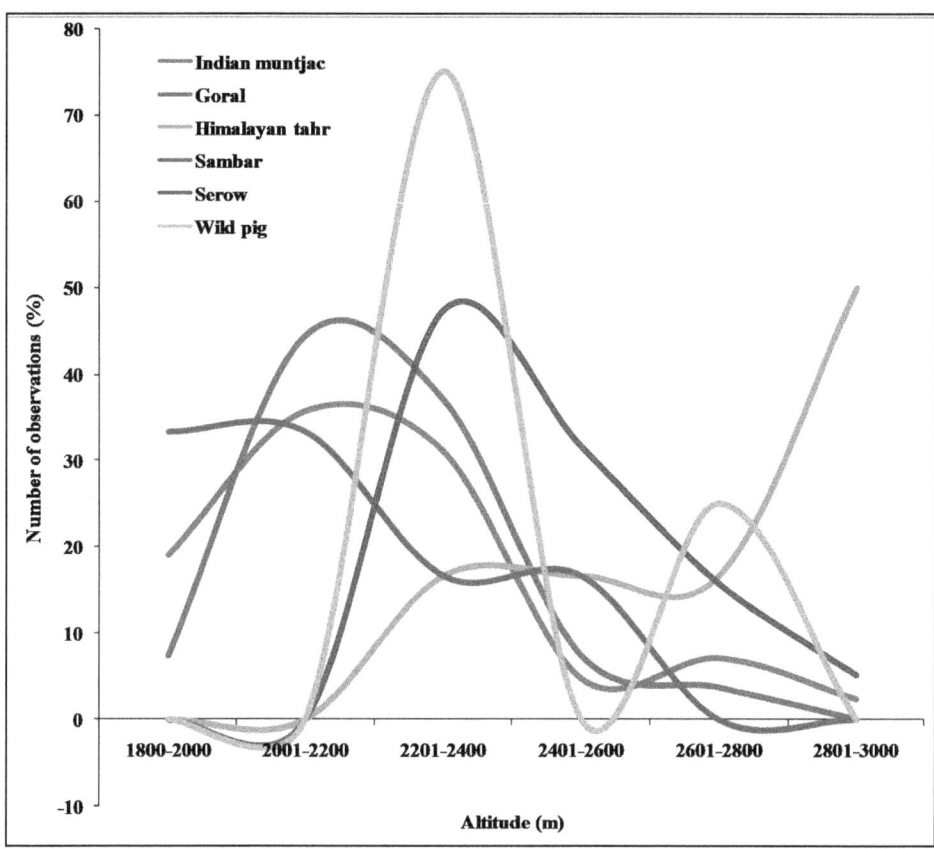

Figure 16.2: Percentage of Observations of the Presence of Ungulate Species by Altitude in Kumaun Himalaya.

Shrub Cover

Significant differences were not observed in use of shrub cover ($F = 7.67$, $p > 0.175$) by different mammal species. Goral and Serow showed high degree of overlapping in use of shrub layer. Sometimes they were seen in the areas having very low or nil shrub cover. Sambar used shrub cover >50 per cent and Green (1987) also always found Sambar distributed in the areas having > 25 per cent shrub cover in Kedarnath Sanctuary during his studies. Indian muntjac was also not found in low shrub cover.

Ground Cover

All mammal species studied used significantly different ground cover ($F = 14.18$, $p < 0.01$). Goral and Himalayan tahr were strongly negatively correlated (rs = -0.885, $p < 0.01$) and therefore both were dissimilar to each other with respect in use of ground cover. Indian muntjac and Sambar showed similar ranges of ground cover but not significantly correlated.

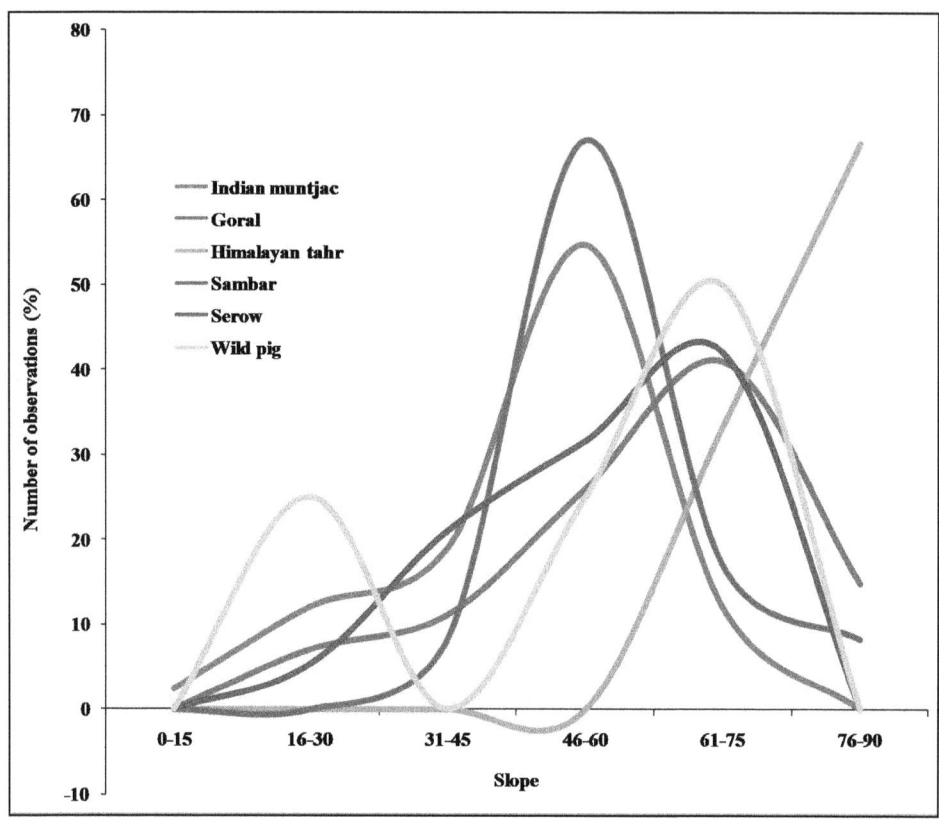

Figure 16.3: Percentage of Observations of the Presence of Ungulate Species by Slope in Kumaun Himalaya.

Aspect of Slope

No significant difference was observed in occurrence of different species with respect to aspects (F = 4.03, p > 0.544). None of them were found significantly correlated. All the surveyed sites were having different aspects so the encounters were more or less dependent on it. However, 33.33 per cent of records of Indian muntjac and Sambar separately were on north-west aspect while Goral (33.33 per cent), Himalayan tahr (66.66 per cent) and Serow (42.1 per cent) preferred dry south-east aspect.

Similarity between Sites in Mammal Species Composition

Clustering of oak patches was accomplished to know the assemblage of mammal species composition on the basis of presence/absence at all the surveyed sites in Kumaon. Overall 10 groups of sites had been formed. The sites Kilbery, Pandavkholi, Sitlakhet, Jageshwer and Munsiary were similar in mammal species composition while Gager, Mukteshwer, Jilling, Gasi, Dhakuri, Wachham, Mechh and Duku showed similar composition. The Pindari formed a separate cluster as it

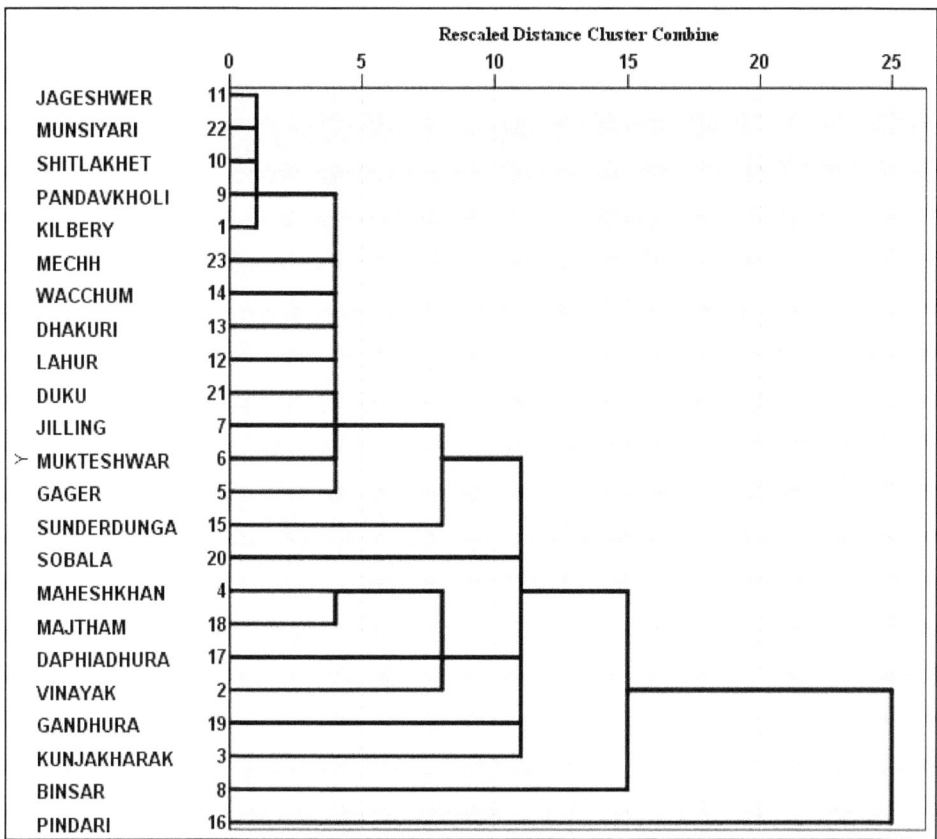

Figure 16.4: Hierarchial Cluster Analysis of different Oak Patches by using 18 Mammal Species Sighted in these Areas.

had maximum number of mammal species (Figure 16.4) such as Musk deer, Serow, Blue sheep, Himalayan tahr, Himalayan Black bear, Goral, Indian muntjac, Leopard, Yellow throated martin, Hanuman Langur, Wild pig and Red giant flying squirrel.

Discussion

There existed no information on the status and abundance of mammal species from this region. This is the first study conducted to document the current status and distribution of the major mammalian community of Kumaon, Central Himalaya. Our main aim was to cover the areas of Oak patches of Kumaon falling within the limits of middle altitude (1500-2500m), but we extended our range in the areas having continuity in the forest up to sub alpine zone (≥3500m).

The mammalian fauna of Kumaon were found to be comparable with part of central Himalaya and northwestern limits of Nepal and Garhwal Himalaya respectively. All species studied in this paper were recorded over the wide range of altitudes and habitats studied. They differed considerably in habitat selection. The distribution of mammal species was not uniform throughout the Kumaon Himalaya. Out of 23 surveyed locations, we documented 18 mammal species with unequal species composition and abundance. The representation mainly included major mammalian communities such as ungulates, primates, large cats, goat antelope, bear, canids, flying squirrel, mountain sheep, porcupine, and mouse hare.

Sites covered during the surveys varied considerably in terms of their mammal species composition due to site characteristics or specific requirements of each mammal species. Marked variation was also observed in their abundance in different locations of Kumaon. Species like Hanuman langur, Rhesus macaque, Indian muntjac, Goral, Sambar, Yellow throated martin and Leopard were found to distribute throughout Kumaon. Out of these species Goral, Indian muntjac, Hanuman langur and Yellow throated martin emerged as generalist species and their representation was in more locations of Kumaon than other species. The species like Mouse hare, Porcupine, Red fox and Red giant flying squirrel were founded at a few locations and their encounters were by chance in low numbers. The species like Blue sheep, Himalyan tahr and Musk deer, which required specific habitat condition were found in Pindari and Sobala (AWS) which were located in remote areas where no human habitation existed.

Maximum group size was recorded for Himalayan tahr while minimum for Serow. In addition to it Shrestha (2006) found the Himalayan tahr in big group size as much as 45 individuals. Results showed that Indian muntjac, Goral, Serow and Sambar are essentially solitary species. Green (1987) also found Goral, Serow and Sambar as a solitary species in Kedarnath Wildlfe Sanctuary. Sambar are predominantly solitary in tropical and subtropical regions of the Indian subcontinent with mean group sizes ranging between 1.4 and 2.4 (Schaller, 1967; Berwick and Jordan, 1971; Eisenberg and Lockhart, 1972; Seidensticker, 1976; Mishra, 1982; Tumang, 1983; Saxton, 1984). Serow was sighted four times and all sightings were of single animal. Dang (1962) reported that Serow may form pairs and Schallar (1977) mentions that they can occur in small family groups. Like a true goat, Himalayan tahr always lives in herd and is always encountered in herd. Goral is more gregarious than any other ungulate species except tahr. It was observed in pair and with juveniles several times and Schallar (1977) states that males are often solitary, pairing with females during the rut. Indian muntjac was also always encountered as a single animal and Menon (2014) also states that Indian muntjac lives in single or pair.

Ecological Separation

Ecological separation explains how species can coexist even though they may have extensive overlap in their ecological requirements (May, 1973). The highest similarity was found between Indian muntjac and Serow in habitat use, *i.e.* oak forest with close canopy while Himalayan tahr used open grassy and rocky slope. But Serow and Himalayan tahr were always found on high altitude while Indian muntjac and Sambar were encountered on relatively lower altitude at different places of Kumaon.

Goral and Serow were found in low shrub cover and ground cover while Indian muntjac and Sambar were found in high shrub cover and ground cover area (Schaller, 1967; Sankar, 1994; Ramesh, 2010). Blue sheep and Musk deer were sighted only once and on very high altitude in Pindari region. Blue sheep was sighted in a herd of 7-8 animals in alpine meadows at the 3600m altitude. Musk deer was sighted in oak-mixed forest at 3200m altitude. Tree cover was low and shrub cover was high. Green (1987) also found Musk deer at low tree cover and high shrub cover but only at the time of resting.

Binsar Wildlife Sanctuary and oak forests of Askot Wildlife Sanctuary were placed in separate cluster and both are protected areas. While Kilbery, Kunjakharak and Vinaiyak also having a good number of mammal species but all are unprotected areas earlier. Now based on our study and recommendation on biodiversity, the forest department of Uttarakhand has declared this area as Naina Devi Himalayan Bird Conservation Reserve. Scats of Himalayan Black Bear and Leopard also were found in Kilbery and Vinaiyak. Pindari region formed a separate cluster by having maximum number of mammal species and different composition and also having many endangered mammal species like Musk deer, Blue sheep, Serow and Himalayan tahr but these areas are unprotected. Snow leopard is also found here, as locals told and pug mark was also sighted at Pindari glacier at 3500m altitude. Presence of Snow leopard and other endangered mammal species makes the Pindari region a valuable place in terms of rich biodiversity. On the other hand, this region touches the boundary of the Nanda Devi Biosphere Reserve, aiding a corridor for animal species to Pindari region. As the mammals roam in large areas and shift also seasonally so a large area should be protected in Pindari region so the animals can disperse freely.

Acknowledgements

This research was supported by Ministry of Environment and Forest, India. Locals and villagers are acknowledged for their help and support during data collection. Yasser Arafat is acknowledged for proving photographs of mammals of the study area.

References

Bagchi, S., Namgail, T. and Ritchie, M.E., 2006. Small mammalian herbivores as mediators of plant community dynamics in the high-altitude arid rangelands of Trans-Himalaya. *Biol. Conserv.* 127(4): 438-442.

Basu, P.K., 2004. Siwalik mammals of the Jammu Sub-Himalaya, India: an appraisal of their diversity and habitats. *Quaternary International.* 117 (1): 105-118.

Berwick, S.H. and Jordon P.A., 1971. First report of the Yale Bombay Natural History Society studies of wild ungulates at the Gir forest, Gujarat, India. *J. Bombay Nat. Hist. Soc.* 68: 412-423.

Conservation International, 2013. Biodiversity Hotspots. Himalaya. http://www.biodiversityhotspots.org/xp/Hotspots/himalaya/

Dinerstein, E., 1980. An ecological survey of the Royal Karnali-Bardia Wildlife Reserve, Nepal. Part III: ungulate populations. *Biol. Conserv.*, 18: 5-38.

Eisenberg, J.F. and Lockhart M., 1972. An ecological reconnaissance of Wilpattu National Park, Ceylon. *Smithson. Contr. Zool.*, 101: 1-118.

Fowler, J. and Cohen L., 1986. *Statistics for Ornithologists.* BTO Guide No. 22.

Gauch, H.G., Jr., 1989. *Multivariate Analysis in Community Ecology.* Cambridge University Press, New York, USA.

Green, M.J.B., 1987. Ecological separation in Himalayan ungulates. *J. Zool. London.* (B) 1: 693-719.

Guisan, A. and Zimmerman, N.E., 2000. Predictive habitat distribution models in ecology. *Ecol. Model.* 135: 147–186.

Hirzel, A.H., Helfer, V. and Metral, F., 2001. Assessing habitat suitability models with a virtual species. *Ecol. Model.* 145: 111–121.

Khoshoo, T.N., 1991. Conservation of biodiversity in Biosphere. In T.N. Khoshoo and M. Sharma Vikas (eds.) *Indian Geosphere-Biosphere.* Publishing House, New Delhi, 178-233.

Legris, P. and Meher-Homji, V.M., 1968. Floristic elements in the Vegetation of India. *Proc.of the Symposium on Recent Advances in Tropical Ecology*, 2: 536-543.

May, R.M., 1973. On the theory of niche overlap. *Theor. Popul. Biol.* 5: 297-332.

Menon, V., 2014. *Indian Mammals: A field guide.* Hachette India Press. Pp: 528.

Mishra, C., Madhusudan M.D. and Dutta A., 2007. Mammals of the high altitudes of western Arunachal Pradesh, eastern Himalaya: an assessment of threats and conservation needs. *Oryx*, 40(1): 1-7.

Mishra, H.R., 1982. *The ecology and behaviour of chital (Axis axis) in the Royal Chitwan National Park, Nepal.* Ph.D. thesis: University of Edinbergh.

Pandit, M.K., Sodhi, N.S., Koh, L.P., Bhaskar, A. and Brook, B.W., 2007. Unreported yet massive deforestation driving loss of endemic biodiversity in Indian Himalaya. *Biodivers. Conserv.* 16 (1): 153-163.

Ramesh, T., 2010. Prey selection and food habits of large carnivores: tiger *Panthera tigris*, Leopard *Panthera pardus* and dhole *Cuon alpinus* in Mudumalai Tiger Reserve, Tamil Nadu. Ph.D. Thesis, Saurashtra University, Gujarat, Rajkot

Sankar, K., 1994. The ecology of three large sympatric herbivores (chital, sambar, nilgai) with special reference for reserve management in Sariska Tiger Reserve, Rajasthan. *Ph.D. Thesis*, University of Rajasthan, Jaipur.

Saxton, J., 1984. *The Feeding Ecology of Sambar (Cervus unicolor niger) in the Monsoon.* Honours Projects: Department of Applied Biology, University of Cambridge

Schaller, G.B., 1967. *The deer and the tiger.* Chicago: University of Chicago Press.

Schaller, G.B., 1977. *Mountain Monarchs: Wild Sheep and Goats of the Himalaya.* Chicago: University of Chicago Press.

Seidensticker, J.C., 1976. Ungulate population in Chitwan Valley, Nepal. *Biol. Conserv.* 10: 183-210.

Seoane, J., Bustamante, J. and Díaz-Delgado, R., 2004. Competing roles for landscape, vegetation, topography and climate in predictive models of bird distribution. *Ecol. Model.* **171:** 209–222.

Shrestha, B., 2006. Status, Distribution and Potential Habitat of Himalayan Tahr (*Hemitragus jemlahicus*), and Conflict Areas with Livestock in Sagarmatha National Park, Nepal. *Nepal Journal of Science and Technology.* 7: 27-33.

Sinha, R.K., 1999. State of Biodiversity in India. *Global Biodiversity*. Published by Ina Shree Publishers. Chap 2: 25-74.

Tumang, K.H., 1983. The status of the tiger (Panthera tigris tigris) and its impact on principalprey population in the Royal Chitwan National Park, Nepal. *Ph.D. Thesis*, Michigan State University.

UNEP, 1992. Convention on Biological Diversity, Text and Annexes, Montreal; Secretariat of Convention on Biological Diversity.

Zar, J.H., 2010. *Biostatistical Analysis*, 5th edn. Pearson publishers, p. 960.

Chapter 17

On Two New Species of *Balantidium* Claparede and Lachmann, 1858 emend. Stein, 1867 from *Rana* (*Dicroglossus*) *cyanophlyctis* Schneider and First Report of Two Species from Kashmir Valley, India

Kanwar Narain, M.K. Raina and P.L. Koul

Introduction

In the present study two new species from *Rana* (*Dicroglossus*) *cyanophlyctis* Schneider of Kashmir India and also two species of the genus reported for the first time from Kashmir India.

Materials and Methods

The host *Rana* (*Dicroglossus*) *cyanophlyctis* Schneider was collected from Srinagar and other adjoining areas. The gut of the pithed host was kept in saline (0.65 per cent). Rapid fixation and staining techniques were used to show nuclei and other structures of protozoans in temporary preparations. Mayer's albumen (Weesner, 1960) used for adhesion. Staining done by using Oelafields Haematoxylin (BOH) and Eosin. Dehydration in grades of alcohol, clearing in Xylene and mounting in Canada balsam. A modified version of Fernandez-Galiano's (1976) wet Silver technique was used.

Balantidium cyanophlycti Shete and Krishnamurthy, 1984 (Figures 17.1 and 17.2, Plates 17.1–17.4)

Host: *Rana (Dicroglossus) cyanophlyctis* Schneider

Locality: Kashmir valley

Infection Loci: Rectum and occasionally in intestine

Prevalence: 453 out of 521 frogs examined were infected.

Description

Balantidiwn cyanophlycti is cylindrical, elongated and measures 60–100.6 ($\bar{x}$ 80.0) µm x 26.0–40.5 ($\bar{x}$ 35.0) µm. The length of the body is generally more than twice the breadth of the body. Both the ends are bluntly rounded. Both the margins of the body are almost parallel to each other (Figure 17.2). The body is covered with uniform coat of closely set cilia arranged in longitudinal rows. The somatic cilia measure 4–6 µm. in length. The cilia at the anterior end are longer and measure 8–9.5 µm. The number of kineties on either side varies from 26–32 (Plate 17.2).

The vestibule is located anteriorly and is nearly equal to the breadth of the body and measures 24.5–39.5 ($\bar{x}$ 33.5) µm in length (Plate 17.3). It is typically V-shaped, narrow and almost of same width throughout its length. The vestibule is provided with adoral row of long cilia. The true oral membranelle are absent.

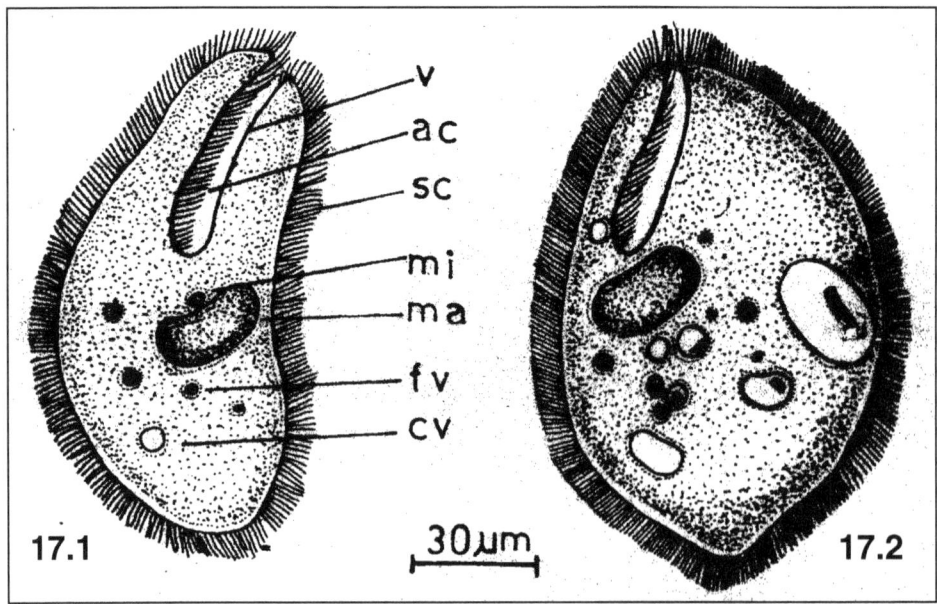

Figures 17.1 and 17.2: Line Drawings Showing Morphology of *Balantidium cyanophlycti* Shete and Krishnamurthy, 1984.

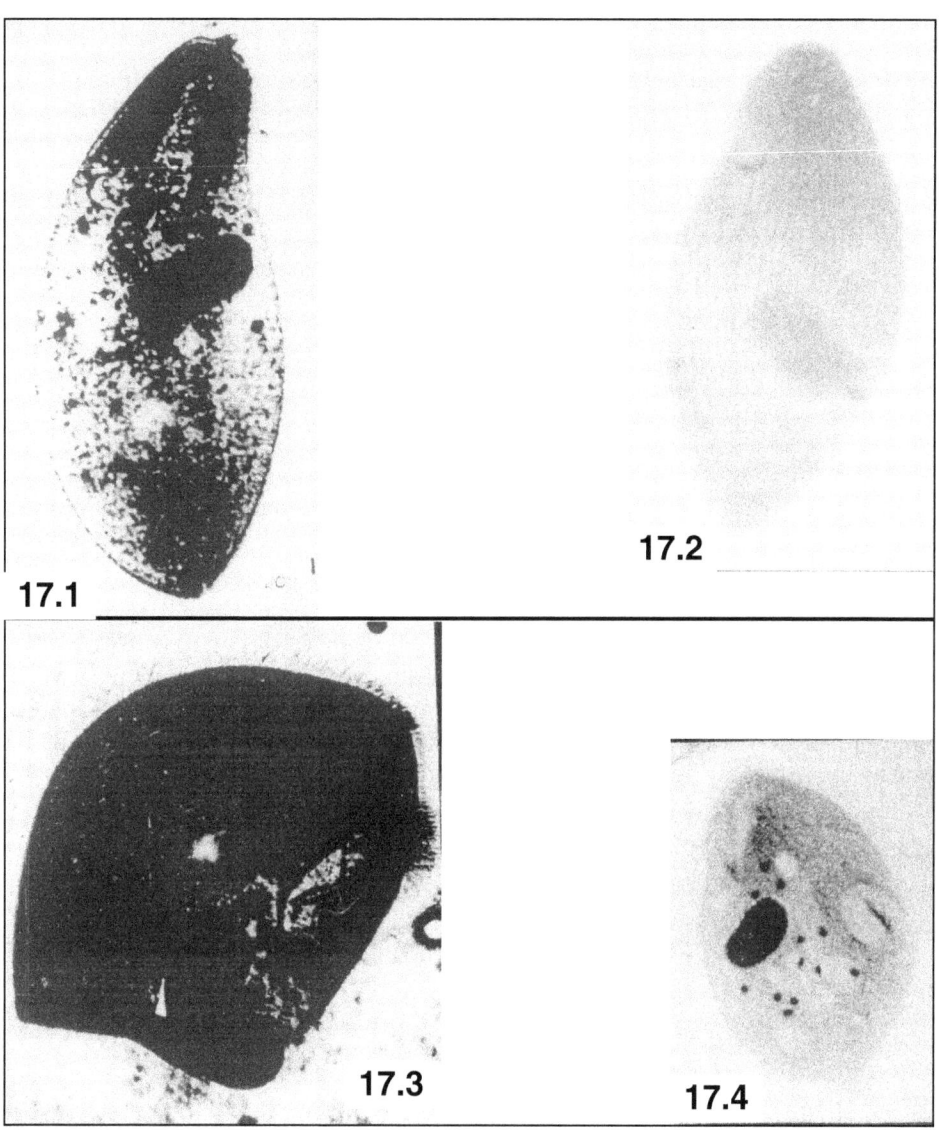

Plates 17.1–17.4: Showing Morphology of *Balantidium cyanophlycti* Shete and Krishnamurthy, 1984 from the rectum of *Rana* (*Dicroglossus*) *cyanophlyctis* Schaeider of Kashmir Valley.

Plate 17.1: Ciliate treated with Rio-Hortega's $AgNO_3$ solution; Plate 17.2: An individual treated with pyridinted ammonical silver carbonate solution showing infraciliature; Plate 17.3: Specimen squashed under controlled cover slip pressure in order to demonstrate morphology of the verstibulum and micro- and maconuclei. Treated as above. Plate 17.4: Specimen stained with Delafield's haematoxylin and Eosin.

The cytoplasm is not clearly marked into ectoplasm or endoplasm. However, endoplasm has dense granular appearance. The macronucleus is more or less reniform in contour and measures 12.6–24.6 ($\bar{x}$ 18.2) µm x 5–10.5 ($\bar{x}$ 8.0) µm. The micronucleus lies adjacent to the macronucleus and measures 3–5 µm (Plate 17.3). In fresh specimens, 2 contractile vacuoles are invariably present, one before and the other behind the macronucleus.

Balantidium ganapatii Shete and Krishnamurthy, 1984 (Figures 17.3 and 17.4; Plates 17.5–17.8)

Host: *Rana* (*Dicroglossus*) *cyanophlyctis* Schneider

Locality: Kashmir valley

Infection Loci: Intestine and occasionally in rectum

Prevalence: 317 out of 521 frogs examined were infected.

Description

The body of *Balantidium ganapatii* is cylindrical, extremely elongated about 4-6 times as long as broad and slightly attenuated at both the ends. The ciliate measures 134–203.5 ($\bar{x}$ 180) µm x 23.3–38 ($\bar{x}$ 33.0) µm. Under live condition the colour of the ciliate ranges from opaque grey to very faint shades of grey. The ciliate moves by rotating on its axis and swims in both forward as well as backward direction. The body is covered with a uniform coat of longitudinally arranged rows of cilia, which are very closely set. The cilia measure 3.5–7.0 µm in length. The number of kineties ranges from 90–101 (Plate 17.8). At the anterior end the cilia are about twice as long as the rest of the body cilia.

The vestibule is located anteriorly and is slightly asymmetrical in position (Plate 17.7). It extends into the cytoplasm upto about one-fifth to one-sixth of the body length. The vestibule measures 25–40.5 ($\bar{x}$ 32.5) µm in length. It is typically sac-shaped being narrow anteriorly and wider posteriorly. The vestibule is equipped with adoral row of long cilia which beat vigorously and independently. True buccal membranelles are absent.

The cytoplasm is distinctly divided into clear ectoplasm and alveolar endoplasm. The macronucleus is oval in contour and measures 19.0–27.5 ($\bar{x}$ 26.0) µm x 7–10 ($\bar{x}$ 8) µm. The micronucleus lies adjacent to the macronucleus (Plate 17.8). The position of the macronucleus varies considerably. In the present study some specimens had two macronuclei which were located anteriorly (Figure 17.4). There are two contractile vacuoles which are distinctly seen in live specimen (Plate 17.5). One of them is located in the middle of the body while the other lies in the hind part of the body.

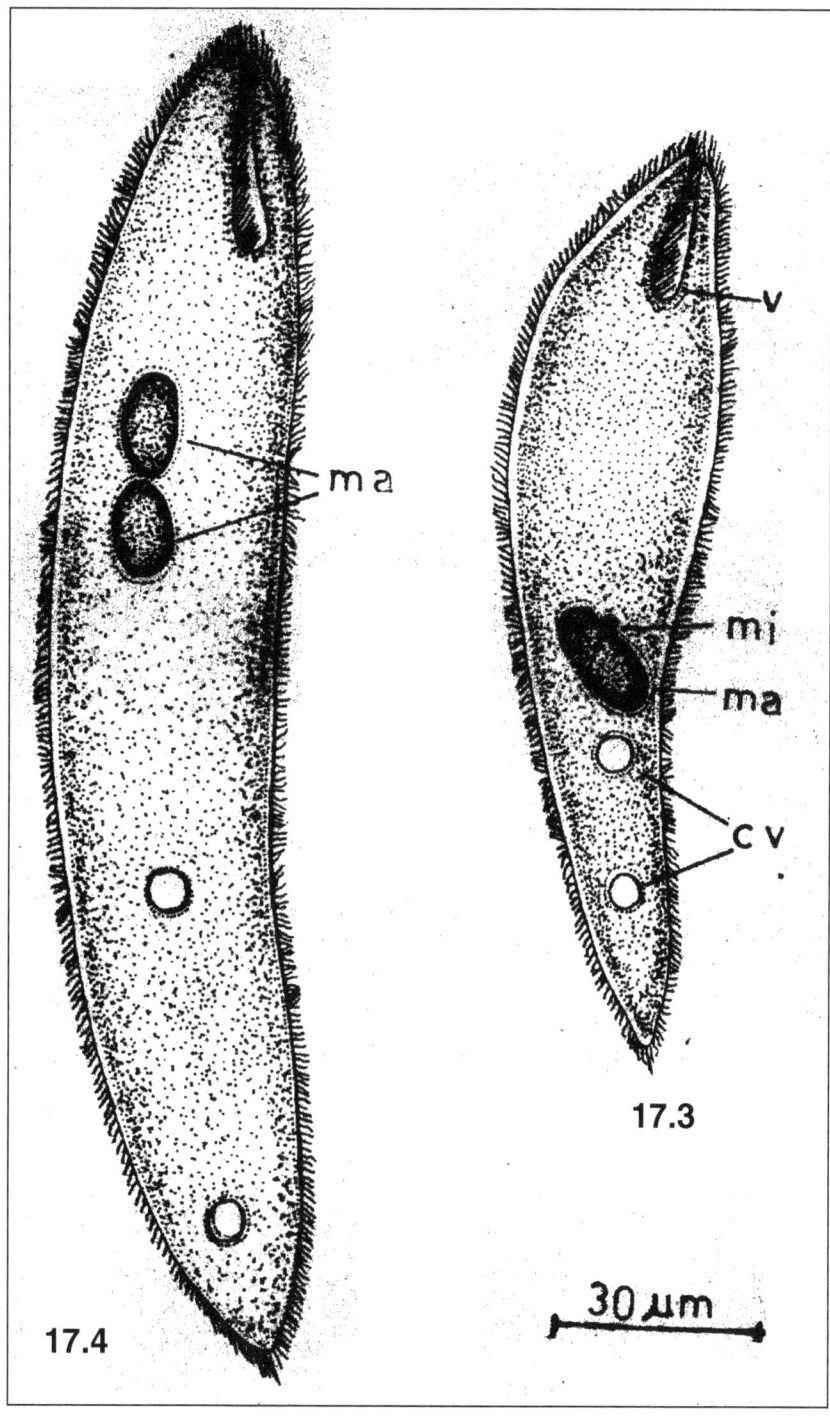

Figure 17.3 and 17.4: Line Drawings Showing Structure of *Balantidium ganapatii* Shete and Krishnamurthy, 1984.

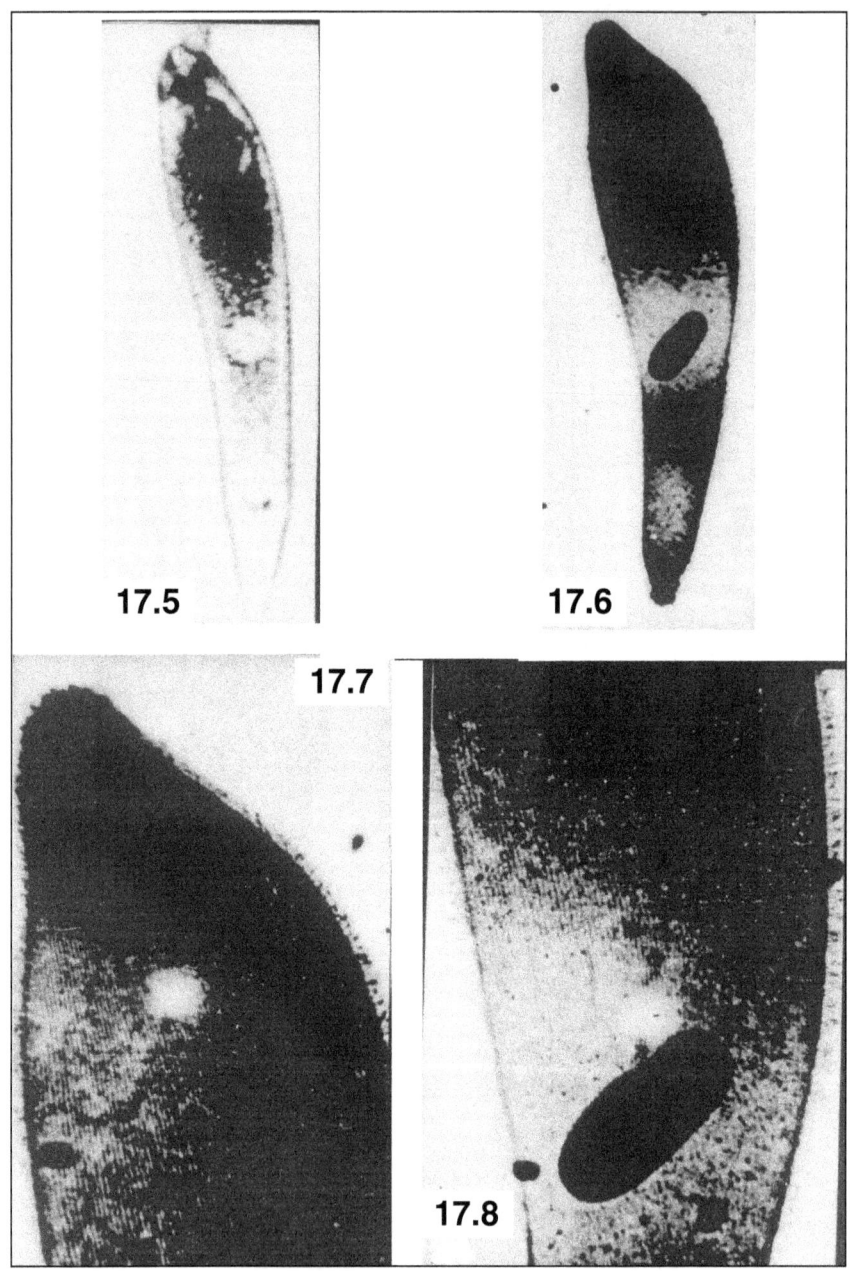

Plates 17.5–17.8: Showing Structure of
Balantidium ganapatii Shete and Krishnamurthy, 1984.

Plate 17.5: Unstained ciliate in phase contrast. Note the position of two contractile vacuoles. Plate 17.6: General morphology. Specimen treated with pyridinated ammonical silver carbonate solution. Plate 17.7: Anterior end of the ciliate treated as above, showing vestibulum and infraciliature. Plate 17.8: Middle region of the ciliate treated above, showing macro- and micronuclei and infraciliature of the body.

Balantidium neotigrinae n.sp. (Figure 17.5; Plates 17.9 and 17.10)

Host: *Rana (Dicroglossus) cyanophlyctis* Schneider

Locality: Kashmir valley

Infection Locus: Rectum

Prevalence: 116 out of 521 frogs examined were infected.

Description

Balantidium neotigrinae n.sp. is an ovoid ciliate. The anterior end of the body is attenuated whereas, the posterior end of the body is dome-shaped (Plate 14.9). The ciliate is markedly large and measures 210–250.5 ($\bar{x}$ 230) µm x 115–147.5 ($\bar{x}$ 135.2) µm. The length of the body is more than 1½ to 2 times the width. The body is covered with a uniform coat of holotrichous cilia which are longitudinal arranged in parallel rows. The number of kineties varies from 44–56 on either side and converge on the vestibulum.

The vestibulum is extremely elongated, 'V'-shaped, set asymmetrically and is tucked into the cytoplasm beyond the median line. The vestibule is provided with conspicuously longer adoral cilia. The true oral membranelles are absent. The vestibulum measures 110.7–155.3 ($\bar{x}$ 130) µm in length While swimming, the ciliate rotates around its body axis.

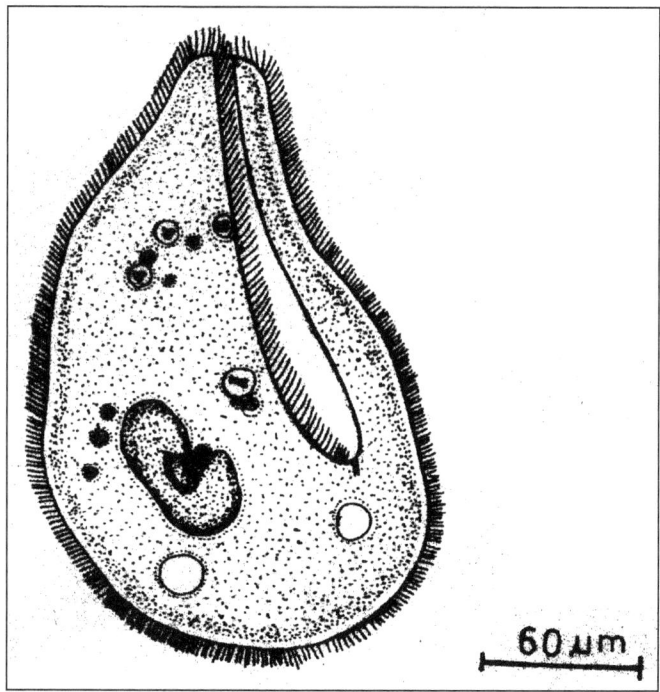

Figure 17.5: Line Drawing of *Balantidium neotigrinae* n.sp.

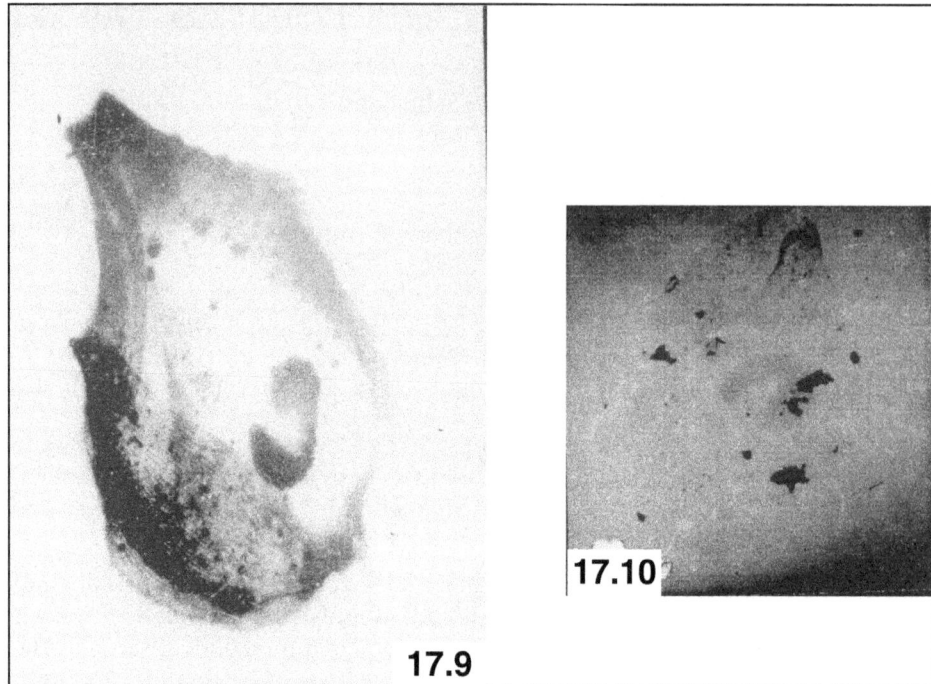

Plates 17.9 and 17.10: *Balantidium neotigrinae* **n.sp. Recovered from the Hind Gut of *Rana (Dicroglossus) cyanophlyctis* of Kashmir Valley.**

Plate 17.9: Specimen treated with pyridinated silver carbonate solution.
Plate 17.10: Haematoxylin and eosin stained specimen.

The cytoplasm is clearly defined into outer ectoplasm and inner endoplasm which is densely granular. The macronucleus is characteristically 'C'-shaped (Plate 17.9) and measures 35.4–67.0 ($\bar{x}$ 50) μm x 12–20.2 ($\bar{x}$ 15) μm. The micronucleus lies a little distance away from the macronucleus and measure 4-5 μm. In fresh specimens 2–4 contractile vacuoles are present and are irregularly distributed in the body.

Balantidium neogiganteum n.sp. (Figure 17.6; Plates 17.11–17.13)

Host: *Rana (Dicroglossus) cyanophlyctis* Schneider

Locality: Kashmir valley

Infection Locus: Rectum

Prevalence: 80 out of 521 frogs examined were infected

Description

B. neogiganteum n. sp. is an ovoid ciliate. The arterial end of the body is attenuated and is often turned slightly to one side, the posterior end is dome-shaped (Plate 17.11). The ciliate is large and measures 151–207 ($\bar{x}$ 183.3) μm x 100–137 (

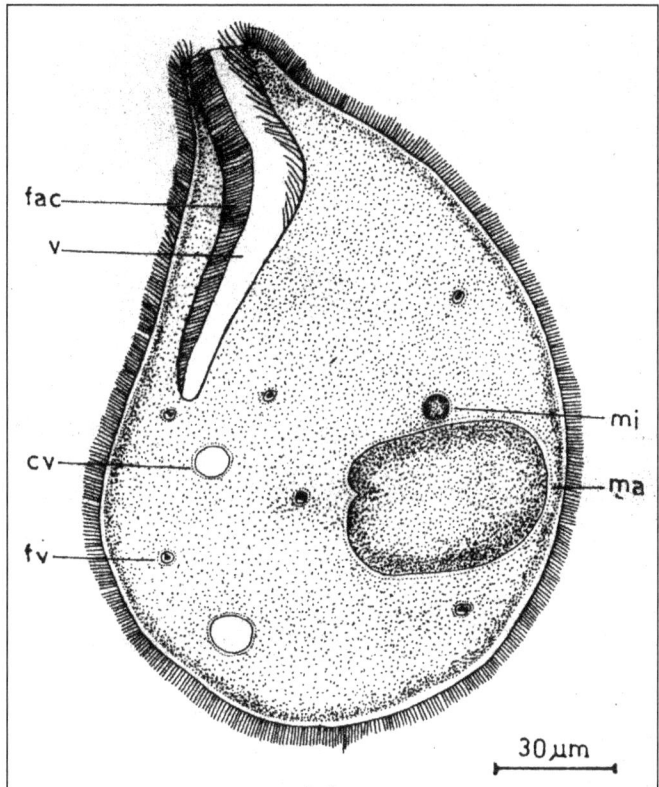

Figure 17.6: Line Drawing of *Balantidium neogiganteum* n.sp.

$\bar{x}$ 125) µm. The length of the body is one and a half time the width. The body is covered with a uniform coat of holotrichous cilia which are longitudinally arranged in parallel rows.

The vestibulum is elongated tube-like, of almost uniform diameter, obliquely set and is tucked into the interior of the cytoplasm at least as far back as the midline of the body. The vestibulum measures 95.2–110 ($\bar{x}$ 104.7) µm. The right lip of the vestibulwn seems to carry membranelles consisting of fused adoral cilia.

The cytoplasm is differentiable into outer ectoplasm and inner densely granular endoplasm. The macronucleus is massive and is broadly ovoid. In some specimens the macronucleus is notched at one of the ends (Figure 17.6). The micronucleus is situated adjacent to the macronucleus and measures 5–7 µm in diameter. Two contractile vacuoles are visible in live ciliates.

Balantidium sp. (Plate 17.14)

Host: *Rana (Dicroglossus) cyanophlyctis* Schneider

Locality: Kashmir valley (Shalimar only)

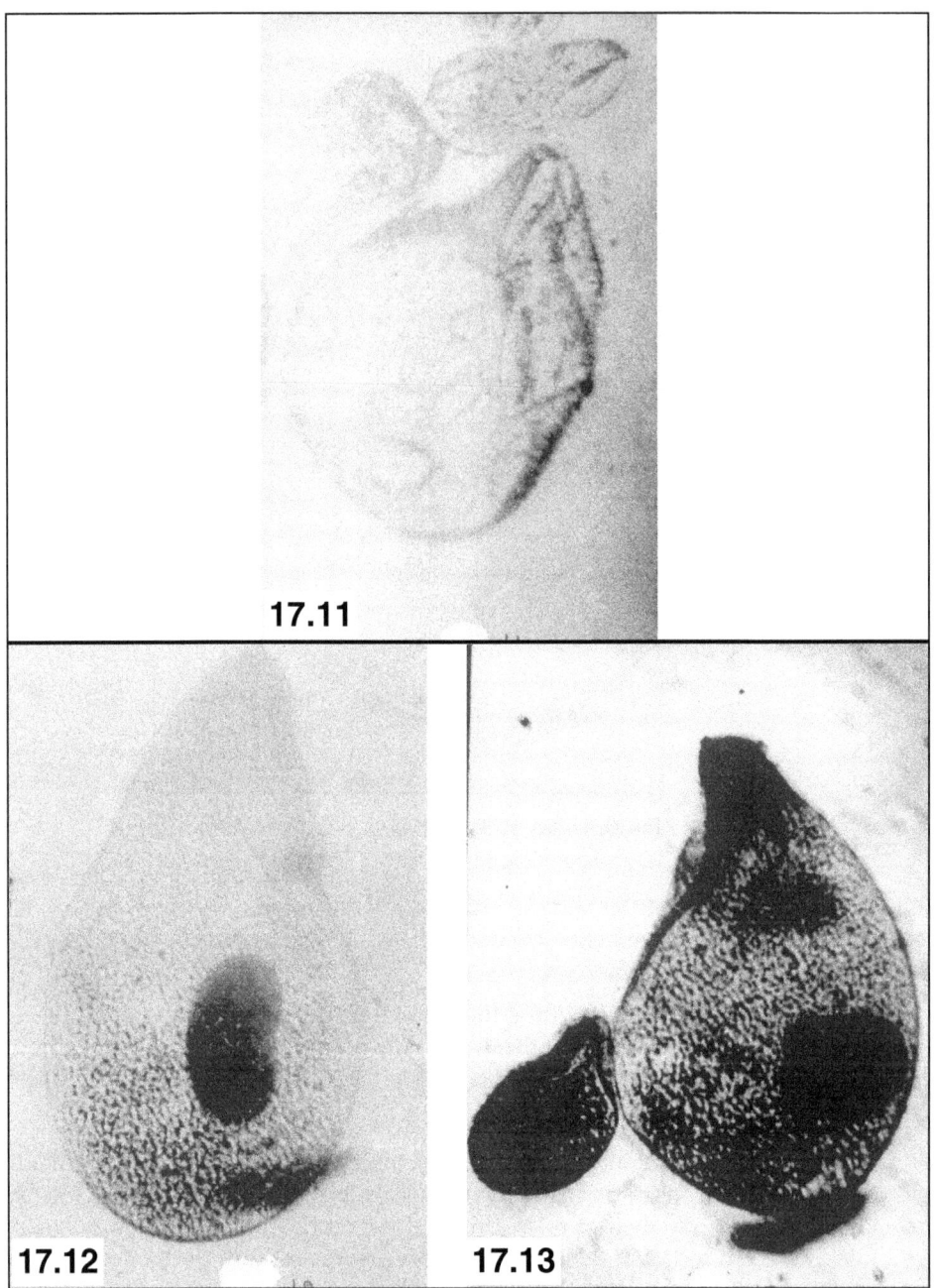

Plates 17.11–17.13: Showing Morphology of *Balantidium neogiganteum* n.sp. Infecting Rectum of *Rana (Dicroglossus) cyanophlyctis* of Kashmir Valley.

Plate 17.11: Unstained ciliate; Plates 17.12 and 17.13: Specimen stained with Delamater's basic fuchsin.

Note the size difference between *B. neogiganteum* n.sp. and *B. cyanophlycti* Shete and Krishnamurthy, 1984.

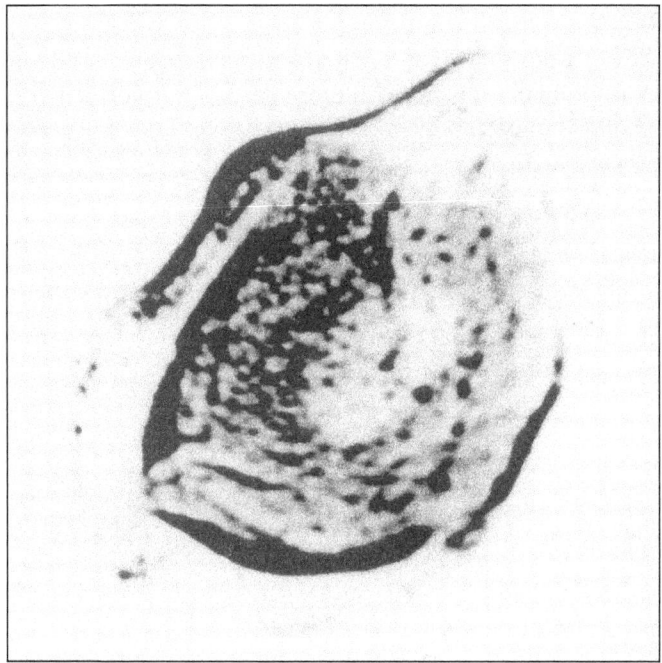

Plate 17.14: Unstained Specimen of *Balantidium* sp. Recovered from the Hind Gut of *Rana (Dicroglossus) cyanophlyctis*.

Infection Locus: Rectum

Prevalence: 1 out of 521 frogs examined was infected

Description

Two individuals of this ciliate were encountered only once and were conspicuous because of their considerably large size. Only one specimen was measured (dimensions 293 x 195.3 µm), the other was destroyed while studying it live. This ciliate is oval, bluntly pointed at anterior end and dome-shaped posteriorly. The dorsal side of the body has a characteristic fold at posterior end (Plate 17.14). The body is covered with a uniform coat of holotrichous cilia. It rotates on its axis during swimming and squeezes its elastic body between various obstacles.

The vestibulum is elongated, 'U'-shaped or tubular with uniform diameter and reaches upto the midline of the body. The vestibulum is slightly sigmoid in its disposition and measures 165.2 µm in length. The vestibulum is provided with conspicuously larger adoral cilia which beat independently. True membranelles are absent.

The cytoplasm is clearly differentiated into outer ectoplasm and inner, coarsely granular endoplasm. The macronucleus is ovoid to slightly reniform located in the middle of the body and measures 90.15 x 45.0 µm. The contractile vacuoles were not discernible.

Discussion

Earl (1973) recognized 59 species of genus *Balantidium* as valid which includes 17 species from amphibian hosts from different parts of the world. In India 27 species of genus *Balantidium* have been recorded from different hosts. These are:

1. *B. coli* Malmsten, 1857
2. *B. duodeni* Stein, 1867
3. *B. elongatum* Stein, 1867
4. *B. gracile* Bezzenberger, 1904
5. *B. giganteum* Bezzenberger, 1904
6. *B. helenae* Bezzenberger, 1904
7. *B. rostundum* Bezzenberger, 1904
8. *B. testudinis* Chagas, 1911
9. *B. depressum* Ghosh, 1921
10. *B. ranarum* Ghosh, 1921
11. *B. ovatum* Ghosh, 1922
12. *B. knowlesii* Ghosh, 1925
13. *B. coli* var. *bovis* Cooper and Gulati, 1926
14. *B. amygdallii* Bhatia and Gulati, 1927
15. *B. bicavata* Bhatia and Gulati, 1927
16. *B. rhesum* Ghosh, 1929
17. *B. sushilli* Ray, 1932
18. *B. rayi* Pal and Dasgupta, 1980
19. *B. tylototritonis* Pal and Dasgupta, 1980
20. *B. aurangabadensis* Shete and Krishnamurthy, 1984
21. *B. corlissi* Shete and Krishnamurthy, 1984
22. *B. cyanophlycti* Shete and Krishnamurthy, 1984
23. *B. ganauatii* Shete and Krishnamurthy, 1984
24. *B. megastomae* Shete and Krishnamurthy, 1984
25. *B. mininucleatum* Shete and Krishnamurthy, 1984
26. *B. ranae* Shete and Krishnamurthy, 1984
27. *B. tigrinae* Shete and Krishnamurthy, 1984

The perusal of the literature revealed that almost all authors have used morphological criteria such as shape and size of the body, macronucleus and vestibule for delineating various species of the genus *Bolantidium* (Bezzenberger, 1904; de Mello, 1932, Hegner, 1934; Earl, 1973 and Shete and Krishnamurthy, 1984).

Table 17.1: Showing a Comparative Characteristics (in microns, µm) of *Balantidium neotigrinae* n.sp. with its Related Species

Sl.No.	Particulars	B. neotigrinae n.sp.	B. elongatum Stein, 1867	B. helenae Bezzenberger, 1904	B. tigrinae Shete and Krishnamurthy, 1984
1.	Length of the body (L_b)	210–250.5 (230)	208–297 (Stain) 90–124 (Bhatia and Gulati)	45–175	73.1–108.7 (92.4)
2.	Width of the body (W_b)	115–147.5 (135.2)	69–130 (Stain) 39–53 (Bhatia and Gulati)	30–62	36.6–52.2 (45.3)
3.	Length of the vestibule (L_v)	110.7–155.3 (130)	—	—	22.0–56.4 (38.7)
4.	Length of the macronucleus (L_{ma})	35.4–67.0 (50)	—	31.37	25.1–35.5 (30.6)
5.	Width of the macronucleus (W_{ma})	12–20.2 (15)	—	10–18	7.3–12.5 (9.6)
6.	W_b/L_b per cent	58.7	36.7–43.7	54	49.0
7.	L_{ma}/L_b per cent	21.7	—	—	33.1
8.	L_v/L_b per cent	56.5	—	—	38.0
9.	W_{ma}/L_{ma} per cent	30	—	—	31.4

B. cyanophlycti was first recorded by Shete and Krishnamurthy in 1984 from rectum of *Rana cyanophlyctis* in Aurangabad, Maharashtra. These authors have given its dimensions as 51.2–104.6 ($\bar{x}$ 82.3) μm × 27.1–41.8 ($\bar{x}$ 33.6) μm. The population of *B. cyanophlycti* in the present investigation falls well within the range of measurements given by Shete and Krishnamurthy (1984). However, Shete and Krishnamurthy reported the number of kineties of the species to be 51 on either side of the body, in contrast to this, the number of kineties in the present investigation ranged from 26–32. The present authors do not feel this difference in the number of kineties to be of any specific importance because Krascheninnikow (1962, cf. Zaman, 1978) also reported the variable number of kineties in different individuals of *B. coli* ranging from 36–106.

B. ganapatii was first recorded by Shete and Krishnamurthy in 1984 from *Rana cyanophlyctis* in rectum of Aurangabad, Maharashtra. These authors gave the dimensions of this ciliate as 123.3–250.8 ($\bar{x}$ 186.6) μm × 21.9–47 ($\bar{x}$ 30.7) pm whereas, in the present study these ciliates measure 134–203.5 ($\bar{x}$ 180) μm × 23.3–38 ($\bar{x}$ 33) μm. In the present study *B. ganapatii* were recovered mostly from small intestines and occasionally in rectum of the hosts in contrast to Shete and Krishnamurthy (1984) who recovered it only from the rectum. Besides, Shete and Krishnamurthy (1984) were not able to demonstrate the kineties of *B. ganapatii* using Klein's dry silver impregnation technique. In the present study the kineties were easily demonstrated by using a modified version of Fernandez-Galianots (1976) wet silver technique. This technique also clearly demonstrated the micronucleus (Figure 17.1) which Shete and Krishnamurthy (1984) failed to demonstrate.

B. neotigrinae n.sp. has a length-width proportion (W_b–L_b per cent) of 43.7 per cent. In this character it comes somewhat closer to *B. elongatum* Stein, 1867 (36.7–43.7 per cent) and *B. helenae* (Bezzenberger, 1904 (54 per cent)). However, in *B. neotigrinae* n.sp. the macronucleus is extremely characteristic being in the form of 'C'-shaped loop, a character reported only in *B. tigrinae* Shete and Krishnamurthy, 1984 so far. However, in *B. tigrinae* the length of vestibulum is small (22–56.4 μm; $\bar{x}$ 38.7 μm) in contrast to the present form, *B. neotigrinae* n.sp. in which the vestibulum extends beyond the midline of the body and measures 110.7–155.3 ($\bar{x}$ 130) μm. Besides, *B. neotigrinae* n.sp. is considerably larger and measures 210–250.5 ($\bar{x}$ 230) μm × 115–147.5 ($\bar{x}$ 135.2) μm as compared to *B. tigrinae* which measures 73.1–108.7 ($\bar{x}$ 92.4) μm × 36.6–52.2 ($\bar{x}$ 45.3) μm.

B. neogiganteum n.sp. has a characteristic undulating membrane on the right lip of the vestibulum. In this character the present form resembles *B. sushilii* Ray, 1932 and *B. giganteum* Bezzenberger, 1904. According to Earl (1973 and 1988 personal communication) ciliates that look like *Balantidium* but have oral membrane instead of free adoral cilia belong to the genus *Dilleria*. However, Corliss (1979) in his encyclopedic work on ciliated protozoa did not recognize this genus. Therefore, in the present investigation, the present form with an undulating membrane is tentatively kept under the genus *Balantidium*. In *B. sushilii* the body is considerably

Table 17.2: Showing Comparative Characteristics of *Balantidium neogiganteum* n.sp. with its Related Species (Measurements are in microns, µm)

Sl.No.	Particulars	*B. neotigrinae* n.sp.	*B. sushilii* Ray, 1932	*B. giganteum* Bezzenberger, 1904
1.	Length of the body (L_b)	151–207 (183.3)	150–319.44	205
2.	Width of the body (W_b)	100–137 (125)	35–65	133
3.	Length of the vestibule (L_v)	95.2–110 (104.7)	–	–
4.	Length of the macronucleus (L_{ma})	42.0–61 (55)	–	–
5.	Width of the macronucleus (W_{ma})	31–45.3 (38)	–	–
6.	W_b/L_b per cent	68.19	–	–
7.	L_{ma}/L_b per cent	30	–	–
8.	L_v/L_b per cent	57.1	–	–
9.	W_{ma}/L_{ma} per cent	69.09	–	–

Table 17.3: Showing a Comparative Study of Five Species of *Balantidium* Reported in the Present Work (Measurement is given in microns, μm)

Sl.No.	Particulars	B. cyanophlycti	B. ganapatti	B. neotigrinae n.sp.	B. neogiganteum n.sp.	B. sp.
1.	Length of the body (L_b)	(60–100.6) 80	134–203.5 (180)	210–250.5 (230)	151.0–2.7 (183.3)	293.0
2.	Width of the body (W_b)	(26.0–40.5) 35	23.3–38.0 (33)	115–147.5 (135.2)	100–137 (125)	195.3
3.	Length of the vestibule (L_v)	(24.5–39.5) 33.5	25–40.5 (32.5)	110.7–155.3 (1330)	95.2–110 (104.7)	165.2
4.	Length of the macronucleus (L_{ma})	(12.6–24.6) 18.2	19.0–27.5 (26)	35.4–67.0 (50)	42.0–61 (55)	90.15
5.	Width of the macronucleus (W_{ma})	(5.0–10.5) 8.0	7–10 (8)	12–20.2 (15)	31–45.3 (38)	45.0
6.	W_b/L_b per cent	43.7	18.3	58.7	68.19	66.65
7.	L_{ma}/L_b per cent	22.7	14.4	21.7	30	30.76
8.	L_v/L_b per cent	41.8	18.5	56.5	571	56.38
9.	W_{ma}/L_{ma} per cent	43.9	30.7	30	69.09	49.91
10.	Reference	Shete and Krishnamurthy, 1984	Shete and Krishnamurthy, 1984	Present study	Present study	Present study

elongated and measures 150–319.44 pm x 35–65 µm whereas, *B. neogiganteum* n.sp. measures 151–207 µm x 100–137 µm. Besides, a knob-like structure is present at the anterior end of *B. sushilii* which is not the case with *B. neogiganteum* n.sp. *B. neogiganteum* n.sp. shows some relationship with *B. giganteum* in overall shape and size of the body. However, in *B. giganteum* the macronucleus is reniform to oval whereas in *B. neogiganteum* n.sp., the macronucleus is massive and broadly ovoid. Besides, the vestibulum of *B. giganteum* is large, broad and does not extend to the middle of the body whereas, in *B. neogiganteum* n.sp. the vestibulum is elongated, narrow tube like and reaches at least as far back as the middle line of the body.

Besides, the above mentioned species which are reported in the present investigation, only 2 specimens of a unique species of *Balantidium* were also encountered. However, these forms were not assigned to a new species because these ciliates were encountered only once and could not be studied in greater detail. These ciliates are considerably large (293 x 195.3 µm) and have a characteristics fold at their posterior ends, a unique character not reported so far.

Acknowledgement

The authors are thankful to Smt. Usha Zutshi (Ex-Principal) for the help provided during preparation of this paper.

Key to the Lettering

AC: Adoral Cilia

CV: Contractile vacuole

FAC: Fused adoral cilia

FV: Food vacuole

MA: Macronucleus

MI: Micronucleus

SC: Somatic Cilia

V: Vestibulum/Vestibule.

References

Bezzen Berger, E., 1904. Uber infusorien aus asiatischen Anuren. *Arch. Protistenk*, 3: 138–174.

Bhatia, B.L. and Gulati, A.N., 1927. On some parasitic Ciliates form Indian frogs, toads, earthworms and cock roaches. *Arch. Protistenk.*, 57: 85–120.

Claparede and Lachmann, 1858–61 Etudes sur les infusoires et les rhizopodes, Geneve. (Extrait des toms V, VI et VII des Memoires de P'instutnt Genevois.).

Cooper, H. and Gulati, A.N., 1926. On the occurrence of *Isospora* and *Balantidium* in cattle. *Mem. Depart. Agric. India, Vet.*, 3: 191–193.

Corliss, J.O., 1979. The Ciliated Protozoa: Characterization, Classification and Guide to Literature, 2nd edn. Pergamon Press, Oxford, New York, Toronto, Sydney, Paris and Frankfurt, 455 pp.

Earl, P.R., 1973. The Ciliate families Balantidae, Difleriidae n.fam. and Nathellidae. *Revista De Biologia*, 9(1–4): 175–185.

Fernandez-Galiano, D., 1976. Silver impregnation of citiated protozoa: Procedure yielding good results with the Pyridinated Silver-carbonate method. *Trans. Microscop. Soc.*, 95: 557–560.

Ghosh, E., 1921. Infusoria from the environment of Calcutta. *Bull. Carmichael Med. Coll.*, 2: 6–17.

Ghosh, E., 1922a. On a new ciliate, *Balantidium blattarum* sp.nov., intestinal parasite in the common cockroach (*Balatta Americana*). *Parasitology*, 14: 15–16.

Ghosh, E., 1922b. On a new ciliate *Balantidium ovatum*, sp.nov., an intestinal parasite in the common cockroach (*Blatta americana*). *Parasitology*, 14: 371.

Ghosh, E., 1925c. On a new ciliate *Balantidium knowlesii*, a coelomic parasite in Culicoides peregrines. *Parasitology*, 17: 189.

Egner, R.W., 1934. Specificity in the genus *Balantidium* based on size and shape of body and Macronucleus, with description of six new species. *Amer. J. Hyg.*, 19: 38–67.

Krasheninnikow, S. and Jeska, E.L., 1961. Agar diffusion studies on the species specificity of *Balantidium coli, B.caviae* and *B. wenrichi. Immunology*, 4: 282–288.

Pal, N.L. and Das Gupta, B. 1980. Observations on two new species of *Balantidium* in the Indian Salamander, *Tylototriton verrucosus* (Caudata : Salamandridae). *Proc. Zool. Soc. Calcutta*, 31(1–2): 47–51.

Shete, S.G. and Krishnamurthy, R., 1984. Observations on the rectal ciliates of the genus *Balantidium* Claparede and Lachmann, 1858 from Indian Amphibians *Rana tigrina* and *R. cyanophlyctis. Arch. Protistenk.*, 128: 179–194.

Stein, F., 1867. Der Organismus de infusionsthiere nach eigenen Forschungen in Systematischer Reihenofolge bearbeitet–II.(Aligemeines U. Heterotricha.) Leipzig.

Weesner, F.M., 1960. *General Zoological Microtechniques*. The Williams and Wilkins Company Baltimore, U.S.A., 230 pp.

Zaman, V., 1978. Balantidium coli. In: *Parasitic Protozoa* Vol. 2, (Ed.) J.P. Kreier. Academic Press, London, 633–653 pp.

Chapter 18

A Review on Diversity and Distribution of Nematodes Associated with Paddy in West Bengal, India

Viswa Venkat Gantait, Suresh Mandal, Paromita Roy and Soumendranath Chatterjee

ABSTRACT

West Bengal is one of the major rice producing state in India. More than 15 per cent of our national production of rice is contributed solely by the state. Among invertebrates, besides insects, nematodes cause major damages to this important cereal crop. The estimated average loss due to nematodes is about 15 per cent reaching up to 90 per cent in severe cases. These noxious pests ask for our much deserved attention so that control measures may be taken in proper way and in proper time against these hidden enemies. The systematic index along with the diversity and distribution of 129 nematode species under 53 genera and 27 families belonging to 4 orders, described and reported so far from different districts of West Bengal are presenting in this article. It will provide a valuable database regarding these pests which will help to take control measures against them. This will help to improve the production of this valuable crop which will ultimately be helpful to improve the GNP of our country.

Keywords: Nematode, Diversity, Distribution, Paddy, West Bengal.

Introduction

Paddy is one of the major agricultural and economic crops being cultivated globally. Different products *e.g.* rice, hay, rice husk, rice bran, rice bran oil *etc.* which are either obtained directly or extracted from this plant, help the human civilization in various ways. The year 2004 was designated as the **'International Year of Rice'** by United Nations considering importance of rice as the major staple food of the world and its role in alleviating malnutrition and poverty. Processed rice feed not only more than 60 per cent of the world human population but other animals and birds too (Sandhu, 2014). The milled rice has been consumed 496.2 million tonnes worldwide in 2015-16, increasing about 1 per cent every year; expected world rice utilization in 2016-17 will be 503.4 million tonnes (FAO, 2016).

In India, paddy is grown in more than 25 per cent of the cultivated lands belonging mostly to the eastern and southern regions (Sandhu, 2014). Contributing about 20 per cent to world paddy, India is lagging behind only to China. About 15 per cent to our national production is contributed solely from West Bengal which is highest among the 29 states. But the major obstacles in rice production are countless pests ranging from virus (*e.g.* Tungro Virus) to vertebrates (*e.g.* Rodents). Among the invertebrates, insects and nematodes are very much important. The estimated average loss due to nematodes is about 15 per cent reaching up to 90 per cent if infested with plant parasitic nematodes like *Aphelenchoides besseyi* (White Tip Nematode), *Ditylenchus angustus* (Stem Nematode), *Heterodera oryzae* (Cyst Nematode), *Hirschmanniella oryzae* (Rice Root Nematode), *Meloidogyne graminicola* (Root Knot Nematode) etc. (*Animal Resources of India*, 1991). These pests ask for our much deserved attention as our knowledge of their existence and dynamics are new to the ecological system. Ufra disease of rice caused by Stem Nematode *Tylenchus angustus* (= *Ditylenchus angustus*) was reported for the first time by Butler in 1913 in East Bengal (Now Bangladesh). In West Bengal nematode researches got pace much later, in the 1970s. Afterwards, Baqri *et al.* (1974-1990) dedicatedly contributed a lot to the research on soil free living and plant parasitic nematodes associated with paddy in West Bengal. Except those works, only very few have so far been done regarding the subject.

In this context, 72 species belonging to the order Dorylaimida, 49 species under the order Tylenchida and 4 species under each of the orders Aphelenchida and Mononchida, described and reported till to date associated with paddy are compiled in the present work. Systematic index along with diversity and distribution of those species are presented here under. District-wise distribution of those species is also presented graphically. It will provide a valuable database to the rice-producers as well as researchers which will help to take control measures against those noxious pests, resulting in improvement of rice production in the state. This will ultimately be helpful to improve the Gross National Production (GNP) of our country.

Systemic Index of Nematode Species

Phylum NEMATODA (Rudolphi, 1808) Lankester, 1877

Order DORYLAIMIDA Pearse, 1942

Suborder DORYLAIMINA Pearse, 1936

Superfamily ACTINOLAIMOIDEA Thorne, 1939

Family ACTINOLAIMIDAE Thorne, 1939

Subfamily NEOACTINOLAIMINAE Thorne, 1967

Genus *Egtitus* Thorne, 1967

    1. *Egtitus elaboratus* (Cobb, 1906) Thorne, 1967

Syn. *Neoactinolaimus elaboratus* (Cobb, 1906) Heyns and Argo, 1969

Genus *Neoactinolaimus* Thorne, 1967

    2. *Neoactinolaimus* sp.

    3. *N. thornei* Chaturvedi and Khera, 1979

Superfamily BELONDIROIDEA Thorne, 1939

Family BELONDIRIDAE Thorne, 1939

Subfamily BELONDIRINAE Thorne, 1939

Genus *Axonchium* Cobb, 1920

    4. *Axonchium amplicolle* Cobb, 1920

Genus *Belondira* Thorne, 1939

    5. *Belondira neortha* Siddiqi, 1964

    6. *B. nepalensis* Siddiqi, 1964

Subfamily DORYLAIMELLINAE Jairajpuri, 1964

Genus *Dorylaimellus* Cobb, 1913

    7. *Dorylaimellus deviatus* Baqri and Jairajpuri, 1968

    8. *D. discocephalus* Siddiqi, 1964

    9. *D. indicus* Siddiqi, 1964

    10. *D. projectus* Heyns, 1962

Subfamily SWANGERIINAE Jairajpuri, 1964

Genus *Paraoxydirus* Jairajpuri and Ahmad, 1979

    11. *Paraoxydirus gigas* (Jairajpuri 1964) Jairajpuri and Ahmad, 1979

Superfamily DORYLAIMOIDEA de Man, 1876

Family APORCELAIMIDAE Heyns, 1965

Subfamily APORCELAIMINAE Heyns, 1965

Genus *Aporcelaimellus* Heyns, 1965

    12. *Aporcelaimellus chauhani* Barqi and Khera, 1975

    13. *A. coomansi* Baqri and Khera, 1975

    14. *A. heynsi* Baqri and Jairajpuri, 1968

    15. *A. tropicus* Jana and Baqri, 1981

Family DORYLAIMIDAE de Man, 1876

Subfamily DORYLAIMINAE de Man, 1876

Genus *Dorylaimus* Dujardin, 1845

16. *Dorylaimus innovatus* Jana and Baqri, 1982
17. *D. stagnalis* Dujardin, 1845
18. *D. thornei* Andràssy, 1969

Genus *Ischiodorylaimus* Andràssy, 1969

19. *Ischiodorylaimus novus* Baqri and jana, 1986
20. *Ischiodorylaimus* sp.

Subfamily LAIMYDORINAE Andràssy, 1969

Genus *Calodorylaimus* Andràssy, 1969

21. *Calodorylaimus andrassyi* Baqri and Jana, 1982
22. *C. simplex* Baqri and Jana, 1982
23. *Calodorylaimus* sp.

Genus *Laimydorus* Siddiqi, 1969

24. *Laimydorus baldus* Baqri and Jana, 1982
25. *L. distinctus* Dey and Baqri, 1986
26. *L. finalis* Thorne, 1892
27. *L. oryzae* Dey and Baqri, 1986
28. *L. siddiqii* Baqri and Jana, 1982

Subfamily THORNENEMATINAE Siddiqi, 1969

Genus *Opisthodorylaimus* Ahmad and Jairajpuri, 1982

29. *Opisthodorylaimus cavalcantii* (Lordello, 1955) Carbonell and Coomans, 1986

Genus *Sicaguttur* Siddiqi, 1971

30. *Sicaguttur coomansi* (Baqri and Jana, 1980) Carbonell and Coomans 1986

Syn. *Medalinema coomansi* Baqri and Jana, 1980

31. *S. sartum* Siddiqi, 1971

Genus *Thornenema* Andràssy, 1959

32. *Thornenema conura* Dey and Baqri, 1986
33. *T. mauritianum* (Williams, 1959) Baqri and Jairajpuri, 1969
34. *T. novum* Dey and Baqri, 1986
35. *T. nodicaudatum* Dey and Baqri, 1986
36. *T. pseudosartum* Carbonell and Coomans, 1987
37. *T. shamimi* (Baqri and Jana, 1980) Carbonell and Coomans, 1987

Syn. *Jairajpuria shamimi* Baqri and Jana, 1980
Family NORDIIDAE Jairajpuri and Siddiqi, 1964
Subfamily PUNGENTINAE Siddiqi, 1969
Genus *Lenonchium* Siddiqi, 1965
  38. *Lenonchium oryzae* Siddiqi, 1965
Family QUDSIANEMATIDAE Jairajpuri, 1965
Subfamily DISCOLAIMINAE Siddiqi, 1969
Genus *Discolaimium* Thorne, 1939
  39. *Discolaimium andrassyi* Baqri and Khera, 1976
Genus *Discolaimoides* Heyns, 1963
  40. *Discolaimoides bulbiferous* (Cobb, 1906) Heyns, 1963
Subfamily QUDSIANEMATINAE Jairajpuri, 1965
Genus *Eudorylaimus* Andràssy, 1959
  41. *Eudorylaimus* spp.
Genus *Takamangai* Yeates, 1967
  42. *Takamangai confusa* (Thorne, 1939) Andràssy, 1991
Syn. *Thonus confusus* Jana and Baqri, 1982
Superfamily LONGIDOROIDEA Thorne, 1935
Family LONGIDORIDAE (Thorne, 1935) Meyl, 1961
Subfamily LONGIDORINAE Thorne, 1935
Genus *Paralongidorus* Siddiqi, Hooper and Khan, 1963
  43. *Paralongidorus citri* (Siddiqi, 1959) Siddiqi *et al.*, 1963
Family XIPHINEMATIDAE Dalmasso, 1969
Subfamily XIPHINEMATINAE Dalmasso, 1969
Genus *Xiphinema* Cobb, 1913
  44. *Xiphinema insigne* Loos, 1949
Superfamily TYLENCHOLAIMOIDEA Filipjev, 1934
Family LEPTONCHIDAE Thorne, 1935
Subfamily BELONENCHINAE Thorne, 1964
Genus *Basirotyleptus* Jairajpuri, 1964
  45. *Basirotyleptus basiri* Jairajpuri, 1964
  46. *B. minimus* Jana and Baqri, 1982
Subfamily LEPTONCHINAE Thorne, 1935
Genus *Proleptonchus* Lordello, 1955
  47. *Proleptonchus clarus* Timm, 1964

48. *P. indicus* Siddiqi and Khan, 1964

Subfamily TYLEPTINAE Jairajpuri, 1964

Genus *Tyleptus* Thorne, 1939

49. *Tyleptus projectus* Thorne, 1939
50. *T. variabilis* Jairajpuri and Loof, 1964

Family MYDONOMIDAE Thorne, 1964

Subfamily CALOLAIMINAE Goseco, Ferris and Ferris, 1976

Genus *Miranema* Thorne, 1939

51. *Miranema gracile* Thorne, 1939

Subfamily MYDONOMINAE Thorne, 1964

Genus *Dorylaimoides* Thorne and Swanger, 1936

52. *Dorylaimoides arcuatus* Siddiqi, 1964
53. *D. arcuicaudatus* Baqri and Jairajpuri, 1969
54. *D. constrictoides* Goseco, Ferris and Ferris, 1976
55. *D. constrictus* Baqri and Jairajpuri, 1968
56. *D. elaboratus* Siddiqi, 1965
57. *D. indicus* Jairajpuri, 1965
58. *D. leptura* Siddiqi, 1965
59. *D. micoletzkyi* (de Man, 1921) Thorne and Swanger, 1936
60. *D. pakistanensis* Siddiqi, 1964
61. *D. parateres* Siddiqi, 1964
62. *D. parvus* Thorne and Swanger, 1936
63. *D. paulbuchneri* Meyl, 1956
64. *D. teres* Thorne and Swanger, 1936

Genus *Morasia* Baqri and Jairajpuri, 1969

65. *Morasia bengalensis* Jana and Baqri, 1982

Family TYLENCHOLAIMIDAE Filipjev, 1934

Subfamily TYLENCHOLAIMINAE Filipjev, 1934

Genus *Discomyctus* Thorne, 1939

66. *Discomyctus cephalatus* Thorne, 1939
67. *D. elongatus* Dhanachand and Jairajpuri, 1980

Genus *Tylencholaimus* de Man, 1876

68. *Tylencholaimus obscurus* Jairajpuri, 1965
69. *T. pakistanensis* Timm, 1964

70. *T. paradoxus* Loof and Jairajpuri, 1968

Suborder NYGOLAIMINA Ahmad and Jairajpuri, 1979

Superfamily NYGOLAIMOIDEA Thorne, 1935

Family NYGOLAIMIDAE Thorne, 1935

Subfamily NYGOLAIMINAE Thorne, 1935

Genus *Laevides* Heyns, 1968

71. *Laevides imphalus* Ahmad and Jairajpuri, 1980
72. *L. paraaquaticus* (Paetzold, 1958) Ahmad and Jairajpuri, 1982

Order TYLENCHIDA Thorne, 1949

Suborder CRICONEMATINA Siddiqi, 1980

Superfamily CRICONEMATOIDEA [Taylor, 1936 (1914)] Geraert, 1966

Family CRICONEMATIDAE [Taylor, 1936 (1914)] Thorne, 1949

Subfamily HEMICRICONEMOIDINAE Andràssy, 1979

Genus *Hemicriconemoides* Chitwood and Birchfield, 1957

1. *Hemicriconemoides cocophillus* (Loos, 1949) Chitwood and Birchfield, 1957
2. *H. mangiferae* Siddiqi, 1961

Subfamily MACROPOSTHONIINAE Skarbilovich, 1959

Genus *Mesocriconema* Andràssy, 1965

3. *M. crenata* (Loof, 1964) Andràssy, 1965

Syn. *Macroposthonia crenata* (Loof, 1964) De Grisse and Loof, 1965

4. *M. onoense* (Luc, 1959) Loof and de Grisse, 1989

Syn. *M. onoensis* (Luc, 1959) de Grisse and Loof, 1965

5. *M. ornata* (Raski, 1958) Loof and de Grisse, 1989

Syn. *M. ornata* (Raski, 1958) de Grisse and Loof, 1965

Superfamily TYLENCHULOIDEA (Skarbilovich, 1947) Raski and Siddiqi, 1975

Family PARATYLENCHIDAE (Thorne, 1949) Raski, 1962

Subfamily PARATYLENCHINAE Thorne, 1949

Genus *Gracilacus* Raski, 1962

6. *Gracilacus janai* Baqri, 1979
7. *Gracilacus* sp.

Genus *Paratylenchus* Micoletzky, 1922

8. *Paratylenchus dianthus* Jenkins and Taylor, 1956

Suborder HEXATYLINA Siddiqi, 1980

Superfamily SPHAERULARIOIDEA (Lubbock, 1861) Poinar, 1975

Family NEOTYLENCHIDAE Thorne, 1941

Subfamily NEOTYLENCHINAE thorne, 1941

Genus *Hexatylus* Goodey, 1926

9. *Hexatylus* sp.

Suborder HOPLOLAIMINA Chizhov and Berezina, 1988

Superfamily DOLICHODOROIDEA (Chitwood in Chitwood and Chitwood, 1950) Siddiqi, 1986

Family TELOTYLENCHIDAE Siddiqi, 1960

Subfamily TELOTYLENCHINAE Siddiqi, 1960

Genus *Tylenchorhynchus* Cobb, 1913

10. *Tylenchorhynchus divittatus* Siddiqi, 1961
11. *T. mashhoodi* Siddiqi and Basir, 1959
12. *T. nudus* Allen, 1955

Superfamily HOPLOLAIMOIDEA (Filipjev, 1934) Paramonov, 1967

Family HETERODERIDAE Filipjev and Schuurmans Stekhoven, 1941

Subfamily HETERODERINAE Filipjev and Schuurmans Stekhoven, 1941

Genus *Heterodera* Schmidt, 1871

13. *Heterodera oryzicola* Rao and Jayaprakash, 1978

Family HOPLOLAIMIDAE Filipjev, 1934

Subfamily HOPLOLAIMINAE Filipjev, 1934

Genus *Hoplolaimus* van Daday, 1905

14. *Hoplolaimus columbus* Sher, 1963
15. *H. indicus* Sher, 1963

Subfamily ROTYLENCHOIDINAE Whitehead, 1958

Genus *Helicotylenchus* Steiner, 1945

16. *Helicotylenchus crenacauda* Sher, 1966
17. *H. digitatus* Siddiqi and Husain, 1964
18. *H. dihystera* (Cobb, 1893) Sher, 1961
19. *H. egyptiensis* Tarjan, 1964
20. *H. exallus* Sher, 1966

Syn. *H. abunaamai* Siddiqi, 1972

21. *H. indicus* Siddiqi, 1963
22. *H. microcephalus* Sher, 1966
23. *H. minzi* Sher, 1966

24. *H. pseudorobustus* (Steiner, 1914) Golden, 1956

25. *H. retusus* Siddiqi and Brown, 1964

Family MELOIDOGYNIDAE (Skarbilovich, 1959) Wouts, 1973

Subfamily MELOIDOGYNINAE Skarbilovich, 1959

Genus *Meloidogyne* Göeldi, 1892

26. *Meloidogyne graminicola* Golden and Birchfield, 1965

Family PRATYLENCHIDAE Thorne, 1949

Subfamily HIRSCHMANNIELLINAE Fotedar and Handoo, 1978

Genus *Hirschmanniella* Luc and Goodey, 1964

27. *Hirschmanniella gracilis* (de Man, 1880) Luc and Goodey, 1964

28. *H. mucronata* (Das, 1960) Luc and Goodey, 1964

29. *H. oryzae* (van Breda de Haan, 1902) Luc and Goodey, 1964

Subfamily PRATYLENCHINAE Thorne, 1949

Genus *Pratylenchus* Filipjev, 1936

30. *Pratylenchus brachyurus* (Godfrey, 1929) Filipjev and Schuurmans Stekhoven, 1941

31. *P. coffeae* (Zimmermann, 1898) Filipjev and Schuurmans Stekhoven, 1941

32. *P. indicus* Das, 1960

33. *P. loosi* Loof, 1960

34. *P. pratensis* (de Man, 1880) Filipjev, 1936

35. *P. scribneri* Steiner 1943

36. *P. thornei* Sher and Allen, 1953

37. *P. zeae* Graham, 1951

Family ROTYLENCHULIDAE (Husain and Khan, 1967) Husain, 1976

Subfamily ROTYLENCHULINAE Husain and Khan, 1967

Genus *Rotylenchulus* Linford and Oliveira, 1940

38. *Rotylenchulus reniformis* Linford and Oliveira, 1940

Suborder TYLENCHINA Chitwood in Chitwood and Chitwood, 1950

Infraorder ANGUINATA Siddiqi, 2000

Superfamily ANGUINOIDEA Nicoll, 1935

Family ANGUINIDAE Nicoll, 1935 (1926)

Subfamily ANGUININAE Nicoll, 1935 (1926)

Genus *Ditylenchus* Filipjev, 1936

39. *Ditylenchus acutus* (Khan, 1965) Fortuner and Maggenti, 1987

Syn. *Nothotylenchus acutus* Khan, 1965

40. *D. angustus* (Butler, 1913) Filipjev, 1936

41. *D. mirus* Siddiqi, 1963

42. *Ditylenchus* spp.

Infraorder TYLENCHATA Siddiqi, 2000

Superfamily TYLENCHOIDEA Örley, 1880

Family TYLENCHIDAE Örley, 1880

Subfamily BOLEODORINAE Khan, 1964

Genus *Basiria* Siddiqi, 1959

43. *Basiria tumida* (Colbran, 1960) Geraert, 1968

Subfamily DUOSULCINAE Siddiqi, 1979

Genus *Malenchus* Andràssy, 1968

44. *Malenchus* sp.

Subfamily TYLENCHINAE Örley, 1880

Genus *Filenchus* Andràssy, 1954

45. *Filenchus filiformis* (Bütschli, 1873) n. Comb.

Syn. *Tylenchus filiformis* Bütschli, 1873

46. *F. goodeyi* (Das, 1960) Goodey, 1963

47. *Filenchus* sp.

Genus *Tylenchus* Bastian, 1865

48. *Tylenchus davainei* Bastian, 1865

49. *Tylenchus* sp.

Order APHELENCHIDA Siddiqi, 1980

Suborder APHELENCHINA Geraert, 1966

Superfamily APHELENCHOIDEA Fuchs, 1937

Family APHELENCHIDAE Fuchs, 1937

Subfamily APHELENCHINAE Fuchs, 1937

Genus *Aphelenchus* Bastian, 1865

1. *Aphelenchus avenae* (Bastian 1865) Chaturvedi and Khera 1979

Superfamily APHELENCHOIDOIDEA (Skarbilovich, 1947) Siddiqi, 1980

Family APHELENCHOIDIDAE (Skarbilovich, 1947) Paramonov, 1953

Subfamily APHELENCHOIDINAE Skarbilovich, 1947

Genus *Aphelechoides* Fischer, 1894

2. *Aphelenchoides besseyi* Christie, 1942

3. *A. pusillus* (Thorne, 1929) Filipjev, 1934

4. *A. subtenuis* (Cobb, 1926) Steiner and Buhrer, 1932

Order MONONCHIDA Jairajpuri, 1969

Suborder MONONCHINA Kirjanova and Krall, 1969

Superfamily MONONCHOIDEA (Filipjev, 1934) Clark, 1961

Family MONONCHIDAE (Filipjev, 1934) Chitwood, 1937

Genus *Mononchus* Bastian, 1865

1. *Mononchus aquaticus* Coetzee, 1968

Genus *Miconchus* Andràssy, 1958

2. *Miconchus dalhousiensis* Jairajpuri, 1969

Family MYLONCHULIDAE Jairajpuri, 1969

Genus *Mylonchulus* (Cobb, 1916) Pennak, 1953

3. *Mylonchulus brachyuris* (Bütschli, 1873) Altherr, 1954

4. *M. mulveyi* (Jairajpuri, 1970) Jairajpuri and Khan, 1982

The diversity and distribution of all the species in different districts of West Bengal are given as follows.

Table 18.1: District-wise Distribution of Nematodes Associated with Paddy in West Bengal

Sl.No.	Species	Family	Locality	References
		Order – DORYLAIMIDA		
1.	Egtitus elaboratus	Actinolaimidae	Darjeeling (Kalimpong-I, Khairibari and Kurseong blocks)	Baqri and Dey, 1991
2.	Neoactinolaimus sp.	Actinolaimidae	Coochbehar; Darjeeling (Kalimpong-I, Khairibari and Kurseong blocks); Jalpaiguri	Baqri and Dey, 1991; Baqri and Ahmad, 2000
3.	N. thornei	Actinolaimidae	Burdwan (Borsul, Jamalpur, Memari blocks)	Baqri et al., 1983
4.	Axonchium amplicolle	Belondiridae	Jalpaiguri	Baqri and Ahmad, 2000
5.	Belondira neortha	Belondiridae	Jalpaiguri	Baqri and Ahmad, 2000
6.	B. nepalensis	Belondiridae	Darjeeling (Kalimpong-I, Khairibari and Kurseong blocks)	Baqri and Dey, 1991
7.	Dorylaimellus deviatus	Belondiridae	Burdwan (Borsul, Jamalpur, Memari blocks); Coochbehar; Malda	Baqri et al. (1983, 1991, 2000)

Contd...

8.	D. discocephalus	Belondiridae	Burdwan (Borsul, Jamalpur, Memari blocks)	Baqri et al., 1983
9.	D. indicus	Belondiridae	Burdwan (Borsul, Jamalpur, Memari blocks); Coochbehar; Jalpaiguri; Malda; West Dinajpur (Balurghat, Hilli, Islampur and Kumarganj blocks)	Baqri et al. (1983, 1990, 1991); Baqri and Ahmad, 2000
10.	D. projectus	Belondiridae	Malda	Baqri and Ahmad, 2000
11.	Paraoxydirus gigas	Belondiridae	Darjeeling (Kalimpong-I, Khairibari and Kurseong blocks); Malda; West Dinajpur (Balurghat, Hilli, Islampur and Kumarganj blocks)	Baqri and Das, 1990; Baqri and Dey, 1991; Baqri and Ahmad, 2000
12.	Aporcelaimellus chauhani	Aporcelaimidae	West Dinajpur (Balurghat, Hilli, Islampur and Kumarganj blocks)	Baqri and Das, 1990
13.	A. coomansi	Aporcelaimidae	Burdwan (Borsul, Jamalpur, Memari blocks)	Baqri and Jana, 1981
14.	A. heynsi	Aporcelaimidae	(Maliyara) Bankura; Burdwan (Jamalpur block); Coochbehar; Darjeeling (Kalimpong-I, Khairibari and Kurseong blocks); Jalpaiguri; Malda; West Dinajpur (Balurghat, Hilli, Islampur and Kumarganj blocks)	Baqri and Jana, 1981; Baqri and Das, 1990; Baqri and Dey, 1991; Baqri and Ahmad, 2000; Ghosh and Manna, 2008
15.	A. tropicus	Aporcelaimidae	Burdwan (Memari block); North 24 Pargana (Duttapukur, Rautara)	Baqri and Jana, 1981; Ghosh and Manna, 2008
16.	Dorylaimus innovatus	Dorylaimidae	Burdwan (Rasulpur); Midnapore; North 24 Pargana (Rautara)	Baqri and Jana, 1982; Ghosh and Manna, 2008
17.	D. stagnalis	Dorylaimidae	Malda	Baqri and Ahmad, 2000
18.	D. thornei	Dorylaimidae	West Dinajpur (Balurghat, Hilli, Islampur and Kumarganj blocks)	Baqri and Das, 1990
19.	Ischiodorylaimus novus	Dorylaimidae	Burdwan	Baqri and Jana, 1986
20.	Ischiodorylaimus sp.	Dorylaimidae	Burdwan (Borsul, Jamalpur, Memari blocks)	Baqri et al., 1983
21.	Calodorylaimus andrassyi	Dorylaimidae	Burdwan	Baqri and Jana, 1982

Contd...

Table 18.1–*Contd...*

Sl.No.	Species	Family	Locality	References
22.	C. simplex	Dorylaimidae	Coochbehar; Malda	Baqri et al. (1991, 2000)

23.	*Calodorylaimus* sp.	Dorylaimidae	Jalpaiguri	Baqri and Ahmad, 2000
24.	*Laimydorus baldus*	Dorylaimidae	Bankura (Jalanpur); Coochbehar (Chakchaka); Darjeeling (Kalimpong-I, Khairibari and Kurseong blocks); North 24 Pargana (Rautara); Purulia (Purulia Town); West Dinajpur (Balurghat, Hilli, Islampur and Kumarganj blocks)	Baqri and Dey, 1991; Baqri and Das, 1990; Ghosh and Manna, 2008
25.	*L. distinctus*	Dorylaimidae	Coochbehar	Baqri *et al.*, 1991
26.	*L. finalis*	Dorylaimidae	Coochbehar; Darjeeling (Kalimpong-I, Khairibari and Kurseong blocks)	Baqri and Dey, 1991
27.	*L. oryzae*	Dorylaimidae	Darjeeling (Kalimpong-I, Khairibari and Kurseong blocks)	Baqri and Dey, 1991
28.	*L. siddiqii*	Dorylaimidae	Coochbehar (Arabindanagar, Pundibari); Jalpaiguri; Malda; North 24 Pargana (Rautara); Purulia	Baqri *et al.* (1991, 2000); Ghosh and Manna, 2008
29.	*Opisthodorylaimus cavalcantii*	Dorylaimidae	Darjeeling (Kalimpong-I, Khairibari and Kurseong blocks)	Baqri and Dey, 1991
30.	*Sicagutuur coomansi*	Dorylaimidae	Burdwan (Borsul, Jamalpur, Memari blocks); Jalpaiguri	Baqri *et al.* (1983, 2000)
31.	*S. sartum*	Dorylaimidae	Burdwan (Borsul, Jamalpur, Memari blocks); Midnapore (Digha); South 24 Pargana (Narendrapur)	Baqri *et al.*, 1983; Ghosh and Manna, 2008
32.	*Thornenema conura*	Dorylaimidae	West Dinajpur	Dey and Baqri, 1986
33.	*T. mauritianum*	Dorylaimidae	Burdwan (Borsul, Jamalpur, Memari blocks); Coochbehar; Jalpaiguri; Malda	Baqri *et al.* (1983, 1991, 2000)
34.	*T. novum*	Dorylaimidae	Darjeeling (Kalimpong-I, Khairibari and Kurseong blocks)	Baqri and Dey, 1991
35.	*T. nodicaudatum*	Dorylaimidae	Darjeeling (Kalimpong-I, Khairibari and Kurseong blocks)	Baqri and Dey, 1991
36.	*T. pseudosartum*	Dorylaimidae	Malda	Baqri and Ahmad, 2000
37.	*T. shamimi*	Dorylaimidae	West Dinajpur (Balurghat, Hilli, Islampur and Kumarganj blocks)	Baqri and Das, 1990

Contd...

Table 18.1–*Contd...*

Sl.No.	Species	Family	Locality	References

38.	Lenonchium oryzae	Nordiidae	Burdwan (Jamalpur block); Malda; West Dinajpur (Balurghat, Hilli, Islampur and Kumarganj blocks)	Baqri and Das, 1990; Baqri and Ahmad, 2000
39.	Discolaimium andrassyi	Qudsianematidae	Darjeeling (Kalimpong-I, Khairibari and Kurseong blocks); West Dinajpur (Balurghat, Hilli, Islampur and Kumarganj blocks)	Baqri and Das, 1990; Baqri and Dey, 1991
40.	Discolaimoides bulbiferous	Qudsianematidae	Coochbehar; West Dinajpur (Balurghat, Hilli, Islampur and Kumarganj blocks)	Baqri and Das, 1990; Baqri et al., 1991
41.	Eudorylaimus spp.	Qudsianematidae	Coochbehar; Darjeeling (Kalimpong-I, Khairibari and Kurseong blocks)	Baqri and Dey, 1991
42.	Takamangai confusa	Qudsianematidae	Coochbehar	Jana and Baqri, 1982
43.	Paralongidorus citri	Longidoridae	Burdwan (Borsul, Jamalpur, Memari blocks)	Baqri et al., 1983
44.	Xiphinema insigne	Xiphinematidae	Darjeeling (Kalimpong-I, Khairibari and Kurseong blocks)	Baqri and Dey, 1991
45.	Basirotyleptus basiri	Leptonchidae	Jalpaiguri; Malda	Baqri and Ahmad, 2000
46.	B. minimus	Leptonchidae	Darjeeling	Jana and Baqri, 1981
47.	Proleptonchus clarus	Leptonchidae	Burdwan (Borsul, Jamalpur, Memari blocks); Coochbehar; Darjeeling (Kalimpong-I, Khairibari and Kurseong blocks); Jalpaiguri; Malda	Baqri et al. (1983, 1991); Baqri and Dey, 1991; Baqri and Ahmad, 2000
48.	P. indicus	Leptonchidae	West Dinajpur (Balurghat, Hilli, Islampur and Kumarganj blocks)	Baqri and Das, 1990
49.	Tyleptus projectus	Leptonchidae	Jalpaiguri	Baqri and Ahmad, 2000
50.	T. variabilis	Leptonchidae	West Dinajpur (Balurghat, Hilli, Islampur and Kumarganj blocks)	Baqri and Das, 1990
51.	Miranema gracile	Mydonomidae	Malda	Baqri and Ahmad, 2000
52.	Dorylaimoides arcuatus	Mydonomidae	West Dinajpur	Baqri and Das, 1990
53.	D. arcuicaudatus	Mydonomidae	Burdwan (Borsul, Jamalpur, Memari blocks); Malda	Baqri et al. (1983, 2000)
54.	D. constrictoides	Mydonomidae	Malda	Baqri and Ahmad, 2000
55.	D. constrictus	Mydonomidae	Coochbehar	Baqri et al., 1991
56.	D. elaboratus	Mydonomidae	Burdwan (Borsul, Jamalpur, Memari blocks)	Baqri et al., 1983

Contd...

Table 18.1–*Contd...*

Sl.No.	Species	Family	Locality	References

Sl.No.	Species	Family	Locality	References
57.	D. indicus	Mydonomidae	Jalpaiguri	Baqri and Ahmad, 2000
58.	D. leptura	Mydonomidae	Malda; West Dinajpur (Balurghat, Hilli, Islampur and Kumarganj blocks)	Baqri and Das, 1990; Baqri and Ahmad, 2000
59.	D. micoletzkyi	Mydonomidae	Malda	Baqri and Ahmad, 2000
60.	D. pakistanensis	Mydonomidae	Coochbehar	Baqri et al., 1991
61.	D. parateres	Mydonomidae	Coochbehar (Tufangung)	Baqri and khera, 1979
62.	D. parvus	Mydonomidae	Burdwan (Borsul, Jamalpur, Memari blocks)	Baqri et al., 1983
63.	D. paulbuchneri	Mydonomidae	Malda	Baqri and Ahmad, 2000
64.	D. teres	Mydonomidae	West Dinajpur (Balurghat, Hilli, Islampur and Kumarganj blocks)	Baqri and Das, 1990
65.	Morasia bengalensis	Mydonomidae	Burdwan (Borsul, Jamalpur, Memari blocks)	Baqri et al., 1983
66.	Discomyctus cephalatus	Tylencholaimidae	Coochbehar; Darjeeling (Kalimpong-I, Khairibari and Kurseong blocks); Jalpaiguri	Baqri and Dey, 1991; Baqri and Ahmad, 2000
67.	D. elongatus	Tylencholaimidae	West Dinajpur (Balurghat, Hilli, Islampur and Kumarganj blocks)	Baqri and Das, 1990
68.	Tylencholaimus obscurus	Tylencholaimidae	Coochbehar	Baqri et al., 1991
69.	T. pakistanensis	Tylencholaimidae	Burdwan (Borsul, Jamalpur, Memari blocks); Darjeeling Jalpaiguri; Malda; West Dinajpur (Balurghat, Hilli, Islampur and Kumarganj blocks)	Baqri et al. (1983,1990, 1991, 2000)
70.	T. paradoxus	Tylencholaimidae	Darjeeling (Kalimpong-I, Khairibari and Kurseong blocks)	Baqri and Dey, 1991
71.	Laevides imphalus	Nygolaimidae	West Dinajpur (Balurghat, Hilli, Islampur and Kumarganj blocks)	Baqri and Das, 1990
72.	L. paraaquaticus	Nygolaimidae	Jalpaiguri	Baqri and Ahmad, 2000
		Order – TYLENCHIDA		
73.	Hemicriconemoides cocophillus	Criconematidae	Burdwan (Borsul, Jamalpur, Memari blocks); Darjeeling; Jalpaiguri	Baqri et al. (1983, 1991, 2000)
74.	H. mangiferae	Criconematidae	West Bengal	Q. H. Baqri, 1999
75.	Mesocriconema crenata	Criconematidae	West Dinajpur (Hilli, Islampur blocks)	Baqri and Das, 1990
76.	M. onoense	Criconematidae	Burdwan (Borsul, Jamalpur, Memari blocks)	Baqri et al., 1983

Contd...

Table 18.1–*Contd...*

Sl.No.	Species	Family	Locality	References

77.	*M. ornata*	Criconematidae	Burdwan (Borsul, Jamalpur, Memari blocks); Coochbehar; Darjeeling; Jalpaiguri; Malda; West Dinajpur (Hilli, Islampur blocks)	Baqri *et al.* (1983, 1990, 1991); Baqri and Ahmad, 2000
78.	*Gracilacus janai*	Paratylenchidae	Burdwan (Borsul, Jamalpur, Memari blocks); Jalpaiguri	Baqri *et al.* (1983, 2000)
79.	*Gracilacus* sp.	Paratylenchidae	Coochbehar	Baqri *et al.*, 1991
80.	*Paratylenchus dianthus*	Paratylenchidae	Burdwan (Borsul, Jamalpur, Memari blocks)	Baqri *et al.*, 1983
81.	*Hexatylus* sp.	Neotylenchidae	Malda	Baqri and Ahmad, 2000
82.	*Tylenchorhynchus divittatus*	Telotylenchiidae	Malda	Baqri and Ahmad, 2000
83.	*T. mashhoodi*	Telotylenchiidae	Birbhum; Burdwan (Memari block); Jalpaiguri; Malda; Murshidabad; Coochbehar; Darjeeling (Kalimpong-I, Khairibari and Kurseong blocks); West Dinajpur (Balurghat, Hilli, Islampur and Kumarganj blocks)	Baqri *et al.*, 1983; Baqri and dey, 1991; Baqri and Ahmad, 2000
84.	*T. nudus*	Telotylenchiidae	Coochbehar	Baqri *et al.*, 1991
85.	*Heterodera oryzicola*	Hetroderidae	Jalpaiguri (Mohitnagar)	Q. H. Baqri, 1999
86.	*Hoplolaimus columbus*	Hoplolaimidae	Burdwan (Borsul, Jamalpur, Memari blocks)	Baqri *et al.*, 1983
87.	*H. indicus*	Hoplolaimidae	Birbhum; Burdwan (Borsul, Jamalpur, Memari blocks); Coochbehar; Darjeeling; Malda; Murshidabad; West Dinajpur (Hilli block)	Baqri *et al.* (1983, 1990); Baqri and dey, 1991; Baqri and Ahmad, 2000
88.	*Helicotylenchus crenacauda*	Hoplolaimidae	Burdwan (Borsul, Jamalpur, Memari blocks); Coochbehar	Baqri *et al.* (1983, 1991)
89.	*H. digitatus*	Hoplolaimidae	Darjeeling (Kalimpong-I, Khairibari and Kurseong blocks); Jalpaiguri	Baqri and dey, 1991; Baqri and Ahmad, 2000
90.	*H. dihystera*	Hoplolaimidae	Coochbehar; Darjeeling (Kalimpong-I, Khairibari and Kurseong blocks); Jalpaiguri	Baqri and dey, 1991; Baqri and Ahmad, 2000
91.	*H. egyptiensis*	Hoplolaimidae	West Dinajpur (Balurghat, Hilli, Islampur and Kumarganj blocks)	Baqri and Das, 1990
92.	*H. exallus*	Hoplolaimidae	Jalpaiguri	Baqri and Ahmad, 2000

Contd...

Table 18.1–*Contd...*

Sl.No.	Species	Family	Locality	References

93.	H. indicus	Hoplolaimidae	Jalpaiguri; Nadia (Krishnanagar); North 24 Pargana (Rautara); West Dinajpur (Balurghat, Hilli, Islampur and Kumarganj blocks)	Baqri and Das, 1990; Baqri and Ahmad, 2000; Ghosh and Manna, 2008
94.	H. microcephalus	Hoplolaimidae	Jalpaiguri; West Dinajpur (Balurghat, Hilli, Islampur and Kumarganj blocks)	Baqri and Das, 1990; Baqri and Ahmad, 2000
95.	H. minzi	Hoplolaimidae	Malda	Baqri and Ahmad, 2000
96.	H. psedorobustus	Hoplolaimidae	Darjeeling (Kalimpong-I, Khairibari and Kurseong blocks)	Baqri and Dey, 1991
97.	H. retusus	Hoplolaimidae	Burdwan (Borsul, Jamalpur, Memari blocks); Malda; West Dinajpur (Balurghat, Hilli, Islampur and Kumarganj blocks)	Baqri et al., 1983; Baqri and Ahmad, 2000
98.	Meloidogyne graminicola	Meloidogynidae	Coochbehar; Darjeeling (Kalimpong-I, Khairibari and Kurseong blocks); Jalpaiguri; Malda; Murshidabad; West Dinajpur	Baqri et al. (1990, 1991, 2000)
99.	Hirschmanniella gracilis	Pratylenchidae	Birbhum; Burdwan (Borsul, Jamalpur, Memari blocks); Coochbehar; Darjeeling (Kalimpong-I, Khairibari and Kurseong blocks); Jalpaiguri; Malda; Murshidabad; Nadia (Jagulia); North 24 Pargana (Rautara); West Dinajpur (Balurghat, Hilli, Islampur and Kumarganj blocks)	Baqri et al. (1983, 1990); Baqri and Dey, (1986,1991); Baqri and Ahmad, 2000; Ghosh and Manna, 2008
100.	H. mucronata	Pratylenchidae	West Bengal	Bala and Khan, 2002
101.	H. oryzae	Pratylenchidae	Birbhum (Bolpur); Burdwan (Borsul, Jamalpur, Memari blocks); Darjeeling (Kalimpong-I, Khairibari and Kurseong blocks); Hooghly (Arambag); Jalpaiguri	Baqri et al., 1983, 1991; Baqri and Ahmad, 2000
102.	Pratylenchus brachyurus	Pratylenchidae	West Bengal	Mukhopadhyay and Haque (1974a, b); Mukherjee and Dasgupta, 1981b; Khan 2005a
103.	P. coffeae	Pratylenchidae	West Bengal	
104.	P. indicus	Pratylenchidae	West Bengal	
105.	P. loosi	Pratylenchidae	West Bengal	
106.	P. pratensis	Pratylenchidae	Malda	Baqri and Ahmad, 2000

Contd...

Table 18.1–*Contd...*

Sl.No.	Species	Family	Locality	References
107.	P. scribneri	Pratylenchidae	West Dinajpur (Balurghat, Hilli, Islampur and Kumarganj blocks); Darjeeling (Kalimpong-I block); Jalpaiguri; Malda	Baqri et al. (1990, 1991, 2000)
108.	P. thornei	Pratylenchidae	Malda	Baqri and Ahmad, 2000
109.	P. zeae	Pratylenchidae	West Bengal	Mukhopadhyay and Haque (1974a, b); Mukherjee and Dasgupta, 1981b; Khan 2005a
110.	Rotylenchulus reniformis	Rotylenchulidae	Burdwan (Borsul, Jamalpur, Memari blocks); West Dinajpur (Balurghat, Hilli, Islampur and Kumarganj blocks)	Baqri et al. (1983, 1990)
111.	Ditylenchus acutus	Anguinidae	West Dinajpur (Balurghat, Hilli, Islampur and Kumarganj blocks)	Baqri and Das, 1990
112.	D. angustus	Anguinidae	Hooghly (Chinsurah)	Chakrabarti et al., 1985; Rao et al., 1986b
113.	D. mirus	Anguinidae	Burdwan (Borsul, Jamalpur, Memari blocks)	Baqri et al., 1983
114.	Ditylenchus spp.	Anguinidae	Coochbehar; West Dinajpur (Balurghat, Hilli, Islampur and Kumarganj blocks)	Baqri et al., 1991; Baqri and Das, 1990
115.	Basiria tumida	Tylenchidae	Coochbehar	Baqri et al., 1991
116.	Malenchus sp.	Tylenchidae	Coochbehar	Baqri et al., 1991
117.	Filenchus filiformis	Tylenchidae	Burdwan (Borsul, Jamalpur, Memari blocks)	Baqri et al., 1983
118.	F. goodeyi	Tylenchidae	West Dinajpur (Balurghat, Hilli, Islampur and Kumarganj blocks)	Baqri and Das, 1990
119.	Filenchus sp.	Tylenchidae	Darjeeling; Coochbehar; Malda	Baqri et al. (1991, 2000)
120.	Tylenchus davainei	Tylenchidae	Burdwan (Borsul, Jamalpur, Memari blocks)	Baqri et al., 1983
121.	Tylenchus sp.	Tylenchidae	Darjeeling	Baqri and Dey, 1991
		Order – APHELENCHIDA		
122.	Aphelenchus avenae	Aphelenchidae	Burdwan (Borsul, Jamalpur, Memari blocks); Coochbehar; Darjeeling; Jalpaiguri; Malda	Baqri et al. (1983, 1991, 2000);
123.	Aphelenchoides besseyi	Aphelenchidae	West Bengal	M. R. Khan, 2001

Contd...

Table 18.1–Contd...

Sl.No.	Species	Family	Locality	References
124.	A. pusillus	Aphelenchidae	Darjeeling	Baqri and Dey, 1991
125.	A. subtenuis	Aphelenchidae	West Dinajpur (Balurghat, Hilli, Islampur and Kumarganj blocks)	Baqri and Das, 1990
		Order – MONONCHIDA		
126	Mononchus aquaticus	Mononchidae	Jalpaiguri; West Dinajpur (Balurghat, Hilli, Islampur and Kumarganj blocks)	Baqri and Das, 1990; Baqri and Ahmad, 2000
127.	Mylonchulus brachyuris	Mononchidae	Darjeeling	Baqri and Dey, 1991
128.	M. mulveyi	Mononchidae	Darjeeling	Baqri and Dey, 1991
129.	Miconchus dalhousiensis	Mononchidae	Darjeeling	Baqri and Dey, 1991

Discussion

The present work deals with a total of 129 species under 53 genera, 38 subfamilies, 27 families, 16 superfamilies, 8 suborders belonging to 4 orders, described and reported so far from different districts of West Bengal. Order Dorylaimida comprises of 72 species under 30 genera, 20 subfamilies, 12 families, 6 superfamilies, 2 suborders; Tylenchida includes 49 species under 18 genera, 16 subfamilies, 11 families, 7 superfamilies, 4 suborders; Aphelenchida contains 4 species under 2 genera, 2 subfamilies, 2 families, 2 superfamilies, 1 suborder; Mononchida includes 4 species under 3 genera, 2 families, 1 superfamily, 1 suborder. The species are arranged systematically following Jairajpuri and Ahmad (1992) in case of order Dorylaimida, Siddiqi (2000) for Tylenchida and Jairajpuri and Khan (1982) for Mononchida. The species viz., *Dorylaimellus indicus, Aporcelaimellus heynsi, Laimydorus baldus, L. siddiqii, Proleptonchus clarus, Tylencholaimus pakistanensis, Mesocriconema ornata, Tylenchorhynchus mashhoodi, Hoplolaimus indicus, Meloidogyne graminicola, Hirchmanniella gracilis, H. oryzae* and *Aphelenchus avenae* are widely distributed in most of the districts. Amongst 53 genera, *Dorylaimoides, Helicotylenchus* and *Pratylenchus* are the most abundant ones. The districts viz., Burdwan, Coochbehar, Darjeeling, Jalpaiguri, Malda and West Dinajpur have been studied well enough but sufficient works have not been done in the remaining districts. A comparison of the number of species described and reported from different districts of West Bengal is diagrammatically represented in Figure 18.1.

Three genera *Eudorylaimus* under the order Dorylaimida and *Hexatylus* and *Malenchus* belonging to the order Tylenchida have been reported without species identification. The genera viz., *Neoactinolaimus, Ischiodorylaimus, Calodorylaimus, Gracilacus, Ditylenchus, Filenchus* and *Tylenchus* have both identified as well as non-identified species. This review work aims to draw attention towards the fact that a major part of paddy fields of this state is yet to be screened thoroughly for the identification and description of more soil free living and plant parasitic nematodes

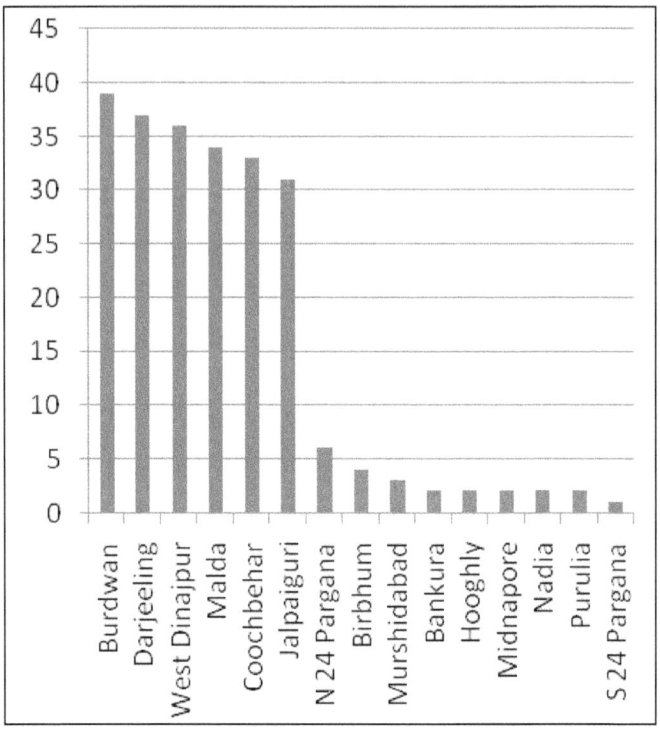

Figure 18.1: Number of Species Reported from different Districts of West Bengal.

associated with the most important cereal crop of our country.

Summary

The present work includes systematic index along with the diversity and distribution of 129 species under 53 genera and 27 families belonging to 4 orders, described and reported so far from different districts of West Bengal. Order Dorylaimida comprises of 72 species under 30 genera, 12 families; Tylenchida includes 49 species under 18 genera, 11 families; Aphelenchida contains 4 species under 2 genera, 2 families; Mononchida includes 4 species under 3 genera, 2 families.

Acknowledgement

Authors are very much thankful to the Director, Zoological Survey of India and the Vice-Chancellor, Burdwan University, West Bengal for their kind co-operation to prepare the paper.

References

Ahmad, N. and Baqri, Q. H. 1987. Nematodes from West Bengal (India) XVIII. Studies on the species of the Subfamily Tylenchorhynchinae (Tylenchorhynchidae: Tylenchida). *Bulletin of the Zoological Survey of India*, **8** (1-3): 135-142.

Ahmad, N., Das, P. K. and Baqri, Q.H. 1984. Evaluation of yield losses in Rice due to *Hirschmaniella gracilis* (de Man, 1880) Luc and Goodey, 1963 (Tylenchida: Nematoda) at Hooghly (West Bengal). *Bulletin of the Zoological Survey of India,* **5** (2 and 3): 85-91.

Annonymous. 1991. Nematoda. *Animal Resources of India: State of the Art,* Zoological Survey of India, pp. 99-113.

Annonymous. 2012. State-wise Area, Production and Productivity of Rice during 2006-07 to 2010-11. *Directorate of Rice Research, Govt. of India, Ministry of Agriculture, Patna (Bihar),* p. 5.

Anwar, S. A., Gorsi, S. and Rauf, C. A. 1993. Nematode diseases of Rice in the Punjub, Pakistan. *Pakistan Journal of Agricultural Research,* **14**, (2 and 3): 184-191.

Bala, S.C. and Khan, M. R. 2002. Occurrence of *Hirschmanniella* species associated with rice in West Bengal and morphometrics of *H. mucronata.* In: *Proceedings of National Seminar on Integrated Pest Management in the Current Century,* 29-30 November 2002 (Eds.: S.K. Mondal, S.S. Ghatak, M. Panda and M.R. Khan) at Department of Agril. Entomology, Bidhan Chandra Krishi Viswavidyalaya, Mohanpur. pp. 310-315.

Baqri, Q. H. 1999. Diversity in Plant and Soil Nematodes of West Bengal (India): An Overview. *Proceedings of the Indian National Science Academy,* **B65** Nos 1 and 2: 1-14.

Baqri, Q. H. and Ahmad, N. 2000. Nematodes from West Bengal (India) XXV. Qualitative and quantitative studies of plant and soil inhabiting nematodes associated with Paddy crop in Malda and Jalpaiguri districts. *Records of the Zoological Survey of India,* **98** (Part-2): 81-91.

Baqri, Q. H., Ahmad, N. and Dey, S. 1991. Nematodes from West Bengal (India) XXIV. Qualitative and quantitative studies of plant and soil inhabiting nematodes associated with Paddy crop in Coochbehar district. *Records of the Zoological Survey of India,* **88** (1): 63-69.

Baqri, Q. H. and Das, P. K. 1990. Nematodes from West Bengal (India) XIX. Qualitative and quantitative studies of plant and soil inhabiting nematodes associated with Paddy crop in West Dinajpur district. *Records of the Zoological Survey of India,* **86** (2): 187-192.

Baqri, Q. H. and Dey, S. 1991. Nematodes from West Bengal (India) XXIII. Qualitative and quantitative studies of plant and soil inhabiting nematodes associated with Paddy crop in district Darjeeling. *Records of the Zoological Survey of India,* **87** (1): 77-81.

Baqri, Q. H. and Jana, A. 1981. On the species of *Aporcelaimellus* Heyns, 1965 from West Bengal (Aporcelaimidae: Nematoda). *Bulletin of the Zoological Survey of India,* **3**(3): 221-225.

Baqri, Q. H. and Jana, A. 1982. Nematodes from West Bengal, (India) XIII. Four new species of Dorylaimidae with a key to the species of *Laimydorus* Siddiqi, 1969 (Dorylaimoidea). *Nematologica,* **28** : 192-205.

Baqri, Q. H and Jana, A. 1986. Nematodes from West Bengal (India) IX. Three new species of the super family Dorylaimoidea (de Man, 1876) Thorne, 1934. *Indian Journal of Helminthology (n.s.)*, **3**(2): 11 -20.

Baqri, Q. H., Jana A., Ahmad, N. and Das, P. K. 1983. Nematodes from West Bengal (India) VIII. Qualitative and quantitative studies of plant and soil inhabiting nematodes associated with Paddy crop in Burdwan district. *Records of the Zoological Survey of India*, **80**: 331-340.

Baqri, Q. H. and Khera, S. 1979. Nematodea from West Bengal (India) IV. Three known and two new species of the Genus *Dorylaimoides* Thorne and Swanger, 1936 (Leptonchidae: Dorylaimida). *Records of the Zoological Survey of India*, **75**: 247-254.

Chakraborti, H.S., Nayak, D.K. and Pal, A. 1985. Ufra incidence in summer rice in West Bengal. *International Rice Research Newsletter*, **10**: 15-16.

Chatterjee, A. and Gantait, V.V. 1999. Plant parasitic nematodes (Tylenchida: Nematoda). *State Fauna Series 3, Fauna of West Bengal*, **Part-11**, Zoological Survey of India: 297-339

Dey, S. and Baqri, Q. H 1986. Nematodes from West Bengal (India) XXI. Five new species of Dorylaimoidea (Dorylaimida: Nematoda). *Indian Journal of Helminthology (n. s.)*, **3**(2): 43-58.

Ghosh, S. C. and Manna, B. 2008. Studies on Nematode parasites associated with paddy crop of West Bengal, India. *Records of the Zoological Survey of India*, Occasional Paper No. **287**: 144 pp.

http://www.fao.org

http://www.ricestat.irri.org:8080/wrs2/entrypoint.htm

Jain, R. K., Khan, M. R. and Kumar, V. 2012. Rice root-knot nematode (*Meloidogyne graminicola*) infestation in rice. *Archives of Phytopathology and Plant Protection*, **45** No. 6: 635-645.

Jana, A. and Baqri, Q. H. 1985. Nematodes from west Bengal (India) XV. In the species of some rare Genera having narrow odontostyle of the Superfamily Dorylaimoidea (Dorylaimida). *Bulletin of the Zoological Survey of India*, **7** (2-3): 297-304.

Khan, M.R. 2001a. White-tip nematode, *Aphelenchoides besseyi* in rice-tuberose cropping system. In: *Proceedings of National Seminar on Frontiers of Crop Management*. 1-3 February 2001 at Visva-Bharati, Sriniketan. pp. 142-144

Khan, M.R. 2005a. Community analyses of plant and soil nematodes in jute- rice cropping system in West Bengal, India. *Pakistan Journal of Nematology*, **23**: 181-186.

Khan, M. R. 2011. Review Article: Nematode pests of crops in West Bengal, India. *Indian Journal of Social and Natural Sciences*, **1** (1): 1-16.

Khan, M. R., Handoo, Z. A., Rao, U., Rao, S. B. and Prasad, J. S. 2012. Observations on the Foliar Nematode, *Aphelenchoides besseyi*, infecting Tuberose and Rice in India. *Journal of Nematology*, **44** (4): 391-398.

Mukherjee, B. and Dasgupta, M. K. 1981b. Soil and plant nematodes of West Bengal, India. *Indian Journal of Nematology*, **9**: 72.

Mukhopadhyay, M. C. and Haque, M. 1974a. Nematodes associated with field crops, fruits and fodder crops in West Bengal. *Indian Journal of Nematology*, **4**: 104-107.

Mukhopadhyay, M. C. and Haque, M. 1974b. Nematodes associated with field crops, fruit trees and fodder crops in West Bengal. *Proceedings of the Indian Science Congress Association*, **61**: 110.

Rao, Y.S., Prasad, J.S. and Panwar, M. S. 1986b. Stem nematode (*Ditylenchus angustus*): a potential pest of rice in Assam and West Bengal, India. *International Nematology Network Newsletter*, **3**(4): 24-26.

Sandhu, J. S. 2014. Status Paper on Rice. *Directorate of Rice Development, Govt. of India, ministry of agriculture, Patna (Bihar)*, 151 pp.

Sen, D., Gantait, V. V. and Sanyal, A. K. 2012. Free living Nematodes (Order Dorylaimida and Tylenchida) of West Bengal, India. *Records of the Zoological Survey of India*, **112** (Part-1): 21-31.

Chapter 19

Observations on Feeding and Breeding Biology of Red-Vented Bulbul *Pycnonotus cafer* (Linnaeus 1766) in Jammu, J&K State, India

B.L. Kaul and A.K. Verma

ABSTRACT

Red-vented bulbul is a common resident bird of Jammu division of the J&K State. The present study deals with the feeding and breeding biology of the bird. It feeds mainly on fruits, petals of flowers, nectar, insects and occasionally small house lizards. It breeds between February and June in and around human habitations in Jammu.

Keywords: Habits, Behaviour, Call, Eggs, Incubation, Hatchling, Nestling, Fledgling and Juvenile.

Introduction

Red-vented bulbul is about 20 cm in length being smaller than a Myna. It is a perky smoke-brown bird with partially crested black head scale-like makings on breast and back, a conspicuous crimson patch below root of tail, and a white rump. Sexes are alike (Plates 19.1 and 19.2). There are seven races (or subspecies). The one under observation is *intermedias* found in North-Western Himalayas (from Jammu and Kashmir to Kumauni). This bird is common in gardens, dry scrub, open forest and cultivated lands.

Plate 19.1: A Breeding Pair of *Pycnonotus cafer* in Courtship.

Plate 19.2: A Single Parent on Watch Duty.

The area of study is in the Jammu region which is spread over the middle Himalayas (Pir Panjal and upper Chenab region, the Shivalik region and the outer plains). The outer plains include the parts of Kathua, Samba and Jammu district. The outer plains and the outer hills are grouped into tropical and subtropical climate regimes and the chief towns are Jammu, Samba, Hirnagar, Kathua, Billawar, Basohli and Akhnoor. These towns experience intense tropical heat in summer. Most of the rainfall comes from July to September. The annual rainfall is almost 150 mm. The winter reason is mild and extends from December to February. Red-Vented Bulbul is a resident bird.

Materials and Methods

The study has been carried out in Jammu district (32.73° N and 74.87° E) to ascertain whether this bird is facing any threat to its existence. The Jammu distinct covers an area of 3097 sq km and lies at a height of 340–410 m from mean sea level (MSL).

For studying the feeding and breeding biology attention has been focused on feeding and foraging habits, breeding season, nest construction, clutch size, egg laying pattern, and incubation period, hatching and feeding the young ones. Incubation period was calculated from the date of laying of the last egg of the clutch to the hatching of that egg. In the same way, nesting period has been recorded from the date of hatching of that egg to hatching out of the last chick up to its fledging time.

Observations

Call

The Red-Vented Bulbul is rather vocal bird and has distinct calls for acts such as greeting and begging as well as two distinctive alarm calls. The bird frequently calls during pair bonding and in mating rituals. According to Salim Ali (1996) "Red-Vented Bulbul has no song as such, but its pugnacity makes it a favourite with fanciers as a fighting bird, and large stakes are wagered on bulbul fights."

Feeding

Red-Vented Bulbul forages in both pairs and large flocks, It eats mostly soft fruits such as bananas, barriers, figs, seeds vegetables and nectar of flowers. It is also not uncommon to find it feasting on insects such as termite swarms and even small reptiles like geckos. The Red-Vented bulbul tends to swallow fruits whole and these seeds travel through its gut completely intact, making it an important agent for dispersing plant seeds.

Breeding

The breeding season of Red-Vented bulbul starts from February and extends up to June. The nest is a shallow cup about 9.5 cm in diameter made from twigs, roots, and other materials such as metal wire and cobwebs, in a bush or tree, 6 to 20 ft above ground. It also builds nests in houses at safe and undisturbed locations.

3 to 4 pale-pinkish eggs with dark red spots are laid in a clutch (Plate 19.3). Two to three broods are raised by a pair, both sharing parental duties in turns.

Plate 19.3: A Live Nest with a Clutch of Eggs.

The eggs start hatching after 14 days. The hatchlings are reddish with yellow beak (Plate 19.4). They are blind. The nestlings and fledglings have black or white

Plate 19.4: A Nest with Hatchlings.

Plate 19.5: Nestlings Begging for Food.

Plate 19.6: A Fledgling.

plumage (Plate 19.5). The fledglings start to move out of the nest for short durations and return (Plate 19.6). The Juveniles look like their parents and leave the nest 3 week after hatching.

The chicks are fed by both parents on insects and on feeding trips wait for the young to excrete, swallowing the faecal sacs produced. Dust storms, heavy rains and predators are the main causes of chick mortality in scrub habitats.

Discussion and Conclusion

The aggressive nature and fruits eating habits of the red-vented bulbul, have unfortunately resulted in it being considered one of the world's worst invasive alien species in areas where it has been introduced

Although not territorial red-vented bulbul, is often considered an aggressive bird. It displaces other birds from their territories and competes directly for food. Thus, in areas where the Red-Vented bulbul has been introduced, this species can have very negative effects on local birds. In addition, the abundance of the Red-Vented bulbul in agricultural areas and gardens, where it destroys flowers, fruits and vegetable and may help spread the seeds of invasive plants has resulted in its reputation as a pest. In fact, the Red-Vented bulbul is now considered to be in the top 100 of the world's invasive alien species. Despite this negative image red-vented bulbul plays positive role in dispersing plant seeds.

Regarding its breeding biology we have observed that it breeds from February to June. It is also reported that red-vented bulbul may breed in some areas round the year but we have not come across any such case in Jammu district. Our study shows that it has up to three broods within a year, with each brood typically containing three to four eggs.

Although there is always danger to the eggs and chicks from the predators and natural causes like unseasonal rains and dust storms etc yet the brood size and breeding over a period of five months goes in its favour to maintain balance of population. Its aggressive nature also intimidates the predators and unwanted neighbors. As of now we do not see any threat to its existence in the Jammu region. It is classified as least concern (LC) on the IUCN Red list.

References

Ali, Salim and Sidney Dillon Ripley, 1998. *Handbook of Birds of India and Pakistan*, Vol. 9. Oxford University Press, New Delhi.

Ali, Salim, 1996. *The Book of Indian Birds* (Revised Edition). Bombay Natural History Society.

Bhatt, D. and Kumar, A., 2001. Foraging ecology of red-vented bulbul *Pycnonotus cafer* in Haridwar, India. *Forktail*, 17: 109–110.

Bird Life International, 2010. *Red-Vented Bulbul* (November).

Brochier, B., Vangeluwe, D. and van den Berg, T., 2010. Alien invasive birds. *Revue Scientifique et Technique de l'OIE*, 29(2): 217–226.

Global Invasive Species Database, 2010. *Red-Vented Bulbul* (November). http://www.issg.org/database/species/ecology.asp?fr= 1 and si= 138 and sts

Goldin, M.R., 2002. *Field Guide to the Samoan Archipelago*. Bess Press Inc., Honolulu, Hawaii.

Grewal, B., Harve, B. and Pfister, O., 2002. *A Photographic Guide to the Birds of India and the Indian Subcontinent*, Princeton University Press, New Jersey, USA.

Harrison, J., 2011. *A Field Guide to the Birds of Sri Lanka*. Oxford University Press, Oxford.

Islam, K. and Williams, R.N., 2000. Re-Vented Bulbul (*Pycnonotus cafer*). In: *The Birds of North America Online*, (Ed.) A. Poole. Cornell Lab of Ornithology, Ithaca. Available at: http://bna.birds.comell.edu/bna/species/520a

IUCN Red List (November, 2010). http://www.iucnredlist.org/

Jerdon, T.C., 1963. *The Birds of India*, Volume 2 Part 1. Military Orphan Press, Calcutta.

Kumar, A., 2004. Acoustic communication in the red-vented bulbul *Pycnonotus cafer*. *Anais da Academia Brasileira de Ciencias*, 76(2): 350–358.

Previous Volumes

— *Volume 1* —

1996, xvi+332p., figs., pls., tab., 25 cm

ISBN 81-7035-156-1

Section I: Fish and Limnology

1. **Melanophore Occurrence in Early Life History Stages (Periodization) of Mahseer, *Tor tor* (Hamilton) and its Role in Identification of the Larvae Inhabiting Jammu Waters of J & K State**

    Y.R. Malhotra & Subash Chander Gupta

2. **Seasonal Variations in Biochemical Composition in Some Freshwater Fishes, Part III, *Channa punctatus* (Bloch)**

    B.N. Pandey, Anupa Sharan, Rumana Perween & M. Kumar

3. **Some Relationships Between Size Structure and Fertility of Rotifer Populations**

    S.S.S. Sarma

4. **Effect of Accessory Pneumectomy on Some Haematological Values of Air Breathing Fish *Clarias batrachus***

    B.D. Joshi

5. Food and Feeding Habits of *Heteropneustes fossils* (Bloch) Inhabiting Gadigarh Stream, Jammu

   S.P.S. Dutta

6. Observations on the Use of Ovaprim for Induced Spawning of Indian Major Carps

   P.K. Roy

7. Aquatic Odonata and Hemiptera of Jammu and their Role in Aquaculture

   Baldev Sharma, Neeru Dhalla & Renu Salaria

8. A Comparative Study of the Renal Organs of Freshwater Teleostean Fish, Part I: Morphology

   B.L. Kaul

9. A Comparative Study of the Renal Organs of Freshwater Teleostean Fish, Part II: Histology

   B.L. Kaul

10. Benthic: Macroinvertebrates as Indicators of Aquatic Environment

    Usha Moza

11. Effect of Mechanical Stress on Early Embryonic Stages of *Tor tor*

    Kuldeep K. Sharma

12. Icthyofauna of the Sector of Kaveri River in Head Region

    M.N. Madyastha and S. Murugan

13. Biology of Indian Belone, *Xenentodon cancila* (Hamilton): A Freshwater Fish from Jammu Waters of J & K State I: Periodization in Life History of *Xenentodon cancila* (Hamilton)

    Subash Chander Gupta & Kuldeep K. Sharma

14. Rotifer Fauna of Devikoppa Tank: Dharwad (Karnataka, India)

    K. Vijay Kumar

15. Terminology of Various Developmental Stages of Fish Larvae Inhabiting Jammu Freshwaters

    Subhash Chander Gupta & Arun Kumar Gupta

16. Inter-Specific Competition in Mixed Culture of Cladocera

    Y.R. Malhotra & Seema Langer

17. Limnology of Farooq Nagar Pond, Jammu Part II: Rotifera

    S.P.S. Dutta & Jyoti Sharma

18. **Relative Population Abundance of Ichthyo-fauna of Lake Mansar**
    Arun K. Gupta, Anil Khajuria, S.C. Gupta & Seema

19. **On the Distribution and Ecology of Some Gastropod Molluscs of the Jammu Province in J & K State**
    Anil K. Verma & P.L. Duda

20. **Macrobenthic Fauna in Relation to Some Environmental Factors in Eutrophicating Lake Mansar, Jammu**
    K. Gupta & Anil Khajuria

21. **Population Structure and Seasonal Succession of Zooplankton of Lake Surinsar, Jammu (India)**
    M.K. Jyoti & H.S. Sehgal

Section II: Wildlife

22. **On the Habitat and Behaviour, Maturity, Size and Sexual Dimorphism in a Population of Freshwater Emydid Turtles of Jammu, J & K State**
    P.L. Duda, Anil K. Verma & D.N. Sahi

23. **Status and Management of Wildlife in Jammu & Kashmir State**
    B.L. Kaul & Indu Kanwal

24. **Eco-geographical Distribution and Present Status of Herptiles in Kashmir Himalayas**
    D.N. Sahi & P.L. Duda

25. **Shrinking Wetlands of India**
    S.K. Chadha

26. **Ecology and Status of Wildlife in Ladakh**
    B.L. Kaul

27. **On the Freshwater Chelonian Fauna of Jammu and Kashmir**
    D.N. Sahi & Anil K. Verma

28. **Threatened Wildlife Habitat in Kashmir Himalayas**
    B.L. Kaul

29. **Role of Gastropods in Trematode Transmission Among Herptiles - Part I: Amphibia A Numerical Analysis**
    Anil K. Verma & P.L. Duda

— Volume 2 —

1999, xvi+231p., fig., plts., 25 cm

ISBN 81-7035-205-3

Section I: Fish and Limnology

1. Role of Thyroid Gland in the Regulation of Metabolic Rate in Fishes with Special Reference to Indian Teleosts

    B.N. Pandey

2. Alcohol Dehydrogenase Isozyme Expression in the Air-breathing Fish, *Clarias batrachus* and *Heteropneustes fossilis* of North Eastern India

    Alka Prakash & Sant Prakash

3. The Ecological Role of Algal Weeds, Charophytes in Particular in Fisheries Water

    Usha Moza

4. Importance of Fish Food Organisms (Live Food) in Aquaculture Practice

    Seem Langer, K. Gupta & R. Gandotra

5. Morphological Studies of Alimentary Canal of Fishes of Lake Mansar

    Arunk K. Gupta, Seema Langer & S.C. Gupta

6. Transgenic Fish: Production and Improvement of Fish Resources

    Anil K. Verma & B.L. Kaul

7. Sewage Fed Fisheries: A Biotechnological Application

    Y.R. Malhotra, Seema Langer & S. Raina

8. The Histopathology of *Pallisentis jagani* and *Pomphorhynchus bulbocolli* Infection in *Channa striatus* and *Schizothorax sinuatus*

    P.L. Kaul & M.K. Rana

9. Female Reproductive System of *Pallisentis jagani*

    P.L. Kaul, M.K. Raina & Usha Zutshi

10. Bacterial Microflora: Their Distribution and Relationship with Fish and Its Environment: A Review

    J.P. Sharma & V.K. Gupta

11. A Comparison of the Feeding Rates of *Streptocephalus torvicornis* and *Chirocephalus diaphanus* (Crustacea : Anostraca) on Rotifers

    S.S.S. Sarma and K.R. Dierckens

12. Population Growth of *Brachionus calyciflorus* Pallas (Rotifera) in Relation to Algal (*Dictyosphaerium chlorelloides*) Density

    S.S.S. Sarma, E.D. Fiogbe & P. Kestemont

13. Ecological Crisis in Lake Mansar Jammu, J & K State

    B.L. Kaul & Anil K. Verma

14. Zooplankton Composition, Abundance and Dynamics in a Lentic Habitat (Kalika Pond, Dhar, M.P.)

    R.K. Dave, M.M. Prakash & N.K. Dhakad

15. Impact of Nutrient Influx on Water Quality Trends of a Vindhyan Lake

    S. Pani & A. Wanganeo

16. Seasonal Variations in Biochemical Composition of Muscle During the Annual Ovarian Cycle of Female *Channa gachua* (Ham.)

    K. Gupta, Sujata Raina, R. Gandotra & S. Langer

17. Effect of Dietary Testosterone Propionate (TP) on the Growth of Common Carp, *Cyprinus carpio* L.

    Y.R. Malhotra, R. Gandotra & K. Gupta

Section II: Wildlife

18. The Common Barn Owl, *Tyto alba stertens* Hartert, 1929: An Effective Bio-Control Agent of Rodent Pests

    P. Neelanarayanan, R. Nagarajan & P. Kanakasabi

19. Morphology of the Male Reproductive Organs in the Indian Saw Back Turtle, Kachuga tecta and Brown Roofed Turtle *Kachuga smithii* from J & K State

    Anil K. Verma, D.N. Sahi & P.L. Duda

20. Preliminary Observations on the Ecology of the Freshwater Soft-Shell Turtles (Family : Trionychidae) of J & K State

    D.N. Sahi, P.L. Duda & Anil K. Verma

21. Impact of Anthropogenic Activities on the Aquatic Birds Population at Bahadur Sagar (Jhabua, M.P.)

    M.M. Prakash & D. Shinde

22. A New Species of Loxogenus (Digenia : Lecithodendriidae) from Rana Cyanophlylctis in Jammu

   P.L. Duda, B.R. Pandoh & A.K. Verma

23. Ecological Notes on the Freshwater and Hard-Shelled Turtles (Family : Emydidae) of Jammu and Kashmir State, India

   P.L. Duda, Anil K. Verma & D.N. Sahi

24. Notes on the Habitat Ecology and Barriers to Dispersal of Some Gastropod Molluscs of J & K State

   P.L. Duda, Anil K. Verma & P.S. Pathania

25. Reprotechnology in Wildlife Conservation

   R.K. Sharma & Manju Sharma

26. Seasonal Variations in Ovarian Weight and the Gonadosomatic Index in the Wall Lizard Hemidactylus Flavivirdis Rupell (Sauria : Gekkonidae) in Jammu

   Bhavana Abrol, Deep N. Sahi, P.L. Duda & Anil K. Verma

27. Impact of Tourism and Development on Biodiversity in Patnitop (J & K State)

   A.K. Parimoo & B.L. Kaul

28. The Guindy National Park : Its History and Physiogeography

   R.K. Menon

— *Volume 3* —

2004, xix+317p., figs., tabls., ind., 25 cm

ISBN 81-7035-327-0

Section I: Fish and Limnology

1. Fish Germplasm Resource Conservation with Special Reference to the Indian Fauna

   A.K. Pandey, S.A. Suryavanshi and P. Das

2. Studies on the Functional Responses of the Mexican Live Bearer Fish *Allotoca meeki* (Godeidae: Cyprinodontiformes)

   Victor M. Peredo-Alvarez, S.S.S. Sarma and S. Nandini

3. A Report on Teratology of *Labeo rohita* (Hamilton): An Important Freshwater Food Fish Inhabiting Lentic and Lotic Environments of Jammu (J&K State)

   Subash C. Gupta, S.P.S. Dutta, N. Sharma and N. Bala

4. **Ecology and Reproductive Plasticity of the Amazonian Cichlid Fishes Introduced to the Freshwater Ecosystems of the Semi-Arid North-Eastern Brazil**

   S. Chellappa and N.T. Chellappa

5. **Influence of *Salvinia* Based Feed on the Growth of *Cyprinus carpio* (L) Fry**

   D.S. Somashekar, M. Venkateshwarlu and M. Shanmugam

6. **Status of Fishes in Protected Areas of Kerala, South India**

   C.R. Biju, K. Raju Thomas and M. John George

7. **Enhancing Recuperation Capacity of River Betwa with the Help of Macrophytes**

   N. Shukla and A. Wanganeo

8. **Competition Between *Ceriodaphnia dubia* and *Moina macrocopa*: A Population Growth Study**

   S.S.S. Sarma and S. Nandini

9. **Diel Variations in Physico-chemical Parameters and Zooplankton in High Altitude Ponds of Bhaderwah (A Himalayan Valley), Jammu and Kashmir, India**

   N. Kumar, Y.R. Malhotra and S.P.S. Dutta

10. **Aquatic Pollution Management by Bio-Remediation Technique: An Overview**

    A. Wanganeo, S. Jagdish, S. Raghuvanshi and R. Wanganeo

11. **Histomorphology of Testes of Some Schizothoracine Fishes (Cyprinidae: Cypriniformes) from Indus River System**

    B.L. Kaul

12. **On Three New Species of *Acanthosentis* Verma et. Datta 1929, from Fishes of Kashmir, India**

    P.L. Kaul and M.K. Raina

13. **Ecological Spectrum, Seasonal and Sex-Wise Distribution, Prevalence, Parasitic Load and Abundance of Various Parasites (Trematoda: Digenea) Infecting Fishes of Jammu Province**

    B.R. Pandoh

14. **Fisheries Sector: A Gendered Perspective**

    Arpita Sharma

Section II: Wildlife

15. Conservation and Future of Saltwater Crocodile, (*Crocodylus porosus* Schneider) in Bhitarkanika, Odisha, India

    S.K. Kar

16. Apoptosis During Follicular Atresia in Mammals

    R.K. Sharma and Neeru

17. Seasonal Fluctuations in Liposomatic and Hepatosomatic Indices of the Common Wall Lizard *Hemidactylus flaviviridis* Ruppell (Reptilia: Gekkonidae) in Jammu

    Bhavna Abrol, D.N. Sahi and Anil K. Verma

18. Habitat Utilisation of Large Herbivores in the Chinnar Wildlife Sanctuary, Kerala

    E.A. Jayson

19. Some Observations on the Population Fluctuation of *Paradistomoides gregarinum* (Tubangui, 1929) in Some Lizards (Reptilia: Sauria) from Jammu Province

    Anil K. Verma and P.L. Duda

20. Ecology of Freshwater Snails (Gastropod Molluscs) in Lake Mansar, Jammu

    K. Gupta and Anil Khajuria

21. Effect of Different Varieties of Mulberry (*Morus* spp.) on the Survival, Growth and Pupation in *Spodoptera litura* (Fabr.) (Lepidoptera: Noctuidae) Larvae

    J.S. Tara and Baldev Sharma

22. Morphometrical Observations on Atresia in *Hemidactylus flaviviridis* Ruppell (Reptilia: Gekkonidae) in Jammu

    Bhavna Abrol and D.N. Sahi

23. Comparison of Changes in Two Wetland Habitats in Salim Ali Bird Sanctuary, Kerala

    K. Seema, R. Sugathan and John George M.

— Volume 4 —

2007, xxii+259p., 12 col. plts., figs., tabls., ind., 25 cm

ISBN 81-7035-517-6

Section I: Fish and Limnology

1. Reproductive Aggression and the Dynamics of Territorial Disputes Between Males of Red Hybrid Tilapia (Osteichthyes : Cichlidae)
   S. Chellapa, A.P.T. Medeiros and N.T. Chellapa

2. Fecundity of *Schizothorax esocinus* (Heckel)
   B.L. Kaul

3. Histochemical Localization of DNA in the Gastrointestinal Tract and Liver of *Nemacheilus kashmirensis*
   Ashok Channa and N.A. Lone

4. Testicular Development of Male Amazonian Red Discus, *Symphysodon discus* Heckel (Perciformes : Cichlidae)
   S. Chellapa, M.R. Câmara and N.T. Chellapa

5. Ichthyofauna of Periyar River (South India) with Special Reference to Distribution of Hill Stream, Endemic and Endangered Species
   K. Raju Thomas, M. John George, C.R. Ajithkumar and C.R. Biju

6. On Genus *Hebesoma* Van Cleave, 1928: A New Species *H. guptai* from the Indian Major Carp *Labeo rohita*
   P.L. Koul and M.K. Raina

7. On Genus *Neoechinorhynchus*, Hamann, 1892 and a New Species *N. ladakhensis* from a Trans-Himalayan Fish *Ptychobarbus*
   P.L. Koul and M.K. Raina

8. Seasonal Variations of Zooplankton from a Drinking Water Reservoir (Valle de Bravo) in Mexico
   S. Nandini, S.S.S. Sarma and Pedro Ramerez-Garcia

9. Seasonal Variations in the Fish Food Organisms (Phytoplankton) from High Altitude Lentic Water Bodies of Bhaderwah (J&K State)
   Neeraj Kumar

10. Conservation and Management of Ecologically Degraded Wular Lake
    Ulfat Jan, G. Mustafa Shah and Bashir A. Mir

11. Sewage Utilization in Agriculture and Pisciculture: An Overview
   A.K. Verma and Surinder Sharma

12. Plankton Ecology of a Paddy Field at Maralia Morh, Miran Sahib, Jammu
   S.P.S. Dutta, S. Kour and M. Khajuria

Section II: Wildlife

13. A Study of Reproductive Biology of Little Bittern, *Ixobrychus minutus minutus* (Linnaeus) in Kashmir
   G. Mustafa Shah, Ulfat Jan, M.F. Fazili and M.K. Raina

14. Large Mammals in the Southern-Western Ghats: A Case Study from the Peppara Wildlife Sanctuary, Kerala
   E.A. Jayson and G. Christopher

15. Management of Wetlands of Kashmir for Waterfowl Populations
   G. Mustafa Shah, Ulfat Jan and Fayaz Ahmad Ahanger

16. Effects of Sublethal Administration of Orthene and Diazinon on Serum Calcium and Inorganic Phosphate Levels, Parathyroid Gland and Calcitonin-Producing C-Cells of *Rattus norvegicus*
   Shaheda P. Rangoonwala, S.A. Suryawanshi and A.K. Pandey

17. Seasonal Variations in Frequency of Atresia in Previtellogenic Follicles in *Hemidactylus flaviviridis* Ruppell in Jammu
   Bhavna Abrol, Deep N. Sahi and A.K. Verma

18. Issues in Patenting of Life Forms
   Sunitha Ninan and Arpita Sharma

19. On Some Aspects of Bio-ecology of the Rhesus Monkey, *Macaca mulatta*
   R.K. Verma and Suman Kapoor

20. A Note on the Unreliability of Pygoferal Spines in Genus *Exitianus* Ball (Homoptera : Cicadellidae)
   Upasna Sharma

Section III: Helminth Parasites of Man and Domestic Animals

21. Gastrointestinal Nematode Parasites of Small Ruminants of Kashmir Valley with Special Reference to Breed, Age and Seasonal Distribution
   Hidayatullah, M.Z. Chishti and Fayaz Ahmed

22. **Epidemiology of Gastrointestinal Helminths in the School Children of District Budgam, Kashmir**

    Showkat A. Wani, Fayaz Ahmed, Showkat A. Zarger, M.Z. Chishti, Hidayat Tak and Zubair Ahmad

23. **Prevalence of Gastrointestinal Helminths in the Primary School Children of District Anantnag, Kashmir and their Effect on the Haemoglobin Status**

    Showkat A. Wani, Fayaz Ahmed, Showkat A. Zarger, M.Z. Chishti, Hidayat Tak and Zubair Ahmad

24. **Microfilarial Periodicity: A Review of the Mechanism**

    Abdul Baqui

— Volume 5 —

2011, xxi+294p., col. plts., figs., tabls., ind., 25 cm

ISBN 978-81-7035-718-6

Section I: Fish and Limnology

1. **Reproductive Biology of *Scomberomorus brasiliensis* (Perciformes: Scombridae)**

    S. Chellappa, J.T.A. Ximenes-Lima, A. Araújo and N.T. Chellappa

2. **Anomalies in *Cirrhinus mrigala* (Ham. Buch.), *Catla catla* (Ham. Buch.) and *Labeo rohita* (Ham. Buch.) Inhabiting Freshwater Environments of Jammu (J&K)**

    Subash C. Gupta and Touseef A. Zargar

3. **Histopathological Alterations in Interrenal and Chromaffin Cells of *Channa punctatus* (Bloch) Exposed to Sublethal Concentration of Carbaryl and Cartap**

    D.K. Mishra, K. Bohidar and A.K. Pandey

4. **Art of Freshwater Prawn, *Macrobrachium rosenbergii*, Culture in Saline Water**

    P.K. Roy

5. **Studies on the Neuroendocrine Regulation of Egg Maturation in the Giant Freshwater Prawn, *Macrobrachium rosenbergii* (de Man)**

    A.K. Pandey and Anjani Kumar

6. **Freshwater Fish Diversity of Tunga and Bhadra Rivers, Western Ghats, Karnataka**

    A. Shahnawaz and M. Venkateshwarlu

7. Population Growth and Demography of *Moinodaphnia macleayi* (King, 1853) (Crustacea : Cladocera) in Relation to Algal (*Chlorella vulgaris* or *Scenedesmus acutus*) Food Density

    S. Nandini and S.S.S. Sarma

8. Bloom Events of Toxin Producing Cyanobacterial Species Associated with Fish Kills in a Tropical Reservoir of Brazil

    N.T. Chellappa and Sathyabama Chellappa

9. Bhoj Wetland: A Review

    Ashwani Wanganeo and Rajni Wanganeo

Secton II: Wildlife

10. Structure and Composition of Birds in the New Amarambalam Tropical Forests of Kerala, Southern Western Ghats, India

    E.A. Jayson

11. An Unusual Breeding Case of the House Sparrow (*Passer domesticus*) at Jammu

    B.L. Kaul and Anil K. Verma

12. Avifauna of Jammu and Kashmir

    O.P. Sharma

13. Responses of Thyroid, Parathyroid, Calcitonin Producing C Cells and Adrenal Cortex of *Rattus norvegicus* to Sublethal Heroin Administration

    S.R. Barai, S.A. Suryawanshi and A.K. Pandey

14. Biodiversity Resources of Bhitarkanika Mangrove Ecosystem: Significance and Threat

    Sudhakar Kar

15. Granulosa: Oocyte Interactions during Folliculogenesis in Mammals

    R.K. Sharma and M.B. Sharma

16. Alterations in Certain Enzyme Activities in Muscles and Nerve Cord of *Periplanata americana* Induced by Sublethal Administration of Cypermethrin, Carbaryl and Monocrotophos

    B.G. Kulkarni, Fatuma A. Mohammed and A.K. Pandey

17. Avifauna of Mansar Wetland (J&K) with Emphasis on their Feeding Ecology

    Deepti Kotwal, Sanjeev Kumar and D.N. Sahi

18. **Penguins: The Amazing Creatures**
    Anil K. Verma and B.L. Kaul

19. **Importance of New Systematics in Parasitological Taxonomy**
    P.L. Koul and M.K. Raina

20. **Insect Pests Associated with the Medicinal Plants in Shivalik Region of Jammu, J&K State**
    Madhu Sudan

— *Volume 6* —

2015, xvi+179p., col. plts., figs., tabls., ind., 25 cm

ISBN 9789351243342

Section I: Fish and Limnology

1. **Influence of aquatic macrophytes on the phytoplankton structure and fish communities in a shallow tropical reservoir**
   N.T. Chellappa, R.K. Oliveira, E.K.R. Pessoa and S. Chellappa

2. **Comparative aspect of seed production of *Macrobrachium gangeticum* (Bate) in natural seawater and brine solution**
   Prasanti Mishra, D. R. Kananjia, K. Bohidar and A. K. Pandey

3. **Inland fisheries in rainfed areas of Jammu (J&K State), India**
   B. L. Kaul and Rajesh Dogra

4. **Water quality assessment of Santhekadur waterbody, Shimoga, Karnataka (India) with relation to its Zooplankton diversity**
   M. Venkateshwarlu, Shahnawaz Ahmed and K. Honneshappa

5. **Limnology of experimental macrophyte floating islands in a high altitude reservoir (Valle de Bravo, Mexico)**
   Pedro Ramirez-Garcia, S. Nandini, S.S.S. Sarma, Martha L. Gayten-Herrara and Victor M. Almeida

6. **Effect of different nutrient media on the relative growth, brood size and population dynamics of *Moina macrocopa***
   Satinder Kour

7. **Fish diversity in a lotic environment, Munawar Tawi of Rajouri, J&K State**
   A.K. Verma, Raheela Mushtaq and Anuradha Gupta

8. Ornamental and food fish diversity of Tunga and Bhadra rivers, Western Ghats, Karnataka India.
    Shahnawaz Ahmed and M. Venkateshwarlu

9. On the Diel fluctuations of some physico-chemical parameters and tropic status of lake Mansar.
    Anil Khajuria

10. Zooplankton diversity of Sader Mouj reservoir Budgam District, Jammu and Kashmir
    Ishrat Bashir, Md. Yousuf, Shahnawaz Ahmad and Shashikanth Majgi

11. A review of the fish parasites of Rion Grande do Norte, Northeastern Brazil
    S. Challappa, E.T.S. Cavaleanti, E.F.S. Costa, G.S. Araujo, J.T.A Ximenes-Lima and N.T. Chellappa

12. On genus Myxidium Butschli 1882 and a new species *Myxidium cholai* from *Puntius chola* (Hamilton)
    Kanwar Narain, M.K. Raina and P.L. Koul

Secton II: Wildlife

13. Urban impacts on the water quality and the macrobenthic invertebrate fauna of river Tawi, J&K State.
    Raheela Mushtaq and Anil K. Verma

14. Feathers are ruffled by winds of change in Kashmir
    B.L. Kaul

15. Avian diversity of Eringole Sacred Grove in the Western ghats of Kerala
    E.A. Jayson

16. Observations on feeding and breeding biology of the Brown Rock Chat *Cercomela fusca* (Blyth) in Jammu, J&K State, India
    B.L. Kaul

17. Status and conservation of endangered Blackbucks, *Antilopa cervicapra* of Odisha, India.
    Sudhakar Kar

18. The breeding biology of Red Wattled Lapwing *Vanellus Indicus Indicus* (Boddaert)
    Vasudha Chaudhari

19. Wetland Avifauna of Kundavada Lake, Davanagere Distt., Karanataka
M.N. Harisha, B.B. Hosetti and Shahnawaz Ahmed

Index

A

Abedus 169
Abiotic Factors 170
Absolute fecundity 13, 96
Acutus (Pareumenes) Pareumenes quadrispinosus ssp. 241, 242
Aedes 176
Aerial ropeway 265
Age-specific reproductive output 122
Age specific survival 124
Agricultural productivity 25
Alderflies 110
Algal cell density 123
Algal density 130
Algal diet 127
Allelopathic interactions 195
Allometric relationship 141
Alluvial grasslands 274
Amphibious fish 139
Anabas testudineus 28
Andaman and Nicobar Islands 38, 136
Annamalai and Cardamom hills 27
Annelida 110
Anopheles larvae 176
ANOVA 276
Anthropogenic activities 78
Anthropogenic stresses 25, 29
Aphelenchida 324
Aquatic ecosystems 29
Aquatic habitat 114
Aquatic organisms 66, 196
Aquatic pollution 31
Arribada 268
Arthropoda 110
Artisanal fisheries 4
Atlantic cod 42
Audio-visuals 269

B

Bagarius bagarius 28, 38
Balantidium 287
Balitora brucei 38
Barilius vagra 37

Barking deer 263
Barrage 30
Batch fecundity 96
Bay of Bengal region 136
Bear 263
Belostoma 168
Belostomatid 167
Belostomatid bugs 168
Bengalensis (Polistes) pareumenes 235
Benthic diversity 136, 147
Benthic fauna 135, 149
Benthic faunal community 136
Benthic invertebrates 106
Benthic studies 144, 146
Bhitarkanika 139
Bhitarkanika National Park 261
Bhitarkanika sanctuary 266
Biodiversity status 25
Biological diversity 106
Biological reserves 40
Biomanipulation 195
Biosphere reserves 40, 41
Biotic factors 197
Birds 261
Birds of prey 228
Bison 263
Bisphenol A 66
Bithinidae 110
Bivalvia 110
Blackbuck habitat 269
Blue bull 263
Blue sheep 283
Botanical garden 265
Brachionus calyciflorus 205
Brackishwater 28
Brahmaputra 27
Brahmaputra system 28

Bray Curtis coefficient 109
Breeding ecology 225
Breeding season 331
Brevirostratus (Eumenes) pareumenes 235, 236, 237, 239, 242

C

Cabomba caroliniana 200
Cameroni (Eumenes) pareumenes fulvipennis 238
Carbohydrates 124
Carps 28
Carrying capacity 85
Catchment area treatment plan (CAT plan) 87
Catfishes 28
Cattle grazing 229
Ceratophyllum 199
Chambal 83
Chaudhria indica 38
Chemically-mediated interactions 196
Chenab region 331
Chilika wetland 268
Chironomidae 111
Chi-square contingency 277
Chi-square test (χ^2) 91
Chlorella 205
Chordata 110
Chromatin nucleolus 17
Chronic abuse 248
Chronic cocaine abuse 256
Ciliate 293
Circular hatchery 183
Cladocerans 207
Clarias batrachus 28
Clinical evidences 248
Clonal cultures 122
Clutch of eggs 332
Coastal waters 3

Index | 355

Cobitoid loach 38
Cochin mangroves 146
Coelomic cavity 11
Coldwaters fish biodiversity 27
Coleoptera 110, 114
Colour description 235
Common carp 33
Community structure 136
Condition factor 9
Conservation 25, 26
Conservation aquaculture 42
Conservation dependent (LR-cd) 34
Conservation programme 40
Conservation status 35
Conservation strategies 44
Contractile vacuole 303
Convention on biological diversity (CBD) 26, 273
Cooum river 173
Copepods 214
Correlation analysis 116
Cortical alveoli stage 96
Corvus splendens 228
Crabs 139, 141
Critically endangered (CR) 34, 38
Crocodile conservation 266
Crossocheilus diplochilus 27
Crustacea 110
Crustaceans 145
Crustacean zooplankton 129
Culex 175, 176
Cultivated lands 329
Culture 182
Cyanobacteria 210
Cyanobacterium 197
Cyanotoxins 210
Cytoarchitecture of testis 255

Cytogenetic abnormalities 32
Cytoplasmic vesicles 96

D

Damming 30
Damsel fly 110
Danio aequipinnatus 37
Data bank 45
Data deficient (DD) 34
Deccan plateau 27
Deforestation 30
Demography experiments 123
Detritus food chain 136
Diatom 197
Dilleria 300
Diplonychus 169
Diptera 110
Diptychus maculatus 27
Diseases 33
Diversity 112
Diversity analysis 112
Diversity index 145, 147
Domestic wastes 129
Dominance 112
Dopamine antagonists 185
Dorylaimida 306
Dragon fly 110

E

East coast river system 28
Echinoderms 140
Ecological niche 169, 261
Ecological separation 282
Ecological significance 211
Eco-tourism 270
Egg feeders 172
Egg hatching 123
Eggs and embryos 44

Elephant 263
Elephant conservation 267
Elephants 267
Endangered (EN) 34, 187
Endangered populations 44
Endocrine disruptors 63
Environmental factors 106
Environmental parameters 113
Ephemeroptera 114
Epizootic ulcerative disease syndrome (EUS) 33
Equilibrium strategists 4
Estuarine crocodile 266
Ethology 234
Exotic fishes 29
Exotic species 33
Experimental manipulation 170
Ex-situ conservation 40, 263
Extinct (EX) 34
Extinct in the Wild (EW) 34

F

FAO/UNDP 266
Farakka 30
Fecundity 13, 91, 96
Fed ad libitum 248
Filiform 11
Fish 1
Fish biodiversity 39
Fish diversity 35
Fishery biology 92
Fishery resources 31
Fish genetic diversity 29
Fish genetic resources 26, 40
Fish germplasm 25, 26, 27, 45
Fishing cat 263
Fish kairomones 198
Fish landings 41

Fish predation 167
Fish reproduction 3, 89
Flatworms 204
Flavobalteata (Montezumia) flavoleptus 236
Fledging time 331
Fledglings 332
Flora and fauna 85
Folliculogenesis 348
Food chain 63, 68
Food vacuole 303
Forage base 167
Foraging and anthropogenic factors 225
Foraging-cum-breeding ground 227
Foraging habits 331
Foraminifera 139, 148
Forest conservation act, 1980 261
Fox 263
Freshwater aquaculture 181
Freshwater catfish 187
Fulvipennis (Eumenes) pareumenes 238
Fulvipennis (Pterochilus) pareumenes brevirostratus 236
Fundulus heteroclitus 32

G

Gamete/embryo bank 43
Ganga 27
Ganga basin 80, 84
Garra gotyla gotyla 27
Garra lamta 27
Gastropoda 110
Gene bank 43, 45
Generation time 127, 130
Genetic management 44
Germplasm resources 30
Gharial breeding 265
Giant squirrel 263
Giant water bugs 168

Gila trout 32
Glyptosternum pectinopterus 27
GnRH-based drugs 185
Golden langur 274
Golden mahseer 33
Golgi 250
Gonadal maturity 9
Gonad development 10, 95
Gonadosomatic index (GSI) 18, 97
Gonoduct 11
Goral 276
Govindsagar reservoir 41
Greater Himalaya 27
Great Lakes of North America 32
Green alga 211
Grey mullet 42
Ground Cover 279
Growth and mortality 93
Gulf of Khambhat 144

H

Habitat alterations 30
Habitat degradation 187
Habitat isolation 115
Habitat modifications 30
Habitats preferences 226
Habitat types 278
Habitat use 171
Haemocytometer 123
Hanuman langur 263
Haridwar city 84
Hatcheries 181
Hatching experiments 183
Hatching systems 182
Hatchlings 332
Hemichordate worm 139
Hemiramphus brasiliensis 4
Hemorrhagic 11

Hemorrhagic appearance 95
Hermetic reaction 202
Heroin administration 247
Heteropneustes fossilis 28
Heterozygosity 31
Himalayan Tahr 273
Himalayan waters 30
Himalayas 27
Hindustan Aeronautics Limited 265
Hirundichythys affinis 4
Histological aspects 18
Histological examinations 96
Holothuroid 139
Homeostatic systems 63
Hooghly (Ganga) river basin 84
Hornbill 263
Horse shoe crabs 139
Human dwellings 228
Human habitations 228
Hydrilla 199
Hydrilla plants 170
Hydrocerius 168
Hyena 263
Hymenoptera 233
Hypophthalmichthys molitrix 33
Hypophysial-gonadal axis 256
Hypothalamo-pituitary-gonadal (HPG) 256

I

Ichthyobiodiversity 28
Ichthyodiversity 39
Immature testes 95
Indian major carps 185
Indian mangroves 135
Indian waters 26
Indices of diversity 109
Induced breeding 185

Induced spawning 181
Indus 27
Indus river system 28
Infochemicals 196
Inland resources 27
Insecta 110
Insects 167
In situ conservation 40
Integrated coastal management plans (ICCPs) 45
Integrated management 86
Integrated pest management (IPM) 116
Interlobular area 250
International year of rice 306
Invasive alien species 334
Isometric growth 93
IUCN red data book of threatened animals 36

J

Jackal 263
Jammu and Kashmir 41
Jammu region 331
Jungle cat 263
Juveniles 334

K

Kashmir Valley 287
Kedarnath wildlfe sanctuary 282
King cobra 263
Kingfishers 263
Kole lands 107
Kole wetlands 225

L

Labeo dero 27
Labeo fimbriatus 36
Laboratory tests 211
Lake Georgia of Africa 31
Lane snapper 4, 90

Lapwings 225
Lapwing's foraging ground 229
Largest river draining 80
Least concern (LC) 35
Length-weight 92
Length-weight relationship 8
Leopard cat 263
Lepidocephalous goalparensis 37
Life history 3, 121
Life history traits 90
Limnogeton 168
Limnology 1
Lipid and protein vitellogenesis 14, 15
Lipid droplets 14, 255
Lipids 124
Lipid vitellogenesis 14
Littoral predatory invertebrates 207
Littoral system 106
Live gene bank 43
Live nest 332
Loktak Lake 33
Lorica morphology 214
Lower risk (LR) 34
Luteinizing
Hormone 247
Lutianus argenti-maculatus 28
Lutjanus synagris 4
Lymnaeidae 110
Lysosomes 250

M

Macrobenthic and meiobenthic diversity 142
Macrobenthic fauna 106, 148
Macronucleus 303
Macrophytes 199
Macroscopic scales 10
Mahakali river 36

Mahseer 41
Major river systems 27
Malabar sole 29
Malacofaunal studies 141
Male brooders 181
Mammalian fauna 274
Management and monitoring 40
Management of watershed 86
Management of wildlife 269
Mangrove ecosystem 136
Mangrove habitats 136
Mangroves 135
Mangroves in India 136
Mangroves of Kerala 146
Man-induced stresses 29
Marine fish 3
Marine fish biodiversity 28
Marine parks 41
Marine protected areas (MPAs) 45
Marine sanctuaries 41
Mastacembalus armatus 37
Mean absolute fecundity 13
Mega biodiversity countries 135
Meiofaunal study 140
Meristic counts 91
Metapenaeus monoceros 28
Microcystis 197
Microenvironmental conditions 170
Micronucleus 303
Microscopic characteristics 13
Migratory birds 263
Mini zoo 263
Mitochondria 250
Mollusca 110
Molluscan fauna 140
Molluscs 139
Morphometric data 213

Mugger crocodile 263
Mugil cephalus 28, 31
Mugil curema 4
Musk deer 283
Mus norvegicus 248
Myriophyllum spicatum 200
Mystus gulio 28

N

Naididae 111
Naina Devi Himalayan Bird Conservation Reserve 283
Nanadankanan 265
Narmada 28
National Bureau of Fish Genetic Resources (NBFGR) 36
Natural ecosystems 167
Near threatened (LR-nt) 34
Nematode 143, 305
Neolissochielus hexagonolepis 27
Neonates 207
Nesting 225, 267
Nesting period 331
Nesting success 229
Nestlings 332
Nest of oogonia 17
Nigerrimus (Pareumenes) Pareumenes 242
Nilgiris 27
Non-indigenous species 32
Non-target organisms 68
Nymphal stages 173

O

Oak Forest 273
Ocelli 168
Oligochaeta 110
Oligochaete 111
Oligoplites palometa 4
Olive Ridley sea turtle 263

Olive Ridley sea turtle conservation 267
Ompok pabda 31, 42
Oncorhynchus apache 32
One way ANOVA 109
Oocytes 14
Oocytes developmental 18
Oral membrane 300
Oreochromis mossambicus 33
Organic wastes 127
Oriental species 242
Orioles 263
Ornamental clownfish 42
Ostracods 148, 211
Otter 263
Ovaries 11
Over-exploitation 31, 187
Over-harvesting 187
Ovigerous lamella 16
Ovipositional Frequency 175

P

Pacific salmon 31
Pacific threadfin 42
Paddy cultivation 114
Paddy fields 105
Palaemon styliferus 28
Pangasius pangasius 28
Pangolin 263, 265
Paramecium caudatum 205
Parthenogenetic females 207, 214
Peafowl 263
Pecking 227
Penaeid prawn 141
Perinucleolus stage 15
Periyar lake-stream system 37
Pesticides pollution 31
Pharmaceuticals 63
Pheromones 207

Physico-chemical parameters 113, 183
Phytoplankton 199
Pichavaram 135
Pine forests 274
Pir Panjal 331
Pisces 110
Pituitary gland extract (PGE) 185
Plankton 199
Plant extracts 178
Plant parasitic nematodes 306
Platyhelminthes 208
Poisonous pollutants 31
Polychaetes 147
Pomadasys corvinaeformis 4
Population dynamics 147, 172
Population growth rate 121
Porcupine 263
Post-Earth Summit Assessment 26
Potter wasp 233
Prawns 139
Predation 170
Predator-prey interactions 170
Predators 229
Predator's niche 176
Predator's performance 175
Prelysosomal vesicular 255
Prey density 178
Prey preference 177
Prey quality 178
Prey species 176
Project tiger 265
Proteins 124
Protozoans 287
Puntius carnaticus 37
Pycnonotus cafer 329
Pyrethroid-K-othrine 174
Python 263

R

Rana (Dicroglossus) cyanophlyctis 287
Ranching 41
Rapid industrialization 83
Rare from Gandaki 36
Rare (R) 34
Ratel 263
Red jungle fowl 263
Redox potential 113, 145
Red-vented bulbul 329
Reproduction 121
Reproductive cycle 94
Reproductive output 214
Reproductive period 18, 97
Reproductive strategies 3, 97
Reptiles 261
Rhesus macaque 263
Rhizophora mucronata 138
Richness 112
River Ganges 77
Rotifer density 207
Rotifers 205
Royal Bengal tiger (*Panthera tigris*) 138

S

Salamanders 214
Sambar 263, 273, 276
Sampling procedure 109
Scenedesmus 129, 202
Schizothorax richardsonii 27, 38
Scomberomorus brasiliensis 4
Sea turtle 267
Secondary metabolites 196
Sedentary hunters 168
Sediment samples 109
Seed production 182
Semiarid region 20
Seminiferous tubules 250
Serow 276
Sertoli cells 250
Serum LH level 249
Sesarmid crabs 141
Sex ratio 9, 93
Sexual maturity 9
Sexual reproduction 207
Sexual strategies 3, 89
Shannon index 109
Shivalik region 331
Shortest lifespan 213
Shrub cover 279
Silonia silondia 28
Siltation 30
Simpson's index 109
Sino-Indian biogeographic realms 274
Spanish mackerel 97
Spawning areas 93
Spawning habitats 41
Spawning period 3, 89
Spearman's rank correlation coefficient 277
Species dominance 109
Spent gonads 11
Spermatocytes 250
Spermatogenic cells 250
Spermatozoa 250
Spirulina platensis 210
Spotted deer 263
Star tortoise 263
Statistical analysis 109
Stress tolerance 173
Striped catfish 187
Sublethal effects 63
Sublethal heroin administration 249
Sundarbans 138
Sunderbans stands 148

Survivorship 121
Sutchi catfish 181
Symbiotic associations 241
Synthetic pesticides 64
Systemic Index of Nematode Species 306

T

Tachysurus spp. 28
Tapti 28
Taxonomic review 233
Temporary wetlands 106
Terrestrial plants 196
Testes 11
Testes and ovaries 95
Testicular morphology 247
Thailand model hatchery 183
Threatened animals 34
Threatened Indian fishes 25
Threatened species 26, 44, 262
Threatened wetlands 226
Threats 225
Tiger 263
Tipulidae 110
Topographic 274
Tor khudree 37
Tor putitora 30
Tor tor 27
Total body length 92
Tree cover 278
Tributaries of river Ganga 82
Trichoptera 114
Triclosan 66
Tropical marine fish 3
Tubificidae 110
Tungro Virus 306
Turbellarians 145, 148, 208
Tylenchida 324

Type of growth 8, 92
Type of spawning 13, 96

U

Ultramicrotome 249
Ungulate species 277

V

Varieties of fish 261
Vascularization 95
Vector control 29
Vertebrates 213
Vertical jar hatchery 183
Vesicular bodies 255
Vespidae 233
Vestibulum 293
Virgin niche 33
Visible oocytes 11
Visual predators 205
Vitologenesis 14
Vulnerable (VU) 34
Vultures 263, 265

W

Wallago attu 28
Wanton destruction 31
Warmwaters 27
Wastewater 83, 129
Wastewater treatment 122
Water 25
Water bugs 167
Water hyacinth 170
Water king 168
Water quality 84, 122, 187
Water sample 174
Watershed 78
Water treatment 129
West coast of India 135
West coast river system 28

Western Ghats 27, 37
Wetlands 106, 114
White bellied sea eagle 263
White tiger cubs 265
Wild boar 263
Wild dog 263
Wildlife 66, 87, 223
Wildlife crime cell 269
Wildlife diversity 261
Wildlife (Protection) Act 1972 261
Wildlife sanctuaries 143
Wild pig 276
Wistar strain 248
Wolf 263

Woodpecker 263
World Fishery 167

X

Xenentodon cancila 37

Y

Yamuna 83
Yellow-Wattled Lapwing 225
Yolk granules 14, 17
Yolk vesicle 14
Yolk vesicle stage 17
Young germ cell 14, 16

Z

Zoological park 265
Zooplankton 63, 195

Lightning Source UK Ltd.
Milton Keynes UK
UKHW052018170220
358876UK00007B/44